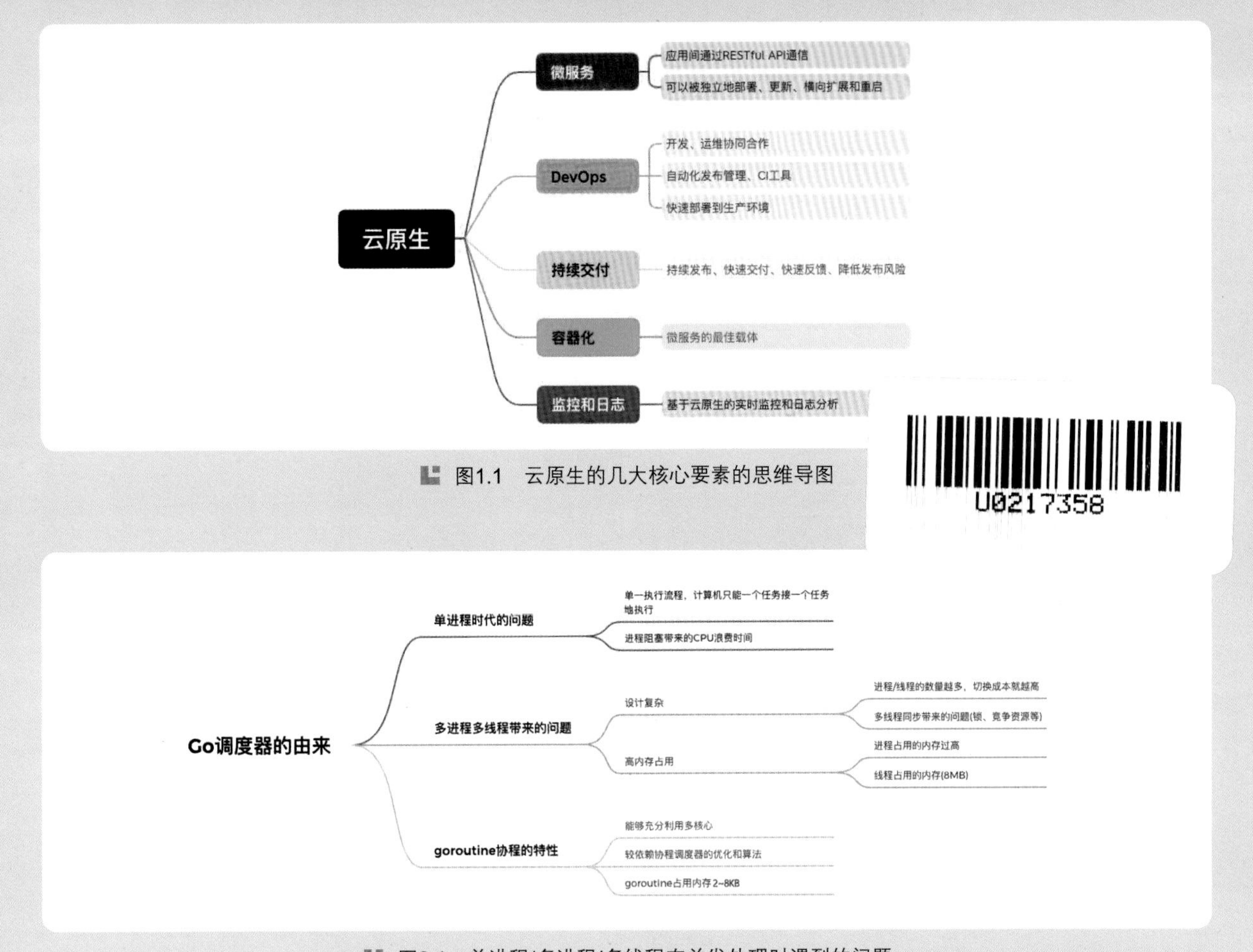

图1.1　云原生的几大核心要素的思维导图

图2.1　单进程/多进程/多线程在并发处理时遇到的问题

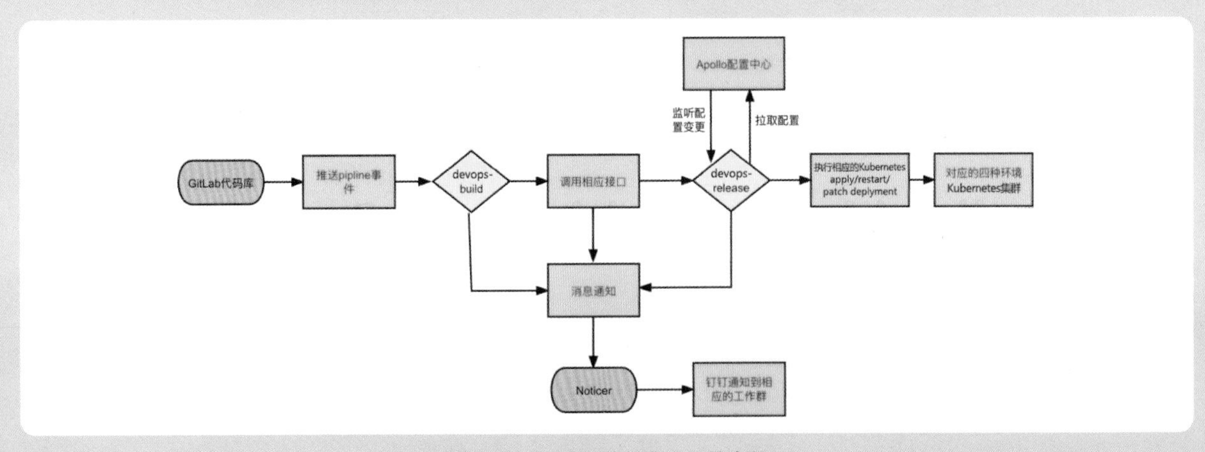

图4.1　CD自动化发版工具流程

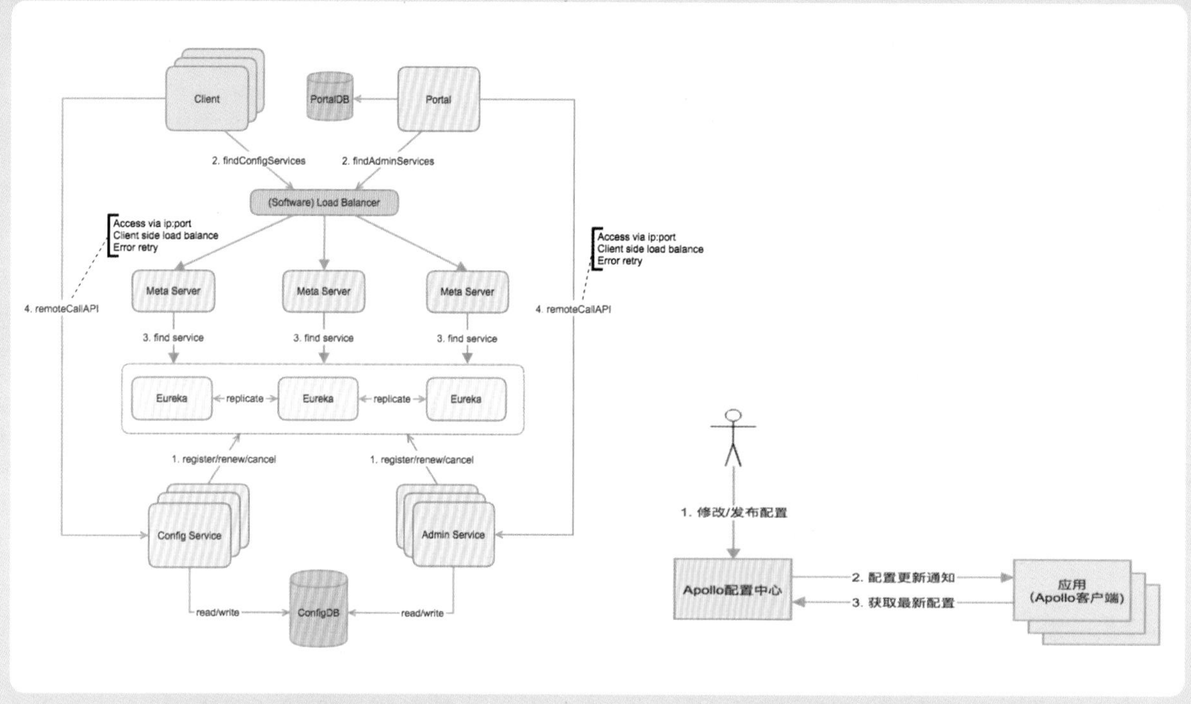

图4.7 Apollo的主架构模块

图4.8 Apollo用户端工作流程

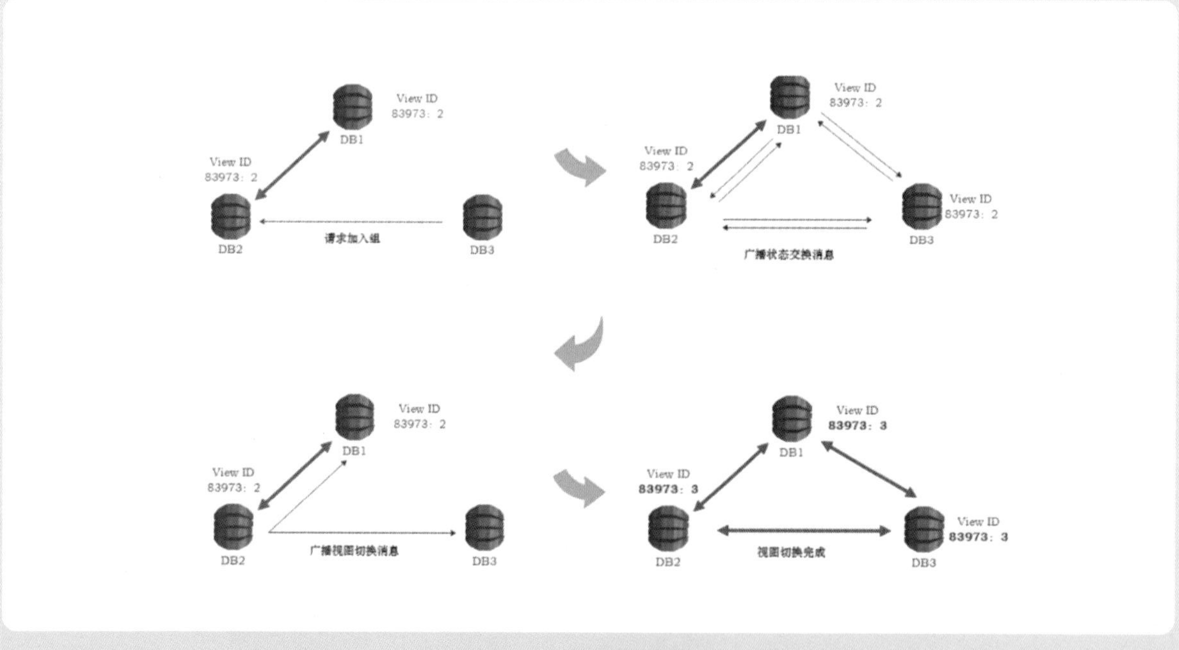

图5.4 MGR视图工作流

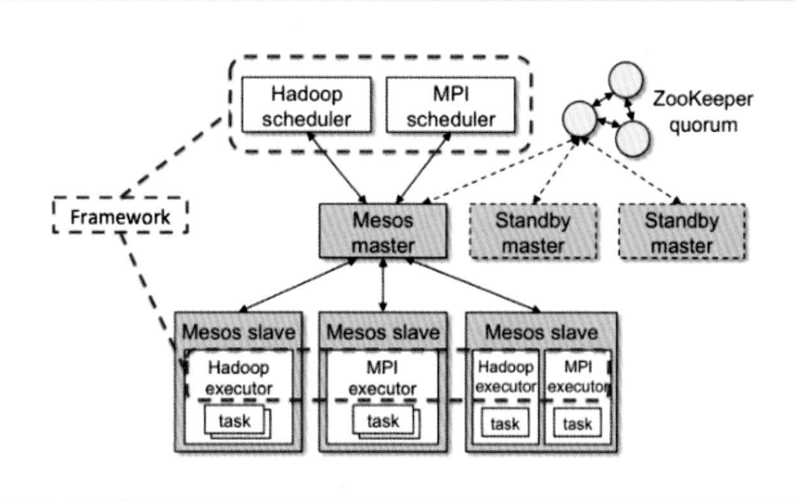

图6.1 Mesos系统架构

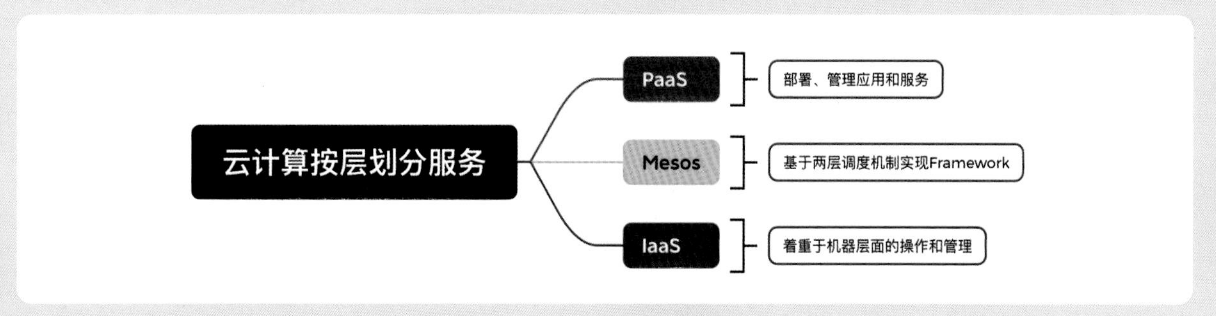

图6.2 Mesos可以实现PaaS服务

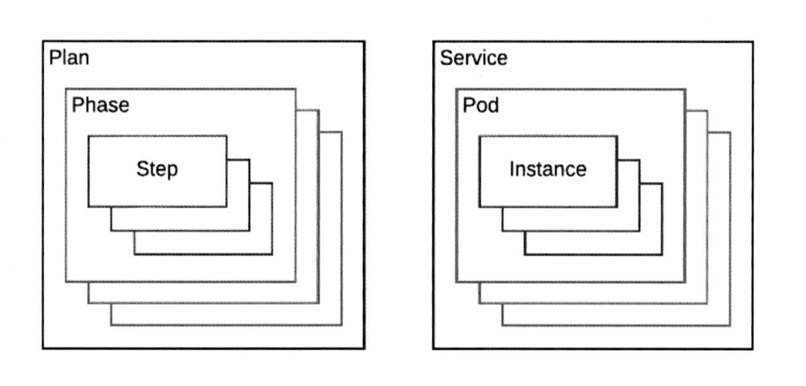

图6.4 DC/OS SDK中的Plan生成对象模型

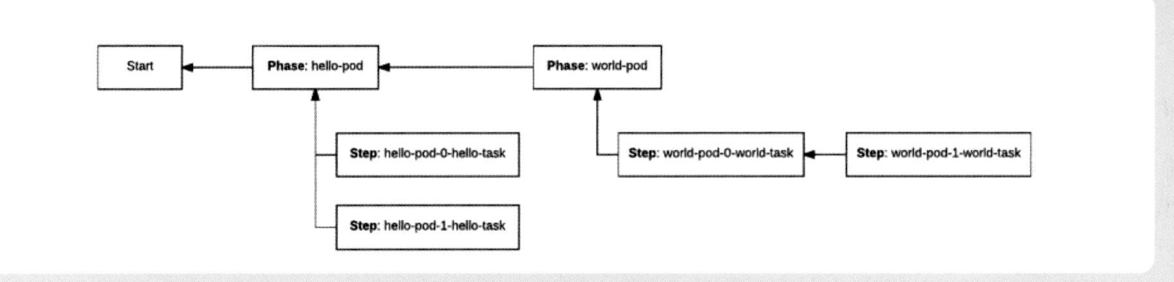

图6.5 名为hello的计划生成实例的阶段演示

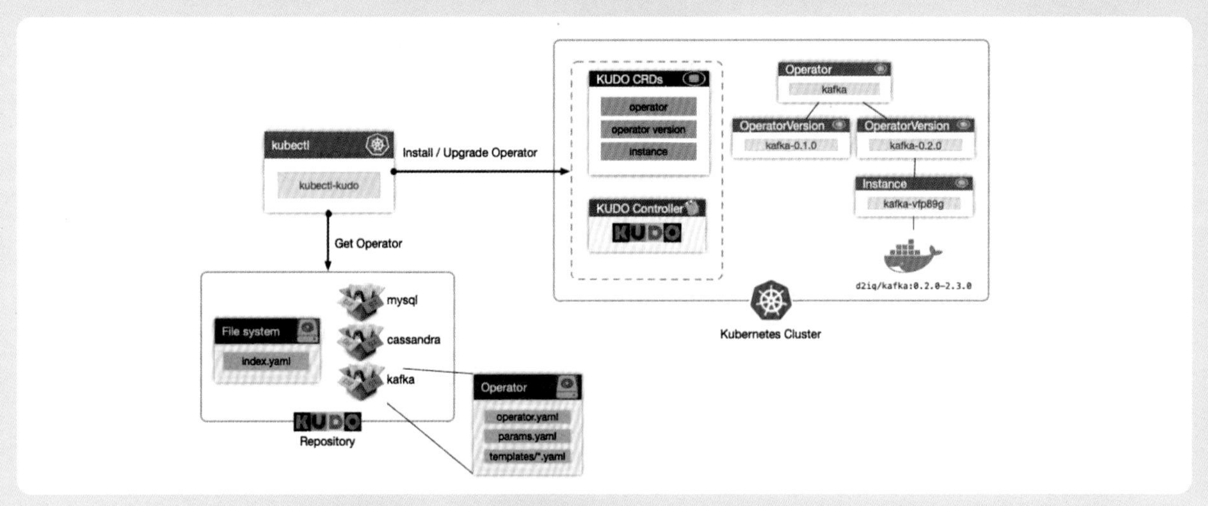

图8.1 KUDO与Kubernetes API进行交互的工作流

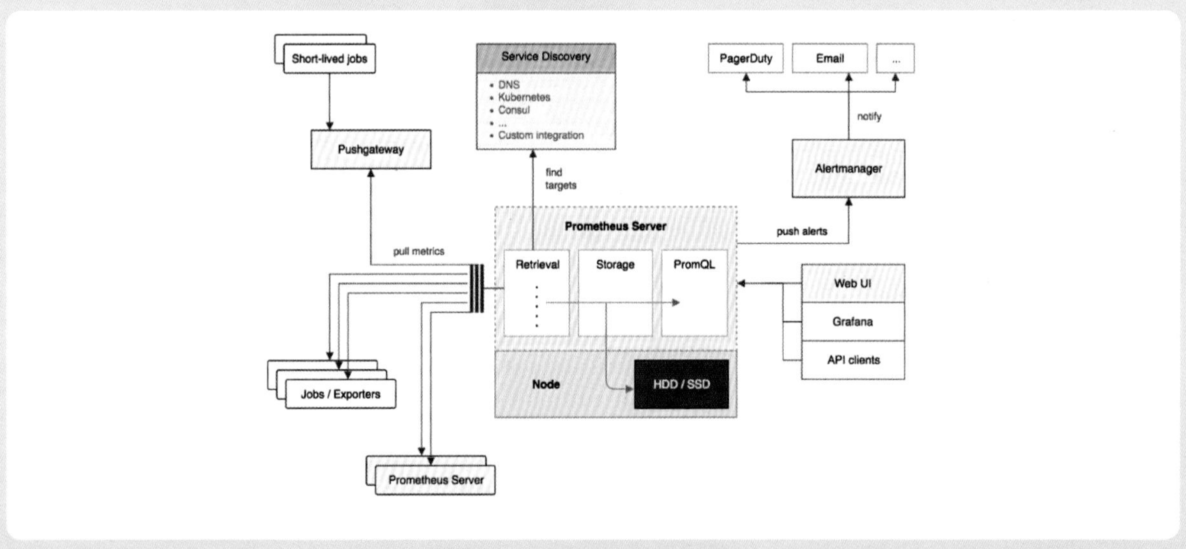

图9.1 Prometheus系统架构

云原生 DevOps

开发实践

■ 余洪春 — 编著

中国水利水电出版社
www.waterpub.com.cn
· 北京 ·

内 容 提 要

这是一本具有实践指导意义的云原生DevOps开发与运维实战宝典，基于作者丰富的实战经验，深入浅出地全面解析了云原生领域的核心要素：Go语言编程、微服务架构设计、持续集成/持续交付（CI/CD）流程、容器化部署技术以及自动化运维策略，并且涵盖了云原生运维的热点内容，如云原生DevOps框架下的CI/CD实践、Operator与Framework的开发应用以及云原生DevOps监控解决方案。

本书内容均紧密贴合实际操作，对一些底层技术选型进行了对比性分析，旨在助力读者不断精进专业技能，掌握云原生DevOps开发运维精髓。本书不仅适合Go开发工程师、高级运维工程师、运维开发工程师、SRE工程师使用，也适合作为高等院校或者培训机构云原生及相关专业的教材和教学参考书。

图书在版编目（CIP）数据

云原生 DevOps 开发实践 / 余洪春编著 . -- 北京：
中国水利水电出版社 , 2025.1. -- ISBN 978-7-5226
-2872-1

Ⅰ. TP311.5

中国国家版本馆 CIP 数据核字第 2024H0F210 号

书　　名	云原生 DevOps 开发实践 YUNYUANSHENG DevOps KAIFA SHIJIAN
作　　者	余洪春　编著
出版发行	中国水利水电出版社 （北京市海淀区玉渊潭南路 1 号 D 座　100038） 网址：www.waterpub.com.cn E-mail：zhiboshangshu@163.com 电话：（010）62572966-2205/2266/2201（营销中心）
经　　售	北京科水图书销售有限公司 电话：（010）68545874、63202643 全国各地新华书店和相关出版物销售网点
排　　版	北京智博尚书文化传媒有限公司
印　　刷	北京富博印刷有限公司
规　　格	190mm×235mm　16 开本　26 印张　641 千字　2 插页
版　　次	2025 年 1 月第 1 版　2025 年 1 月第 1 次印刷
印　　数	0001—3000 册
定　　价	99.80 元

前　言

作者已从事系统运维 /DevOps/ 运维架构师 15 余年，亲历了云计算从物理机时代到虚拟化时代，再到容器化时代的变革。目前，企业应用架构在云原生时代如何重塑是一个越来越现实的问题。云原生技术是一种全新的方法论，这种技术的流行给传统的运维知识体系带来了巨大的冲击。作者的许多读者朋友和学生也经常向作者咨询工作中的困惑。比如，从事系统运维工作 5 年以后应该如何规划自己的职业生涯等。作者想通过本书，分享自己多年来的工作经验和心得，解决读者心中的困惑。

当前，企业应用架构与数字化转型的需求日益凸显，尤其是为客户提供解决方案的比重持续增大，这导致对云原生技术的需求也随之不断攀升。与此同时，所面临的问题愈发复杂，且这些挑战往往缺乏统一且普适的解决方案。本书基于作者多年的实战经验，聚焦于云原生 DevOps 开发这一核心主题，全面系统地阐述了云原生领域中的 CI/CD、Operator 及 Framework 开发实践等内容。这对于具备一定 Linux 服务器管理经验的读者，尤其是希望转型至云原生领域的 DevOps 技术人员而言，具有显著的助益。通过本书中的项目实践及线上环境案例，读者能够清晰地认识并确定云原生环境下 DevOps 的研发方向。我们期望，通过学习本书，读者朋友们能够深刻掌握云原生 DevOps 的精髓，享受工作的乐趣，并不断提升自身的职业技能。这正是作者编写此书的初衷，也是作者所乐见的。

本书内容

本书共 9 章，具体内容如下。

第 1 章，主要介绍企业采用云原生下的 DevOps 进行开发的作用和意义。

第 2 章，介绍 Go 语言在云原生下的基础及进阶，主要涉及工作中应该掌握的 Go 语言的知识点和基础概念。

第 3 章，介绍如何用脚本语言开发云原生 CI SDK。

第 4 章，主要介绍如何利用 Go 语言开发 CD 自动化发版工具。

第 5 章，介绍云原生 MySQL 架构选型，和基于云原生高可用数据库 MGR 集群的原理和搭建流程。

第 6 章，介绍 Mesos SDK 的概念和工作流，以及如何将 MGR 封装成一个 Framework 来运行。

第 7 章，介绍 Kubernetes 的 API 简单源码分析，方便后续理解 Operator 开发工作。

第 8 章，介绍 Kubernetes 下的低代码化 Operator 开发工具 KUDO。

第 9 章，介绍云原生监控 DevOps 实践。

本书特色

1. 书中有几大板块是目前云原生运维方向的热点。例如，基于云原生的 DevOps 中的 CI/CD，Operator/Framework 开发实践、云原生 DevOps 监控实践等，相信能提升读者的工作技能。

2. 书中的所有案例均取自于生产服务器或项目实施，读者无须再花费大量时间进行调研，所有提及的项目方案均可直接对照自己的业务场景有针对性地使用。

3. 本书对许多底层技术选型进行了对比性分析，例如，在云原生 DevOps 监控领域，放弃热门的 InfluxDB，而直接选用较新的 VictoriaMetrics 时序数据库（原因在于，在生产环境中，尤其是在磁盘 I/O 资源有限的情况下，InfluxDB 容易发生 iowait 过高的问题）。这些对比分析对读者在自己的业务场景中做出技术选型，具有一定的参考价值。

本书服务

为避免代码占用篇幅过多，本书对代码进行了简略处理，若想获得全部代码，可关注以下公众号二维码，输入关键词 YYS28721 至公众号后台，获取本书代码文件的下载链接。读者也可以加入本书的读者交流群，在群中分享心得，或提出对本书的建议，以及咨询本书相关的问题等。

人人都是程序猿

致谢

本书能够顺利出版，是作者、编辑和所有审校人员共同努力的结果，在此表示深深的感谢。同时，祝福所有读者在职场一帆风顺。

作　者

2024 年 11 月

目　录

第 1 章　云原生下 DevOps 工作概述

云原生这个词读者应该已经很熟悉了，在很多技术资料或相关参考书上都能看到这个词。那么什么是云原生呢？云原生下的 DevOps 又是如何工作的呢？

1.1　什么是云原生

什么是云原生？其实，云原生借了云计算的东风，没有云计算，就没有云原生，云计算是云原生的基础。随着虚拟化技术的成熟和分布式框架的普及，在容器技术、可持续交付、编排系统等开源社区的推动下，以及微服务等开发理念的带动下，应用云已经是不可逆转的趋势。云计算的 3 层划分，即基础设施即服务（IaaS）、平台即服务（PaaS）、软件即服务（SaaS）为云原生提供了技术基础和方向指引，真正的云化不仅仅是基础设施和平台的变化，应用也需要做出改变，也就是摈弃传统的方法，在架构设计、开发方式、部署维护等各个阶段和方面都基于云的特点进行重新设计，从而建设全新的云化的应用，即云原生应用。

传统应用与云原生应用的区别有以下几点：

（1）本地部署的传统应用往往采用 C++、Java 语言编写，而云原生应用则需要用以网络为中心的 Go 或 Node.js 等新兴语言编写。

（2）本地部署的传统应用可能需要停机更新，而云原生应用则始终是最新版本，需要支持频繁变更、持续交付、蓝绿部署。

（3）本地部署的传统应用无法动态扩展，往往需要冗余资源以抵抗流量高峰，而云原生应用利用云的弹性自动伸缩，通过共享降本增效。

（4）本地部署的传统应用对网络资源（如 IP、端口等）有依赖，有时候代码里面还需要带上 IP 和端口配置，而云原生应用对网络和存储都没有这种限制。

（5）本地部署的传统应用通常通过人工手动运维或 Python 自动化工具来部署，而云原生应用的部署都是自动化的。

（6）本地部署的传统应用通常依赖系统环境，而云原生应用不会硬连接到任何系统环境，而是依赖抽象的基础架构，从而获得良好的移植性。

（7）本地部署的传统应用有些是单体应用，或者强依赖，而基于微服务架构的云原生应用，纵向划分服务，模块化更合理。

可见，要转向云原生应用需要以新的云原生方法开展工作，云原生包括基础架构服务、虚拟化、容器化、容器编排、微服务等方面。幸运的是，开源社区在云原生应用方面做出了大量卓有成效的工作，很多开源的框架和设施可以直接使用，2013 年，Docker 推出并很快成为容器事实标准，在随后围绕容器编排的混战中，于 2017 年诞生的 Kubernetes 很快脱颖而出，而这些技术极大地降低了开发云原生应用的技术门槛。

技术的变革，一定是思想先行，云原生是一种构建和运行应用程序的方法，是一套技术体系和方法论。云原生（CloudNative）是一个组合词：Cloud+Native。其中，Cloud 表示应用程序位于云中，而不是位于传统的数据中心；Native 表示应用程序从设计之初即考虑到云的环境，原生为云而设计，在云上以最佳方式运行，充分利用和发挥云平台的弹性 + 分布式优势。

云原生的几大核心要素的思维导图如图 1.1 所示。

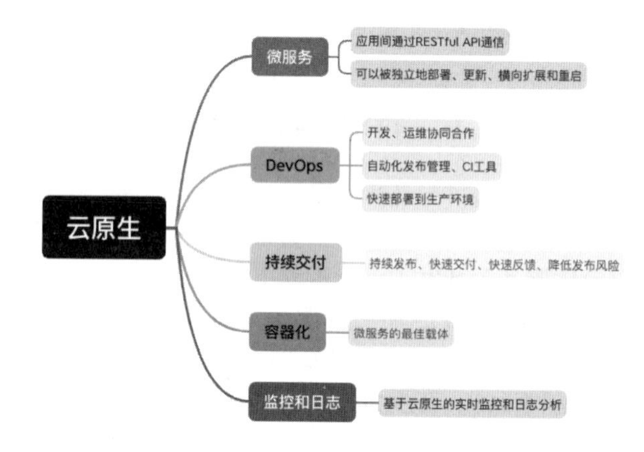

图 1.1　云原生的几大核心要素的思维导图

（1）微服务：微服务解决的是软件开发中一直追求的低耦合 + 高内聚，在未使用微服务架构时，若某个子应用出现问题，则很有可能会影响整个平台或产品的正常使用。而微服务的本质是把一块大饼分成若干块低耦合的小饼，例如，一块小饼专门负责接收外部数据；另一块小饼专门负责响应前台的操作。另外，小饼还可以进一步拆分，例如，负责接收外部数据的小饼可以继续分成多个微小饼，以负责接收不同类型的数据。这样即使有一个小饼出现问题，其他小饼还能正常对外提供服务。

（2）DevOps：这是个组合词，Dev+Ops，即开发和运维组合，实际上 DevOps 还应该包括测试，DevOps 是一种敏捷思维，也是一个方法论，为云原生提供持续交付能力。

（3）持续交付：持续交付是不延时开发、不停机更新、小步快跑、反传统瀑布式开发模型，这要求开发版本和稳定版本并存，其实需要很多流程和工具支撑。

（4）容器化：Docker 是应用最为广泛的容器引擎，在思科、谷歌等公司的基础设施中大量使用，其基于 LXC 技术，容器化为微服务提供实施保障，可起到应用隔离作用，Kubernetes 是容器编排系统，用于容器间的负载均衡和容器的管理。

（5）监控和日志：基于云原生环境的实时监控和日志分析，实时监控和日志分析对于快速检测和解决问题至关重要。现在很多基于云原生的开源工具已经很成熟，如 Prometheus/Alertmanager/Loki/Grafana 等。监控和日志对业务的意义重大，建议读者重视。因此，也可以简单地把云原生理解为：云原生 = 微服务 + DevOps + 持续交付 + 容器化 + 监控和日志。

总而言之，符合云原生架构的应用程序应该是：采用开源堆栈 Kubernetes 进行容器化，基于微服务架构提高灵活性和可维护性，借助敏捷方法、DevOps 支持持续迭代和运维自动化，借助开源组件实现基于容器的应用程序的监控和日志收集，利用云平台设施实现弹性伸缩、动态调度、优化资源利用率。

1.2　云原生下 DevOps 的概念和实践

云原生环境下的 DevOps 是一种融合了云计算与原生应用开发的方法论，其核心目的是加速软件交付流程，同时确保软件的质量与可靠性。在云原生架构中，DevOps 实践与云原生应用的开发、部署紧密相连，以便更好地适应云环境的动态变化与弹性需求。以下是 DevOps 的几个关键概念与实践。

（1）容器化技术：在云原生开发中，容器技术（如 Docker 或 Podman）被广泛采用。该技术可以将应用及其所有依赖项打包到一个固定环境的沙盒中，确保在不同的部署阶段和环境中具有一致性。

（2）编排工具：云原生应用的复杂性需要有效的编排工具，以管理和协调容器的部署、伸缩和运行。Kubernetes 是当前最流行的开源容器编排工具之一。

（3）持续集成 / 持续交付（CI/CD）：DevOps 在云原生环境中强调自动化。通过 CI/CD 流水线，开发人员能够频繁地将代码集成到主干分支，并通过自动化工具实现快速、可靠的部署。

（4）微服务架构：云原生开发倡导使用微服务来构建应用程序。微服务是独立部署的小型服务，通过 API 进行通信，使得应用更加灵活、可维护和可扩展。

（5）监控和日志：在云原生环境中，实时监控和日志分析对于应用的快速检测和解决问题至关重要。使用工具（如 Prometheus 和 Grafana）可以实现对应用性能的实时监控，通过 Loki 可以实现对应用日志的实时采集。

（6）自动化基础设施：通过 Infrastructure as Code（IaC）的实践，云原生 DevOps 可以实现对基础设施的自动化管理，以确保环境的一致性和可重复性。

综合所述，云原生下的 DevOps 强调敏捷性、自动化和弹性，以适应云环境的动态特性，同时提高软件开发交付的效率和质量。

1.3　云原生下 DevOps 开发语言简介

一般来说，在云原生下进行 DevOps 开发工作，作者推荐用 Go 作为首选语言。下面将 Go 语言与现在的热门开发语言进行对比，并介绍 Go 作为云原生 DevOps 开发语言的优势。

（1）语法简单。Go 语言的语法非常简单、直接，学习难度低。Go 语言的设计是为了弥补 C++ 语言的缺陷，去除 C++ 语言中的烦琐与笨重之处，改进 C++ 语言的效率和扩展性，让编程工作变得方便。所以它本身就具有编译时间短，运行效率高，稳定性强，拥有强大的编译检查和完整的软件生命周期工具等优点。

（2）并发编程。Java语言的编码非常烦琐，为了应用设计模式做了大量的冗长设计。而Go语言与Java语言不同，它提供了便利的并发编程方式，简单的Go语句，就可以创建多个goroutine执行并发任务，即Go语言是为并发而生的。

（3）具有强大的标准库。Go语言里包括互联网应用、系统编程和网络编程标准库。Go语言中的标准库基本上是稳定的，网络层、系统层的库非常实用。Go语言的标准库“麻雀虽小，五脏俱全”。虽然有些库还不是很好用，但随着技术的发展和成熟，Go语言的库会越来越成熟和实用。

（4）C语言的理念和Python的特点。C语言的理念是信任程序员，保持语言的小巧，不屏蔽底层且对底层友好，关注语言的执行效率和性能。Python的特点是用尽量少的代码做尽量多的事。而Go语言将C语言和Python语言相结合，兼具二者的优势。

（5）Go语言适合资源密集型的应用。Go语言的并发性能优势使其成为开发高性能、高并发API服务的理想选择。Go语言适合处理资源密集型的应用，如网络服务等；同样的网络服务应用，用Go语言跟Node.js做开发对比，Go语言的处理效率要优于Node.js数十倍；如笔者目前从事的CDN行业及HPC行业，物理机器基本上都超过了10 000台，容器数量基本上也是以万级计算（含产研环境），像这种量级的机器和容器，不管是开发API还是中间件服务，性能是我们考虑的首要条件之一。

（6）Go语言是云计算时代的语言。随着云计算平台的逐渐成熟，应用云已经成为一个不可逆转的趋势。很多公司都选择将基础架构/业务架构云化，如阿里巴巴集团、腾讯公司，都在将公司内部业务全面云化。也就是说，以后所有的技术后期可能都围绕着云来构建。而云目前是朝着云原生架构的方向演进的，云原生架构中有60%以上的具有统治力的云原生项目都是用Go语言构建的。

综上所述，笔者也将Go语言作为云原生下DevOps开发推荐的语言，关于Go语言的使用在后面章节会重点介绍。

1.4 云原生下DevOps技术的价值

优秀的DevOps工程师不应该只从事开发方面的工作，还需要了解业务运营情况、熟悉业务流程，这样才能有针对性地开发监控采集。只有了解了业务线，有针对性地开发与业务相关的平台或工具，才能让其产生更大的价值。

优秀的DevOps工程师应该确保所有系统（包括业务系统）都有监控和警报，以帮助提高生产服务的可用性和保证正常运行时间。对于这一点，有许多云原生开源工具可以采用，如Prometheus/Grafana/Alertmanager/Loki等。有了全面的监控指标数据和应用采集日志，就能在出现问题时快速解决技术故障，以帮助生产系统恢复正常，提升业务的健康度。

DevOps是一种将开发和运维紧密结合的方法论，旨在通过自动化、持续集成和持续交付来加快软件交付速度并确保其可靠性。其核心优势在于能够实现软件的快速且可靠的持续交付，这不仅显著缩短了从开发到部署的周期，还降低了开发与运维团队之间的协作难度，促进了团队间的紧密合作，进而提升了整个交付流程的效率。

DevOps着重于推动开发和运维团队之间的协作与文化变革，它不仅仅关注技术层面的自动化和快速交付，更强调敏捷开发理念的融入。通过注重流程的持续改进、增强团队协作以及推动文化层面的转变，DevOps 致力于打破传统开发与运维之间的壁垒，确保软件产品能够迅速响应市场需求并持续保持高质量。

下面用实际案例来说明云原生下 DevOps 技术的价值。

（1）可以快速投放市场。每天可快速迭代数百个版本。2018 年，作者所在公司的大数据产品团队一周时间迭代了 200 ~ 300 次，每天差不多有几十次的小版本迭代，而且用蓝绿版本迭代，这样频繁地发布也只有通过 DevOps 才能实现。最开始是用虚拟机 + Java 应用的方式直接部署和运维，运维人员经常需要加班交付，因此，做发布和交付的运维人员离职率较高。而用容器来做，运维人员就不用加班到凌晨了。只需两套环境，一套蓝、一套绿，下午做好部署，晚上切换网络就可以了。

（2）降低成本。有 DevOps 工具之后，大数据团队可以减少人工投入，进而减少因停机时间带来的损失。作者在云原生 DevOp 团队工作的时候，对接最多的客户就是某大型汽车制造厂商，他们都愿意做云原生数字化转型，也愿意为云原生 DevOps 方案买单和持续投入，这说明 DevOps 在节约人力成本方面的价值已被客户认可。

（3）节省人力。DevOps 可以让开发者不用做一些低价值的事情，包括安装、部署、配置等都可以用工具来做。此外，DevOps 平台还可以满足整个生命周期管理的需求，从最早的项目管理、需求，到构建代码，再到最后的运维。这就是 DevOps 的价值体现。

DevOps 的未来前景：DevOps 作为一种文化和实践方法，在快速变化的业务环境中将持续发展。大量实践也已经证实，DevOps 对提高软件的生产质量和安全性、缩短软件的发布周期等都有非常明显的促进作用。随着云原生技术、容器化技术的流行和推动，DevOps 有望进一步加速软件交付流程，提高团队协作效率。

1.5　小　　结

本章介绍了云原生及其 DevOps 的工作概述，云原生下 DevOps 开发语言 Go 的特点，以及云原生下 DevOps 技术的价值体现。本章只是一些简单的理论介绍，后续章节会重点介绍具体实践。

第 2 章　Go 语言基础语法及高级进阶

2.1　Go 语言的安装部署及包结构

下面从 Go 语言开发环境的安装及部署、Go 工程的项目组织结构及 Go 工程中的模块管理几方面来依次说明。

2.1.1　Go 语言开发环境的安装及部署

本小节从 Go 语言开发环境的安装及部署开始，以一台 CentOS 7.9 x86_64 的开发机器为例进行说明。

首先从 Go 官网下载较新的源码包，即 1.20 大版本，然后解压缩并 mv 到 /usr/local 目录下，其命令如下：

```
cd /usr/local/src && tar xvf go1.20.6.linux-amd64.tar.gz
mv go /usr/local/
```

接着编辑 /etc/profile 文件，主要定义 GOROOT 及 GOPATH 环境变量。代码如下：

```
export GOROOT=/usr/local/go
export GOBIN=$GOROOT/bin
export GOPATH=/data/home/go
export PATH=$PATH:$GOPATH:$GOBIN:$GOROOT
```

最后不用重启机器而使变量生效，其命令如下：

```
source /etc/profile
```

下面可以写一个简单的程序验证 Go 语言开发环境是否部署成功。建立一个 hello.go 文件，文件内容如下：

```
import "fmt"

func main() {
  fmt.Println("hello,world")
}
```

可以用下列命令执行 hello.go 文件：

```
go run hello.go
```

命令显示结果：

```
hello,world
```

这里查看一下当前 Go 语言的版本，命令如下：

```
go version
```

命令显示结果：

```
go version go1.20.6 linux/amd64
```

以上结果说明，Go 语言开发环境已经部署成功。

> 注：Go语言版本是具备高版本向下兼容特性的。Go语言的未来版本会确保向下的兼容性，不会破坏现有程序，即10年前用Go 1.0写的代码，用10年后的Go 1.20版本，依然可以正常运行。也就是说，较高版本的Go语言能正常处理较低版本的Go语言代码。

至于 Go 语言代码编辑器，可选择 GoLand 或 Visual Studio Code，这里不做过多描述。

2.1.2 Go 工程的项目组织结构

Go 工程为什么要分层？分层有时也称为分模块。简单来说，它会带来三点好处：有利于标准化工作、降低层次依赖和降低整体的复杂度。

其实，不论采用何种编程语言，良好的工程项目组织结构都至关重要，因为这将直接影响项目内部依赖的复杂程度以及项目对外提供 API 等服务的灵活性等。最好在项目初期便制定好项目组织结构约定，甚至可以为其开发脚手架之类的工具（后面章节提到的 devops-build/release 工程项目都是采取统一的脚手架工具）来生成项目模板，让开发者尽量按照统一的规范参与项目。

一个常见的 Go 应用项目（如名为 go-project 的项目）布局通常有以下结构：

（1）cmd。cmd 目录是项目的主干，是编译构建的入口，main 文件通常放置在此处。一个典型的 cmd 目录结构如下：

```
├── app1
│   └── main.go
└── app2
    └── main.go

2 directories, 2 files
```

从上述例子可以看出，cmd 下可以挂载多个需要编译的应用，只需在不同的包下编写 main 文件即可。需要注意的是，cmd 中的代码应该尽量“保持简洁”，main 函数中可能仅仅是参数初始化、配置加载、服务启动的操作。

（2）pkg。pkg 目录中存放的是可供项目内部 / 外部使用的公共代码。例如，用来连接第三方服务的 client 代码等。也有部分项目将该目录命名为 lib，如 consul 项目，所表示的含义其实相同。

（3）internal。internal 包主要用于提供一个项目级别的代码保护方式，存放在其中的代码仅供项目内部使用。具体使用的规则是，.../a/b/c/internal/d/e/f 仅仅可以被 .../a/b/c 下的目录导入，.../a/

b/g 则不允许。internal 是 Go 1.4 版本中引入的特性。

在 internal 内部可以继续通过命名对目录的共享范围做区分。例如，internal/myapp 表示该目录下的代码供 myapp 应用使用；internal/pkg 表示该目录下的代码可以供项目内多个应用使用。

（4）doc。Go 项目按照习惯，一般都由 go-swagger 库来自动生成接口文档，doc 中会保存由 go-swagger 自动生成的接口文档。

（5）go.mod 文件与 go.sum 文件。go.mod 文件与 go.sum 文件是采用 go modules 进行依赖管理所生成的配置文件。go modules 是 Go 1.11 版本中引入的版本管理功能，目前已经是 Go 依赖管理的主流方式，所以此处不再讨论 vendor、dep 等依赖管理方式所生成的目录。

（6）Makefile 文件。Makefile 文件通常存放项目的编译部署脚本。Go 的编译命令虽然简单，但总是手写命令效率还是低，因此使用 Makefile 编写编译部署脚本是工程实践中常见的方式。Makefile 提供了一个非常有效的方法来为应用程序自动化各种任务。例如，可以用 make build 或 make run 执行不同的编译任务。

（7）Dockerfile。项目中 Dockerfile 的作用很明确，就是将当前项目打包成 Docker 镜像的文件。

（8）.gitlab-ci.yaml。.gitlab-ci.yaml 是 GitLab CI 流水线配置文件。

2.1.3 Go 工程中的模块管理

Go 工程中使用模块来管理包依赖的步骤比较简单，具体如下：

（1）开启模块支持。在 go 命令行中使用 modules，而不是在 GOPATH 目录下查找。其命令如下：

```
set GO111MODULE=on
```

（2）配置 Golang 国内代理。

```
export GOPROXY=https://goproxy.cn
```

（3）初始化项目。如果在项目目录下面，就直接执行 go mod init 命令。

```
go mod init [包名]
```

（4）生成项目目录下的文件。执行以上命令，会在项目目录下生成 go.mod 和 go.sum 文件，go.mod 文件中包含当前 Go 工程项目的所有包。

（5）下载模块。可以用 go mod tidy 下载模块，这条语句在这里有两种作用：

1）引用项目需要的依赖增加到 go.mod 文件。

2）去掉 go.mod 文件中项目不需要的依赖。

（6）执行 go build 命令。如果能成功执行 go build 命令，就说明前面的流程都没有问题，即

```
CGO_ENABLED=0 GOOS=linux GOARCH=amd64 go build.
```

或

```
CGO_ENABLED=0 GOOS=linux GOARCH=amd64 go build main.go
```

2.2 Go 语言的函数和指针

Go 语言作为新式语言，具有以下重要特征。

2.2.1 Go 语言的内置函数

Go 语言拥有一些不需要进行导入操作就可以直接使用的内置函数。它们有时可以针对不同的数据类型进行操作，如 len、cap 和 append 函数，或用于系统级的操作，如 panic。因此，它们需要直接获得编译器的支持。下面列举一些在开发工作中常用的内置函数。

（1）append：用来向 arrgy（数组）、slice（切片）中追加元素，返回修改后的 array、slice。

（2）close：主要用来关闭 channel（通道）。

（3）delete：用来从 map（字典）中删除 key 及其对应的 value。

（4）panic：停止常规的 goroutine 执行（panic 和 recover，用来做错误处理）。

（5）recover：允许程序定义 goroutine 的 panic 函数。

（6）make：用来分配内存，返回 Type 本身（只能应用于 slice、map 和 channel）。

（7）new：用来分配内存，主要用来分配值类型，如 int、struct，返回指向 Type 的指针。

（8）cap：capacity 是容量的意思，用于返回某个类型的最大容量（只能用于 slice 和 map）。

（9）copy：用来复制和连接 slice，返回复制的数目。

（10）len：用来求长度，如 string、array、slice、map 和 channel，返回长度。

（11）print、println：底层打印函数，在部署生产环境中建议使用 fmt 包，即使用 fmt.Println 这种用法。

另外，Go 内置了接口 error，该接口只声明了一个方法 Error()，返回值是 string 类型，用以描述错误：

```
type error interface {
    Error() string
}
```

2.2.2 init 函数和 main 函数

1. init 函数

Go 语言的 init 函数主要用于包的初始化。该函数是 Go 语言的一个重要特征，其主要特点如下：

（1）init 函数用于程序执行前做包的初始化，如初始化包中的变量等。

（2）每个包可以拥有多个 init 函数。

（3）包的每个源文件也可以拥有多个 init 函数。

（4）同一个包中多个 init 函数的执行顺序在 go 语言中没有明确的定义（说明）。

（5）不同包的 init 函数按照包导入的依赖关系决定该 init 函数的执行顺序。

（6）init 函数不能被其他函数调用，而是在 main 函数执行之前，自动被调用。

其主要作用如下：

（1）在程序运行前进行注册。

（2）实现 sync.once 功能。

（3）初始化无法在全局上下文中初始化的全局变量。

2. main 函数

Go 语言程序的默认入口主函数，其语法如下：

```
func main() {
    // 函数体
}
```

3. init 函数和 main 函数的异同

相同点：两个函数在定义时不能有任何参数和返回值，且 Go 程序自动调用。

不同点：

（1）init 函数可用于任意包中，且可以重复定义多个。

（2）main 函数只能用于 main 包中，且只能定义一个。

2.2.3 占位符

占位符“_”是 Go 语言中独有的，用来忽略结果。其用法如下：

```
import "database/sql"
import _ "github.com/go-sql-driver/mysql"
```

在 Go 语言中，import 的作用是导入其他包（与 Python 的作用一样），以上代码中的第二个 import 就是不直接使用 mysql 包，只执行这个包中的 init 函数，把 mysql 的驱动注册到 sql 包中，然后在程序中就可以使用 sql 包来访问 mysql 数据库了。

如果占位符“_”出现在代码主体中，表示忽略此变量。

```
package main

import (
    "os"
)

func main() {
    buf := make([]byte, 1024)
    f, _ := os.Open("/Users/yuhongchun/Desktop/text.txt")
    // 如 os.Open，返回值为 *os.File, error，这里用到了占位符，则表示忽略 error 变量
    defer f.Close()
    for {
        n, _ := f.Read(buf)
```

```
        if n == 0 {
            break

        }
        os.Stdout.Write(buf[:n])
    }
}
```

2.2.4　Go 语言的指针

区别于 C 语言中的指针，Go 语言中的指针不能进行偏移和运算，是安全指针。下面介绍 Go 语言中与指针有关的 3 个概念：指针取值、指针地址和指针类型。

1. Go 语言中的指针和指针取值

Go 语言中的函数传参都是值复制，当需要修改某个变量时，可以创建一个指向该变量地址的指针变量。传递数据使用指针，而无须复制数据。Go 语言中的指针操作非常简单，只需记住两个符号：&（取地址）和 *（根据地址取值）。

2. 指针地址和指针类型

每个变量在运行时都拥有一个地址，这个地址代表变量在内存中的位置。在 Go 语言中，将 & 字符放在变量前面表示对变量进行“取地址”操作。Go 语言中的值类型（int、float、bool、string、array、struct）都有对应的指针类型，如 *int、*int64、*string 等。

下面用一个简单的例子进行说明，代码如下：

```
package main

import "fmt"

func main() {
   a := 10
   b := &a
   fmt.Printf("a:%d ptr:%p\n", a, &a)       // a:10 ptr:0x1400011a018
   fmt.Printf("b:%p type:%T\n", b, b)       // b:0x1400011a018 type:*int
   fmt.Println(&b)                          // 0x14000120018

}
```

总结，取地址操作符 & 和取值操作符 * 是一对互补操作符。其中，& 取出地址，* 根据地址取出地址指向的值。

再看一个例子，代码如下：

```
func main() {
    var a *int
```

```
    *a = 100
    fmt.Println(*a)

    var b map[string]int
    b["test"] = 100
    fmt.Println(b)
}
```

报错如下：

```
panic: runtime error: invalid memory address or nil pointer dereference
[signal SIGSEGV: segmentation violation code=0x2 addr=0x0 pc=0x105003bc0]
```

执行上面的代码会引发 panic，这是因为在 Go 语言中对于引用类型的变量，在使用时不仅需要声明，还要为它分配内存空间，否则值就没办法存储。而对于值类型的变量，在声明时不需要分配内存空间，这是因为它们在声明时已经默认分配好了内存空间。要分配内存，就要用到前面提到的 new 函数和 make 函数。Go 语言中的 new 和 make 是两个内置函数，主要用来分配内存。

下面依次介绍 new 函数和 make 函数的用法及差异。

new 是一个内置的函数，其用法如下：

```
func new(Type) *Type
```

在以上代码中，Type 表示类型，new 函数只接收一个参数，这个参数是一个类型；*Type 表示类型指针，new 函数返回一个指向该类型内存地址的指针，这里用实际例子来说明：

```
func main() {
    a := new(int)
    b := new(bool)
    fmt.Printf("%T\n", a)           // *int
    fmt.Printf("%T\n", b)           // *bool
    fmt.Println(*a)                 // 0
    fmt.Println(*b)                 // false
}
```

new 函数不太常用，使用 new 函数得到的是一个类型指针，并且该指针对应的值为该类型的零值。

make 函数也用于内存分配，区别于 new 函数，它只用于 slice、map 以及 channel 的内存创建，而且它返回的类型就是这三个类型本身，而不是它们的指针类型。因为这三个类型就是引用类型，所以没有必要返回它们的指针。make 函数的用法如下：

```
func make(t Type, size ...IntegerType) Type
```

make 函数是无可替代的，在使用 slice、map 和 channel 时，必须使用 make 函数进行初始化，然后才可以对它们进行操作。

下面将前面报错的例子用 make 函数进行修改：

```
func main() {
	b := make(map[string]int, 10)
	b["test"] = 100
	fmt.Println(b) // map[test:100]
}
```

综上所述，new 函数和 make 函数具有以下相同点及区别：

（1）二者都用于内存分配。

（2）make 函数只用于 slice、map 和 channel 的初始化，返回的还是这三个引用类型本身。

（3）new 函数用于类型的内存分配，并且内存对应的值为类型零值，返回的是指向类型的指针。

2.2.5　Go 语言的 goroutine 和 channel

Go 语言的并发实现，主要依靠 goroutine 和 channel。goroutine 与 channel 搭配可以实现基于消息通信的 CSP 模型。下面可以看一个 Go 语言并发的例子，代码如下：

```
package main

import (
  "fmt"
  "time"
)


func HelloWorld() {
  fmt.Println("hello,world")
  time.Sleep(10*time.Second)
}

func main() {
  for i:=0;i<10;i++ {
    go HelloWorld()
  }
  time.Sleep(2*time.Second)
  fmt.Println("The end!")
}
```

上面的代码会并发输出 10 次“hello,world”，正因为有了 goroutine 和 channel，Go 语言并发编程才会如此轻松，后面的章节将重点讲述 goroutine 和 channel 的用法。

2.3 Go 语言的基础概念

本节从最基础的概念讲起，即 Go 语言的变量和常量。

在数据概念中，变量表示没有固定值且可以改变的数。但从计算机系统实现角度来看，变量是一段或多段用来存储数据的内存。

2.3.1 变量

使用变量需要两个过程：声明变量、变量赋值。声明变量也常被称为“定义变量”。变量声明后必须要使用，否则会报错。

定义变量的常用方式如下：

```
var type
```

关键字 var 用于定义变量，与 C 语言不同，类型被放在变量名后面。

这里举例说明下，代码如下：

```
var a int
var b bool
var c string
//或者
var (
    a int
    b bool
    c string
)
```

声明变量和变量赋值可以结合使用，代码如下：

```
var a int = 15
var i = 5
var b bool = false
var str string = "Hello World"
```

（1）简短模式。除关键字 var 以外，Go 语言还可以使用简短的定义变量和初始化语法，用法如下：

```
func main() {
     x := 100
     a,b := 1,2
}
```

使用简短模式给变量赋值比较方便，但是也要注意这种使用方法也是有限制的：

1）只能用在函数内部（即有作用域限制）。

2）不能提供数据类型。

3）定义变量，同时显式初始化。

（2）变量的作用域。Go 语言的变量作用域采用的是词法作用域，表示定义文本段的位置决定了可看见的值范围。Go 语言的变量作用域有以下特点：

1）定义在函数内部的变量为局部变量，只在函数内部可见。

2）定义在代码块 {} 内的变量也是局部变量，出了代码块就消失。

3）定义在代码块外部和函数外部的变量为包变量或者全局变量，它们可以被同一个目录下同一个包的多个文件访问（因为 Go 语言中一个目录下只能定义一个包，但一个包可以分成多个文件）。

4）如果变量的名称以小写字母开头，则其他包不能访问该变量。

5）如果变量的名称以大写字母开头，则其他包可以访问该变量。

6）不同作用域的变量名可以相同，但建议采取名称唯一的方式为变量命名。

事实上，关于变量的作用域问题，用户在后续的开发工作中会经常遇到，初学者在开始写 Go 语言代码时都会遇到这些问题，后续熟练了就会掌握这些要点了。

2.3.2　常量

常量包含不会发生更改的数据。常量的数据类型只能是 boolean、number（int/float/complex）或 string。

常量的定义方式如下：

```
const NAME [TYPE] = VALUE
```

其中，TYPE 基本可以省略，因为常量都是简单数据类型，编译器可以根据值推断出它的数据类型。例如：

```
const PI = 3.14159
```

还可以一次性定义多个常量，代码如下：

```
const a,b,c = "meat", 2, "veg"
const Mon, Tues, Wednes = 1, 2, 3
const (
    Mon, Tue, Wednes = 1, 2, 3
    Thurs, Fri, Satur = 4, 5, 6
)
```

2.3.3　枚举

枚举是一个被命名的整型常数的集合，枚举在日常生活中很常见。例如，“星期”这个词就是一个枚举，星期一、星期二、星期三、星期四、星期五、星期六、星期日就是这个枚举中的成员。

Go 语言并没有明确意义上的枚举类型，但是如果是整型常量，可以使用 iota 常量计数器借助 iota 标识符来实现一组自增常量值，从而实现枚举类型。具体用法如下：

```
package main

func main() {
```

```
    const a = iota                          // a=0
    const b = iota + 3                      // b=3
    const c, d = iota, iota + 3             // c=0,d=3
    const(
            e = iota                        // e=0
            f = iota + 4                    // f=5
            g                               // g=6
    )
    println(a, b, c, d, e, f, g)
}
```

还可以使用类型别名，让常量看起来更直观：根据类型就能明确知道该常量是哪种枚举。代码如下：

```
type Gender = string

const (
   Male   Gender = "男性"
   Female Gender = "女性"
)
```

通过类型别名可以很直观、很明确地知道 Male 和 Female 两个常量是性别枚举。

2.3.4 Go 语言的基本数据类型

Go 语言的基本数据类型见表 2.1。

表2.1 Go语言的基本数据类型

类 型	长 度	默认值	说 明
bool	1	false	
byte	1	0	uint8
rune	4	0	Unicode Code Point, int32
int, uint	4 或 8	0	32 位或 64 位
int8, uint8	1	0	-128~127, 0~255，byte 是 uint8 的别名
int16, uint16	2	0	-32768~32767, 0~65535
int32, uint32	4	0	-21 亿 ~21 亿, 0 亿 ~42 亿，rune 是 int32 的别名
int64, uint64	8	0	
float32	4	0.0	
float64	8	0.0	
complex64	8		
complex128	16		
uintptr	4 或 8		用于存储指针的 uint32 或 uint64 整数
array			数组

续表

类　型	长　度	默认值	说　明
struct			结构体
string		""	UTF-8 字符串
slice		nil	切片，引用类型
map		nil	字典，引用类型
channel		nil	通道，引用类型
interface		nil	接口
function		nil	函数

这里需要注意的是，Go 语言是强制类型转换。除常量及未命名类型外，Go 语言强制要求使用显式类型转换。

隐式转换造成的问题远大于它带来的好处。隐式转换，也称为自动转换，是指在程序执行过程中，编译器自动进行的类型转换。

表达式的格式为 T(x)，其中 T 为变量 x 要转换的最终类型，适用于整数类型与浮点类型互转、字节数组与字符串类型互转、结构体类型转换为接口类型（反之不行）。

使用 strconv 包提供的方法：

- strconv.Atoi(): 将字符串类型转换成整数类型。
- strconv.Itoa(): 将整数类型转换成字符串类型。

2.4　Go 语言的数据结构

下面依次从 string（字符串）、array（数组）、slice（切片）及 struct（结构体）几个方面来简单说明 Go 语言的数据结构。

2.4.1　string

string 是 Go 语言中的基础数据类型，是一个不可变对象，string 的底层是 byte 数组，每个 string 其实只占用两个机器字长：一个指针和一个长度。只不过这个指针在 Go 语言中完全不可见，所以可以这么总结，string 是一个底层 byte 数组的值类型而非指针类型，其结构如下：

```
type stringstruct struct {
    str unsafe.Pointer
    len int
}
```

声明 string 类型变量非常简单，常见的方式有以下两种：

（1）声明一个空字符串后再赋值。

```
var s string
s = "hello world"
```

需要注意的是，空字符只是长度为 0，但不是 nil。不存在值为 nil 的 string。

（2）使用简短变量声明。

```
func main() {
    s := "hello world"        // 直接初始化字符串
}
```

在实际开发工作中，string 用得最多的地方还是配合 Go 语言正则进行业务开发，Go 语言正则会在后续章节中介绍。

在实际项目中，数据经常需要在 string 和字节 []byte 之间切换。

（1）[]byte 转 string，代码如下：

```
func ByteToString(){
  b:=[]byte{'h','e','l','l','o'}
  s:=string(b)
  fmt.Println(s)                //hello
}
```

（2）string 转 []byte，代码如下：

```
func StringToByte(){
  s := "hello"
  b := []byte(s)
  fmt.Println(b)
}
```

需要注意的是，无论是 string 转成 []byte，还是 []byte 转成 string，都将发生一次内存复制，会有一定的性能开销。正因为 string 和 []byte 之间的转换非常方便，所以在某些高频场景中往往会成为性能的瓶颈，如数据库访问、HTTP 请求处理等。需要处理大量字符串时用 []byte，性能会好很多。

实际上，在项目开发需求中，会经常需要将 int 类型转换成 string 类型。前面提到过，Go 语言中的 int 类型转 string 类型是强类型转换，可以用 strconv 包来实现，其语法如下：

```
str := strconv.Itoa(intvar)
```

这里写个简单的例子，代码如下：

```
package main
import (
"fmt"
"strconv"
)

func main() {
      num := 65
      str := strconv.Itoa(num)
```

```
        //str := string(num)
        fmt.Printf("%v, %T\n", str, str) // 65, string
}
```

字符串是字符数组，如果字符串中全是ASCII字符，则直接遍历即可，但如果字符串中包含多字节字符（如中文），则可以用[]rune(str)将其转换后再遍历。

代码如下：

```
package main
import "fmt"

func main() {
    str := "Hello 你好 "
    r := []rune(str)          // 8
    for i := 0; i < len(r); i++ {
        // 输出 unicode 字符
        fmt.Printf("%c", r[i])
    }
}
```

对于字符串的操作，可以用Go语言的几个内置包来处理：

- strings：提供搜索、比较、切分与字符串连接等功能。
- bytes：如果要对字符串的底层字节进行操作，可以使用[]bytes类型转换后进行处理。
- strconv：主要是字符串与其他类型的转换，如整型、布尔型等。

2.4.2 array

Python语言的array（数组）是动态的，可以随需求自动增大数组长度。但Go语言中的数组是固定长度的，数组一经声明，就无法扩大、缩减数组的长度。Go语言数组具备如下特点：

- 数组定义：var a [len]int。例如，var a[5]int，数组长度必须是常量，且是类型的组成部分。一旦定义，长度不能变。
- 长度是数组类型的一部分，因此，var a[5] int和var a[10]int是不同的类型。
- 访问越界，如果下标在数组合法范围之外，则触发访问越界，会产生panic错误。
- 数组是值类型，赋值和传参会复制整个数组，而不是指针。因此改变复制副本数组的值，不会改变数组本身的值。

（1）数组的定义，用法如下：

```
var 数组变量名 [数组数量]T
```

初始化数组时，可以使用初始化列表来设置数组元素的值，用法如下：

```
var arr0 [5]int = [5]int{1, 2, 3}
var arr1 = [5]int{1, 2, 3, 4, 5}
var arr2 = [...]int{1, 2, 3, 4, 5, 6}
var str = [5]string{3: "hello world", 4: "tom"}
```

举例如下：

```
package main

import (
    "fmt"
)

var arr0 [5]int = [5]int{1, 2, 3}
var arr1 = [5]int{1, 2, 3, 4, 5}
var arr2 = [...]int{1, 2, 3, 4, 5, 6}
var str = [5]string{3: "hello world", 4: "tom"}

func main() {
    a := [3]int{1, 2}                   // 未初始化的元素值为 0
    b := [...]int{1, 2, 3, 4}           // 通过初始化值确定数组长度
    c := [5]int{2: 100, 4: 200}         // 使用冒号初始化元素
    d := [...]struct {
        name string
        age  uint8
    }{
        {"user1", 10},                  // 可省略元素类型
        {"user2", 20},                  // 别忘了最后一行的逗号
    }
    fmt.Println(arr0, arr1, arr2, str)
    fmt.Println(a, b, c, d)
}
```

（2）数组的遍历。遍历数组有两种方法：for 循环遍历和 for range 遍历。下面举例说明：

```
package main
import "fmt"

func main() {
   a := [...]struct {
          name string
          age  uint8
   }{
          {"user1", 10},                  // 可省略元素类型
```

```
        {"user2", 20},                      // 别忘了最后一行的逗号
    }

    // 方法 1: for 循环遍历
    for i := 0; i < len(a); i++{
            fmt.Println(a[i])               // {user1 10} {user2 20}

    }

    // 方法 2: for range 遍历
    for index, value := range a{
            fmt.Println(index, value)       //0 {user1 10} 1 {user2 20}
    }
}
```

2.4.3 slice

slice 是 Go 语言中重要的数据结构之一。

slice 是数组的一个引用，因此切片是引用类型。但切片本身是结构体，以值复制传递，其特点有：

（1）切片的长度可以改变，因此切片是一个可变的数组。

（2）切片的遍历方法和数组一样，可以用 len 函数求长度。这个长度表示切片中可用元素的数量，进行读写操作时不能超过该限制。

（3）cap 函数可以求出 slice 的最大扩张容量，且不能超出数组限制。0 <= len(slice) <= len(array)，其中 array 是 slice 引用的数组。

（4）如果 slice == nil，那么 len 函数和 cap 函数的结果都等于 0。

因为数组是固定长度且按值传递，很不灵活，所以在 Go 程序中很少用到数组。然而 slice 无处不在，slice 以数组为基础提供强大的功能。

切片是对数组的一个连续片段的引用，所以切片是一个引用类型（因此更类似于 C/C++ 中的数组类型，或者 Python 中的 list 类型），这个片段可以是整个数组，也可以是由起始和终止索引标识的一些项的子集。需要注意的是，终止索引标识的项不包括在切片内。Go 语言中切片的内部结构包含地址、大小和容量，一般用于快速操作一块数据集合，如果将数据集合比作切糕，切片就是要切的“那一块”，切的过程包含从哪里开始（切片的起始位置）及切多大（切片的大小），容量可以理解为装切片的口袋大小。

那么 len（长度）与 cap（容量）的区别在哪里呢？

首先切片是动态的数组，所以才会有 cap 这个属性。cap 是指底层数组的大小，len 是指可以使用的大小。cap 主要用于在使用 append 函数添加切片的元素时判断是否需要更换底层数组。如果新的长度小于 cap，则不会更换底层数组；否则，Go 会重新申请一个底层数组，复制原来数组

的值，然后把原来的数组丢掉。也就是说，cap 是在数据复制和内存申请的消耗与内存占用之间提供一个权衡。

而 len，则是为了帮助限制切片可用成员的数量，提供边界查询。所以用 make 函数申请空间后，需要注意不要越界（即超过 len 数量），不然会产生 panic 错误。

（1）切片的定义。代码如下：

```
type slice struct {
    array unsafe.Pointer
    len int
    cap int
}
```

（2）切片的初始化。代码如下：

```
var arr = [...]int{0, 1, 2, 3, 4, 5, 6, 7, 8, 9}
var slice0 []int = arr[start:end]
var slice1 []int = arr[:end]
var slice2 []int = arr[start:]
var slice3 []int = arr[:]
var slice4 = arr[:len(arr)-1]       // 去掉切片的最后一个元素
```

可以从 slice 中继续切片生成一个新的 slice，这样能实现 slice 的缩减。对 slice 进行切片的语法如下：

```
slice[x:y]
slice[x:y:z]
```

其中，x 表示从 slice 的第几个元素开始切；y 表示控制切片的长度（y-x）；z 表示控制切片的容量（z-x）。如果没有给定 z，则表示切到底层数组的尾部。注意，截取时要遵循“左闭右开”的原则。

例如，slice[1:3]，表示新的 slice 是从 slice 的 index=1 处开始截取，截取到 index=3 为止，但不包括 index=3 这个元素。因此，新的 slice 是由 my_slice 中的第 2 个元素、第 3 个元素组成的数据结构，长度为 2。

这里举例子说明如下：

```
package main
import "fmt"

var arr = [...]int{0, 1, 2, 3, 4, 5, 6, 7, 8, 9}
var slice0 []int = arr[2:8]
var slice1 []int = arr[0:6]        // 可以简写为 var slice []int = arr[:end]
var slice2 []int = arr[5:10]       // 可以简写为 var slice[]int = arr[start:]
var slice3 []int = arr[0:len(arr)] // var slice []int = arr[:]
var slice4 = arr[:len(arr)-1]      // 去掉切片的最后一个元素
func main() {
    fmt.Printf("arr %v\n", arr)    // arr [0 1 2 3 4 5 6 7 8 9]
```

```
    fmt.Printf("slice0 %v\n", slice0)     // slice0 [2 3 4 5 6 7]
    fmt.Printf("slice1 %v\n", slice1)     // slice1 [0 1 2 3 4 5]
    fmt.Printf("slice2 %v\n", slice2)     // slice2 [5 6 7 8 9]
    fmt.Printf("slice3 %v\n", slice3)     // slice3 [0 1 2 3 4 5 6 7 8 9]
    fmt.Printf("slice4 %v\n", slice4)     // slice4 [0 1 2 3 4 5 6 7 8]
}
```

除了可以使用这些方法来创建切片外，还可以使用 make 函数。代码如下：

```
// 创建一个 length 和 cap 都等于 5 的 slice
slice := make([]int,5)

// 创建一个 length=3, cap=5 的 slice
slice := make([]int,3,5)
```

make 函数比 new 函数多一些操作，new 函数只进行内存分配并做默认的赋零值初始化，而 make 函数可以先为底层数组分配好内存，然后从这个底层数组中再额外生成一个 slice 并初始化。另外，make 函数只能构建 slice、map 和 channel 这 3 种数据结构的数据对象，因为它们都指向底层数据结构，所以都需要先为底层数据结构分配好内存并初始化。

Go 语言中的 slice 依赖于数组，它的底层就是数组，所以数组具有的优点 slice 都有，并且 slice 支持通过 append 函数向其中追加元素，长度不够时会动态扩展。通过再次切片，可以得到更小的 slice 结构，可以迭代、遍历等。另外，虽然 slice 具有数组的所有优点，Go 语言中的数组也不是完全没有优势，其相对于 slice 的优势在于：

（1）数组是值对象，可以进行比较，数组可以作为 map 的映射键。而 slice 无法作为 map 的映射键。

（2）数组有编译安全的检查，可以在开始就避免越界行为。切片是在运行时会出现越界的 panic，越界出现的阶段不同。

（3）数组可以更好地控制内存布局，若用 slice 替换，会发现不能直接在带有 slice 的结构中分配空间，但数组可以。

（4）在访问单个元素时，数组的性能比 slice 好。

（5）数组的长度是其类型的一部分，在特定场景下具有一定的意义。

（6）数组是 slice 的基础，每个数组都可以是一个 slice，但并非每个 slice 都可以是一个数组。如果值是固定大小，可以使用数组来获得较小的性能提升（至少节省 slice 头部占用的空间）。

在两个 slice 之间复制数据，复制长度以 len 小的为准。如果两个 slice 可指向同一底层数组，则允许元素区间重叠。这表示将 src slice 复制到 dst slice 时，如果 src 比 dst 长，就截断；如果 src 比 dst 短，则只复制 src 那部分。copy 函数的返回值是复制成功的元素数量，也就是 src slice 或 dst slice 中最小的那个长度。

copy 函数的语法定义如下：

```
func copy(dst, src []Type) int
```

需要注意以下问题：

（1）copy 函数只能用于 slice，不能用于 map 等任何其他类型。

（2）copy 函数的返回结果为一个 int 类型值，表示复制的长度。

举例说明如下：

```
package main
import (
  "fmt"
)

func main() {
      s := []int{1,2,3,4,5,6,7,8,9}
      s1 := s[5:8] // [6 7 8]
      n := copy(s[4:],s1)      // 在同一底层数据的不同区间复制
      fmt.Println(n,s)

      s2 :=make([]int,6)      // 在不同数组之间复制，s2 的值为 [0 0 0 0 0 0]
      n = copy(s2,s)
      fmt.Println(n,s2)
}
```

输出结果如下：

```
3 [1 2 3 4 6 7 8 8 9]
6 [1 2 3 4 6 7]
```

使用 copy（dst, src [] type）函数时需要注意以下问题：

（1）切片 dst 需要先初始化长度。前面已经提到过，不是定义好类型，就能将 src 完全复制到 dst 的，还需要初始化长度。如果 dst 的长度小于 src 的长度，则复制部分；如果大于，则全部复制过来，只是没占满 dst 的坑位而已；相等时则刚好完全复制过来。

（2）原切片中的元素类型为引用类型时，复制的是引用。由于只复制 slice 中的元素，因此，如果切片中元素的类型是引用类型，则复制的也将是个引用。

向 slice 尾部添加数据，返回新的 slice 对象。这里举个简单的例子进行说明，代码如下：

```
func main() {
    s := make([]int,0,3)
    s1 := append(s,10)
    s2 := append(s1,20,30)

    fmt.Println(s,len(s),cap(s)) // [] 0 3
    fmt.Println(s1,len(s),cap(s)) // [10] 0 3
    fmt.Println(s2,len(s),cap(s)) // [10 20 30] 0 3
}
```

如果超出原 slice.cap 限制，就会重新分配底层数组，即便原数组并未填满。下面举例说明：

```
package main

import (
"fmt"
)

func main() {

	data := [...]int{0, 1, 2, 3, 4, 10}
	s := data[:2]
	fmt.Println(s, data)
	s = append(s, 100, 200)        // 一次 append 两个值，超出 s.cap 限制

	fmt.Println(s,data)
	fmt.Println(s[0], data[0])     // 重新分配底层数组，与原数组无关
	fmt.Println(&s, &data)         // 比对底层数组的起始指针
}
```

输出结果如下：

```
[0 1] [0 1 2 3 4 10]
[0 1 100 200] [0 1 100 200 4 10]
0 0
&[0 1 100 200] &[0 1 100 200 4 10]
```

Go 语言在 append 函数和 copy 函数方面的开销是可预知和可控的，在简单的程序上应用有很好的效果。

在实际的 DevOps 开发工作中，有很多地方需要使用 slice 的组合操作，这样可以将相关的函数封装成一个 common 包（即文件），以方便需要时直接调用。部分代码如下：

```
//求两个 slice 的并集
func UnionJob(slice1, slice2 []string) []string {
	m := make(map[string]int)
	for _, v := range slice1{
		m[v]++
	}

	for _, v := range slice2{
		times, _ := m[v]
		if times == 0{
			slice1 = append(slice1, v)
		}
	}
```

```
        return slice1
}

//求两个 slice 的交集
func IntersectJob(slice1, slice2 []string) []string {
        m := make(map[string]int)
        nn := make([]string, 0)
        for _, v := range slice1{
                m[v]++
        }

        for _, v := range slice2{
                times, _ := m[v]
                if times == 1{
                        nn = append(nn, v)
                }
        }
        return nn
}

//两个 slice 相减
func Diffjob(a, b []string) []string {
        var(
                r []string
                m = make(map[string]struct{}, len(a))
        )

        for _, v := range b{
                m[v] = struct{}{}
        }

        for _, v := range a{
                if _, ok := m[v]; !ok {
                        r = append(r, v)
                }
        }

        return r
}

// InSlice 函数用于判断字符串是否在 slice 中
```

```
func InSlice(items []string, item string) bool {
        for _, eachItem := range items{
                if eachItem == item{
                        return true
                }
        }
        return false
}
```

2.4.4 map

map（字典）是 Go 语言中重要的数据结构之一，是一种无序的基于 key-value 的数据结构。Go 语言中的 map 是引用类型，必须初始化才能使用。

Go 语言中 map 的定义语法如下：

```
map[KeyType]ValueType
```

其中，KeyType 表示键的类型；ValueType 表示键对应的值的类型。

map 类型的变量默认初始值为 nil，需要使用 make 函数来分配内存。语法如下：

```
make(map[KeyType]ValueType, [cap])
```

其中，cap 表示 map 的容量，该参数虽然不是必需的，但是应该在初始化 map 时就为其指定一个合适的容量。在创建时预先准备足够的容器有助于提升性能，减少扩张时的内存分配和重新哈希操作。

1. map 的基本使用

map 中的数据都是成对出现的，其基本使用示例代码如下：

```
func main() {
    s := make(map[string]int, 8)
    s["a"] = 9
    s["b"] = 10
    fmt.Println(s)                     // map[a:9 b:10]
    fmt.Println(s["a"])                // 9
    fmt.Printf("type of a:%T\n", s)   // type of a:map[string]int
}
```

2. 判断某个键是否存在

Go 语言中有一种判断 map 中的某个键是否存在的特殊写法，格式如下：

```
value, ok := map[key]
```

下面举个简单的例子进行说明：

```
func main() {
    scoreMap := make(map[string]int)
```

```
    scoreMap["张三"] = 90
    scoreMap["小明"] = 100
    // 如果key存在，则ok为true，v为对应的值；如果key不存在，则ok为false，v为值类型的零值
    v, ok := scoreMap["张三"]
    if ok {
        fmt.Println(v)
    } else {
        fmt.Println("查无此人")
    }
}
```

3. map 的遍历

Go 语言中使用 for range 遍历 map，由于 map 是无序的，因此每次返回的键值结果都是不一样的。

```
func main() {
      s := make(map[string]int, 8)
      s["a"] = 9
      s["b"] = 10
      s["c"] = 1
      for k,v := range s{
            fmt.Println(k,":",v,"") // map是无序的，所以每次输出的结果都不一样
      }
}
```

输出结果如下：

```
b : 10
c : 1
a : 9
```

在实际开发中使用 Go 语言中的 map 时，经常会遇到如下报错：

```
concurrent map read and map write
```

具体原因为：在 Go 语言中，map 并不是线程安全的，如果一个 goroutine 协程正在对字典进行写操作，那么其他 goroutine 协程不能对该字典进行并发操作（读、写、删除），否则会导致进程崩溃。需要使用 sync.Lock 或 sync.RWLock，推荐用后者，其性能更好。下面举个简单的例子复现，代码如下：

```
package main

import (
     "time"
)
```

```
func main() {
        var m = make(map[string]int)
        go func(){
                for{
                        m["a"] += 1          // 写操作
                        time.Sleep(time.Microsecond)
                }
        }()

        go func(){
                for{
                        _ = m["b"]           // 读操作
                        time.Sleep(time.Microsecond)
                }
        }()

        select{}                             // 阻塞主进程
}
```

下面启用数据竞争来检查 map 的线程安全问题，它会输出详细的检测信息。命令如下（测试文件名为 6.go）：

```
go run -race 6.go
```

输出结果如下：

```
==================
WARNING: DATA RACE
Read at 0x00c00011c060 by goroutine 6:
  runtime.evacuate_fast32()
      /usr/local/go/src/runtime/map_fast32.go:374 +0x38c
  main.main.func2()
      /Users/yuhongchun/repo/workspace/cloudnative/woking/6.go:18 +0x40

Previous write at 0x00c00011c060 by goroutine 5:
  runtime.mapaccess1_faststr()
      /usr/local/go/src/runtime/map_faststr.go:13 +0x40c
  main.main.func1()
      /Users/yuhongchun/repo/workspace/cloudnative/woking/6.go:11 +0x70

Goroutine 6 (running) created at:
  main.main()
      /Users/yuhongchun/repo/workspace/cloudnative/woking/6.go:16 +0xc4
```

```
Goroutine 5 (running) created at:
  main.main()
      /Users/yuhongchun/repo/workspace/cloudnative/woking/6.go:9 +0x74
==================
fatal error: concurrent map read and map write

goroutine 18 [running]:
main.main.func2()
        /Users/yuhongchun/repo/workspace/cloudnative/woking/6.go:18 +0x44
created by main.main
        /Users/yuhongchun/repo/workspace/cloudnative/woking/6.go:16 +0xc8

goroutine 1 [select (no cases)]:
main.main()
        /Users/yuhongchun/repo/workspace/cloudnative/woking/6.go:23 +0xcc

goroutine 17 [sleep]:
time.Sleep(0x3e8)
        /usr/local/go/src/runtime/time.go:195 +0x118
main.main.func1()
        /Users/yuhongchun/repo/workspace/cloudnative/woking/6.go:12 +0x98
created by main.main
        /Users/yuhongchun/repo/workspace/cloudnative/woking/6.go:9 +0x78
exit status 2
```

也可以使用读写锁来解决此问题，将前面的代码更改如下：

```
package main

import (
      "fmt"
      "sync"
      "time"
)

func main() {
      var lock sync.RWMutex
      var m = make(map[string]int)
      go func(){
            for{
                  lock.Lock()
                  m["a"] += 1         // 写操作
```

```
                lock.Unlock()
                time.Sleep(time.Microsecond)
                fmt.Println(m)
            }
        }()

        go func(){
            for{
                lock.RLock()
                _ = m["b"] // 读操作
                lock.RUnlock()
                time.Sleep(time.Microsecond)
            }
        }()

        select{}                // 阻塞主进程，可以按快捷键 Ctrl+C 终止程序
}
```

程序运行以后会一直循环下去，需要按快捷键 Ctrl+C 来终止程序，输出结果如下（输出结果太多，已做截断处理）：

```
map[a:493747]
map[a:493748]
map[a:493749]
map[a:493750]
map[a:493751]
map[a:493752]
map[a:493753]
map[a:493754]
map[a:493755]
map[a:493756]
map[a:493757]
map[a:493758]
map[a:493759]
map[a:493760]
map[a:493761]
map[a:493762]
map[a:493763]
map[a:493764]
map[a:493765]
map[a:493766]
map[a:493767]
```

```
map[a:493768]
map[a:493769]
map[a:493770]
map[a:493771]
map[a:493772]
map[a:493773]
map[a:493774]
map[a:493775]
map[a:493776]
map[a:493777]
map[a:493778]
map[a:493779]
map[a:493780]
map[a:493781]
map[a:493782]
map[a:493783]
^Cmap[a:493827]
map[a:493828]
map[a:493829]
signal: interrupt
```

需要注意的是，在实际业务开发场景中，map的线程安全是经常遇到的问题。

2.4.5 struct

struct也是Go语言中重要的数据结构之一。Go语言中的基本数据类型可以表示对象的基本属性，但是当想表示一个对象的全部或部分属性时，单一的基本数据类型明显已无法满足需求。因此，Go语言提供了一种自定义数据类型，可以封装多个基本数据类型。这种数据类型称为struct（结构体），通过struct可以定义自己的数据类型。

下面使用type和struct关键字来定义结构体，其语法如下：

```
type 类型名 struct {
     字段名 字段类型
     字段名 字段类型
…
}
```

其中，类型名表示自定义结构体的名称，在同一个包内不能重复；字段名表示结构体字段名，结构体中的字段名必须唯一；字段类型表示结构体字段的具体类型。

下面定义一个person结构体，代码如下：

```
type person struct {
    name string
    city string
```

```
    age  int8
}
```

以上代码定义了一个名为 person 的自定义类型，它有 name、city、age 三个字段，分别表示姓名、居住城市和年龄。使用这个 person 结构体就能够在程序中表示和存储人的信息了。

语言内置的基本数据类型用来描述一个值，而结构体用来描述一组值。例如，一个人有名字、年龄和居住城市等，本质上是一种聚合型的数据类型。

1. struct 实例化的方式

struct 有以下 3 种实例化的方式。

（1）用 var 实例化，代码如下：

```
type person struct{
    age   int
    name  string
}

func main()  {
    var  fan person
    fan.age = 10
    fan.name = "fan"
}
```

（2）用 new 关键字实例化，代码如下：

```
type person struct{
    age  int
    name  string
}

func main()  {
    fan := new(person)              // new 返回一个指针
    fmt.Printf("%T\n", fan)         // *main.person
    fan.name = "fan"
    fan.age = 26
}
```

（3）命名初始化。代码如下：

```
func main()  {
    fan := person{
    name: "fan",
    age:  10,
}
```

前两种实例化方式通过“.”来访问结构体的字段（成员变量），如 fan.name 和 fan.age。这里推荐用命名初始化的方式，好处是：在扩充结构字段或调整字段顺序时，不会导致初始化语句出错。

2. 嵌套匿名结构体

当匿名字段是一个 struct 时，那么这个 struct 所拥有的全部字段都被隐式地引入当前定义的 struct 中。示例代码如下：

```
package main
import "fmt"
type Human struct {
     name string
     age int
     weight int
}
type Student struct {
      Human  // 匿名字段，默认 Student 包含 Human 的所有字段
      speciality string
}

func main() {
      mark := Student{Human{"Mark", 25, 120}, "Computer Science"}
                                                          // 初始化一名学生
      fmt.Println("His name is", mark.name)               // 访问相应的字段
      fmt.Println("His age is", mark.age)
      fmt.Println("His weight is", mark.weight)
      fmt.Println("His speciality is", mark.speciality)
      mark.speciality = "AI"                              // 修改对应的备注信息
      fmt.Println("Mark changed his speciality")
      fmt.Println("His speciality is", mark.speciality)
      fmt.Println("Mark become old")
      mark.age = 46
      fmt.Println("His age is", mark.age)
      fmt.Println("Mark is not an athlete anymore")
      mark.weight += 60
      fmt.Println("His weight is", mark.weight)
}
```

程序执行结果如下：

```
His name is  Mark
His age is  25
His weight is  120
His speciality is  Computer Science
```

```
Mark changed his speciality
His speciality is  AI
Mark become old
His age is 46
Mark is not an athlete anymore
His weight is 180
```

Go 语言中通过 struct 来实现面向对象，可以类比 C/Java 语言中的 Class；但 struct 不是 Class，它有自己的特色。

3. 结构体标记

在 Go 语言中，结构体标记（struct tag）也称为结构体注释（struct annotation），它是一种对 Go 结构体中的字段进行元数据附加的机制。这些标记是用反引号 (`) 引起来的键值对，可以在运行时通过反射机制获取并处理。

结构体标记使用 key:"value" 的格式来定义。其中，key 是标记的名称，value 是标记的值。一个结构体字段可以有多个标记，每个标记之间用空格分隔。

struct tag 的应用很广泛，特别是在 json 序列化，或者在数据库 ORM 映射方面。

4. 使用方法和代码示例

下面介绍一个例子，代码如下：

```
type User struct {
   Name string `json:"name"`
   Age int `json:"age"`
}
```

注意：以上结构体中的反引号引起来的内容就是 Go 语言中的 struct tag，接下来看它的作用，如果输出 json 格式，代码如下：

```
u := &User{Name: "xiaohong", Age: "18"}
j, _ := json.Marshal(u)
fmt.Println(string(j))
```

输出结果如下：

```
{"name": "xiaohong","age": 18}
```

如果去掉 struct tag 会输出什么呢？代码如下：

```
type User struct {
  Name string
  Age int
}
u := &User{Name: "xiaohong", Age: "18"}
j, _ := json.Marshal(u)
fmt.Println(string(j))
```

输出如下内容：

```
{"Name": "xiaohong","Age": 18}
```

从输出结果可以看出，加上 struct tag 之后，输出的内容跟着发生了变化。因此 struct tag 通常用于在转换 struct 编码的过程中提供一些转换规则信息。一般来讲，struct tag 都是以 key:"value" 键值对的形式来定义的，如果有多个键值对，可以用空格分隔。代码如下：

```
type User struct {
    ID    int    `json:"id" db:"id"`
    Name  string `json:"name" db:"name"`
    Email string `json:"email" db:"email"`
}
```

5. 空结构体 struct{}

空结构体 struct{} 的宽度是 0，占用 0 字节的内存空间。空结构体具有以下特点。

- 省内存，尤其在事件通信的情况下。
- struct 零值就是本身，读取 close 的 channel 返回零值。
- 可以控制协程并发度。

下面写一个简单的例子。代码如下：

```
package main
import (
     "fmt"
     "unsafe"
)
var s struct{
     A struct{}
     B struct{}
}

func main() {
     // unsafe.Sizeof 用于计算一个数据类型实例需要占用的内存字节数
     fmt.Println(unsafe.Sizeof(s)) // 0
     fmt.Println(unsafe.Sizeof(struct{}{})) // 0
}
```

空结构体 struct{} 实例不占任何内存空间。接下来，再举个实际工作中的例子。代码如下：

```
package main
import ("fmt"
)
```

```
func main() {
	exit := make(chan struct{})
	go func(){
		fmt.Println("hello,world")
		exit <- struct{}{}
	}()

	<-exit
	fmt.Println("end")
}
```

输出结果如下：

```
hello,world
end
```

有时 channel 不需要发送任何数据，只用来通知子协程（goroutine）执行任务，或只用来控制协程并发度。在这种情况下，使用空结构体作为占位符就非常合适了。还可以用 channel+ 空结构体来控制并发度，避免过多的 goroutine 对机器负载产生不良的影响。具体示例如下：

```
package main

import (
	"log"
	"sync"
	"time"
)

func main() {
	var wg sync.WaitGroup
	ch := make(chan struct{}, 4)   // 结合实际情况定义的缓存 channel
	for i := 0; i < 29; i++{
		ch <- struct{}{}
		//fmt.Println(ch)
		wg.Add(1)
		go func(i int){
			defer wg.Done()
			log.Println(i)
			time.Sleep(time.Second * 1)
			<-ch
		}(i)
	}
	wg.Wait()
}
```

6. 利用 mapstructure 库将 map 转为 struct

mapstructure 库用于将通用的 map[string]interface{} 解码到对应的 Go 结构体中，或者执行相反的操作。一般情况下，解析来自多种源头的数据流时，一般事先并不知道这些数据流所对应的具体类型，只有读取到一些字段之后才能做出判断。这时，可以先使用标准的 encoding/json 库将数据解码为 map[string]interface{} 类型，然后根据标识字段利用 mapstructure 库将 map 转为相应的 Go 结构体以便使用。

下面用实例说明：

```
package main
import (
	"fmt"
	"github.com/mitchellh/mapstructure"
)

func main() {
	type Person struct{
		Name string `mapstructure:"person_name"`
		Age  int    `mapstructure:"person_age"`
	}
	input := map[string]interface{}{
		"person_name": "Mitchell",
		"person_age":  91,
	}
	fmt.Printf("inputtype:%T\n", input) // input type:map[string]interface {}
	var result Person
	err := mapstructure.Decode(input, &result)
	if err != nil{
		panic(err)
	}
	fmt.Printf("result type:%T\n", result) // result type:main.Person
	fmt.Printf("%#v", result)
}
```

最后输出结果如下：

```
main.Person{Name:"Mitchell", Age:91}
```

mapstructure 是非常实用的库，很多 Go 工程都会用到，读者应熟悉其用法，以在实际工作中简化代码量。

另外，Go 语言还自带了 3 个容器结构：list（双向链表）、heap（堆）和 ring（圈）。虽然官方标准库中的数据结构还是很少，但是在遇到适合场景时直接使用标准库还是比较方便的。这 3 个数据结构的实现位于 container 库下，可以根据实际业务开发需求来使用。

2.5 函　　数

函数是结构化编程的最小模块单元。Go 语言中的函数声明包含函数名、参数列表、返回值列表和函数体。如果函数没有返回值，则返回值列表可以省略。函数从第一条语句开始执行，直到执行 return 语句或者执行函数的最后一条语句。

函数的语法格式如下：

```
func funcName(input1 type1, input2 type2) (output1 type1, output2 type2) {
  //这里处理逻辑代码
  return value1, value2 //返回多个值
}
```

函数的主要特点如下：

- 关键字 func 用来声明一个函数 funcName。
- 函数可以有一个或者多个参数，每个参数后面带有类型，通过逗号分隔。上面函数的返回值声明了两个变量 output1 和 output2，type1 和 type2 为类型。
- 函数可以返回多个值，用“_”可以省略某个返回值。
- 如果只有一个返回值且不声明返回值变量，那么可以省略包括返回值的括号。
- 如果没有返回值，那么直接省略最后的返回信息。
- 如果有返回值，那么必须在函数的外层添加 return 语句。

下面举个简单的例子说明函数的特点，代码如下：

```
package main
import "fmt"

func add(a,b int) int {
    return a + b
}
func main() {
    sum := add(1,2)
    fmt.Println(sum)
}
```

Go 语言中如何调用函数?

函数在定义后，可以通过调用的方式，让当前代码跳转到被调用的函数中执行，调用前的函数局部变量都会被保存起来不会丢失，被调用的函数运行结束后，恢复到调用函数的下一行继续执行代码，之前的局部变量也能继续访问。

函数内的局部变量只能在函数体中使用，函数调用结束后，这些局部变量都会被释放并且失效。

调用函数的语法如下：

```
返回值变量列表 = 函数名（参数列表）
```

默认情况下，Go 语言使用的是值传递，即在调用过程中不会影响实际参数。

Go 语言的参数传递可以分为按值传递和按引用传递。其中的按值传递，表示传递给函数的是复制的副本，所以函数内部访问、修改的也是这个副本。例如：

```
func main() {
   a,b := 10,20
   min(a,b)
   func min(x,y int) int{}
}
```

在上面程序中调用 min 函数时，首先将 a 和 b 的值复制一份，然后将复制的副本赋值给变量 x、y。因此，在 min 函数内部访问、修改的一直是 a、b 的副本，和原始的数据对象 a、b 没有任何关系。

如果要修改函数外部的数据（上面代码中的 a、b），则需要传递指针。

代码如下：

```
package main
import "fmt"

func main() {
    a := 10
    func_value(a)
    fmt.Println(a)     // 10

    b := &a
    func_ptr(b)
    fmt.Println(*b)    // 11
}

func func_value(x int) int{
    x = x + 1
    return x
}

func func_ptr(x *int) int{
    *x = *x + 1
    return *x
}
```

注意：

- Go 语言默认都是采用值传递，即复制传递。
- 有些值的类型就是指针。例如，slice、map 和 channel 这些语言内置的类型。

2.5.1 匿名函数

匿名函数是指不需要定义函数名的一种函数实现方式。在 Go 语言中，函数可以像普通变量一样被传递或使用，Go 语言支持随时在代码中定义匿名函数。

匿名函数由一个不带函数名的函数声明和函数体组成。其优越性在于可以直接使用函数内的变量，不必声明。

这里举个简单的例子进行说明：

```
package main

import (
    "fmt"
    "math"
)

func main() {
    getSqrt := func(a float64) float64 {
        return math.Sqrt(a)
    }
    fmt.Println(getSqrt(4)) // 2
}
```

在上面的代码中，先定义了一个名为 getSqrt 的变量，初始化该变量的方式和之前的变量初始化有些不同，这里使用了 func，func 是用于定义函数的，但是这个函数和上面说的函数的最大不同就是没有函数名，也就是匿名函数。这里将函数当作变量进行操作。

2.5.2 闭包

函数 A 返回函数 B，最典型的用法就是闭包（closure）。简单地说，闭包就是“一个函数 + 一个作用域环境”组成的特殊函数。这个函数可以访问不是它自己内部的变量，也就是这个变量在其他作用域内，并且这个变量是未赋值的，即等待赋值的。

Go 语言中的闭包是引用了自由变量的函数，被引用的自由变量和函数一同存在，即使已经离开了自由变量的环境也不会被释放或者删除，在闭包中可以继续使用这个自由变量。使用闭包最初的目的是减少全局变量，在函数调用过程中隐式地传递共享变量。 这种方式所带来的结果就是不够直接、不够清晰，所以一般不建议使用闭包。

这里建议用以下公式来总结闭包的定义：

匿名函数 + 引用环境 = 闭包。

一个函数类型就像结构体一样，可以被实例化，函数本身不存储任何信息，只有与引用环境结合后形成的闭包才具有“记忆性”，函数是编译期静态的概念，而闭包是运行期动态的概念。

这里举个简单的例子，代码如下：

```
package main
import "fmt"
func test(x int) func() {
	fmt.Println(&x) // 0x1400011c018
	return func(){
		fmt.Println(&x)} // 0x1400011c018
}

func main() {
	f := test(123)
	//fmt.Println(f)
	f()
}
```

通过代码输出结果应该可以判断，操作的是同一个对象。

再举个 DevOps 开发中实际业务用到的例子，代码如下：

```
package main

import (
	"fmt"
	"log"
	"sync"
	"time"
)
// 实际的业务处理逻辑
func Downlaod_file(file_name string) {
	log.Println(file_name)
	time.Sleep(time.Second * 1)
}

func main() {
	max_concurrency := 4
	control_chan := make(chan struct{}, max_concurrency)
	for i := 0; i < max_concurrency; i++{
		control_chan <- struct{}{}
	}
```

```
    var wg sync.WaitGroup
// 这里举个简单的例子，实际应用场景是上万级别的文件数量切片
// 开启过多 goroutine 协程可能会导致程序崩溃
    file_list := []string{"aaa", "bbb", "cccc", "dddd"}

    for _, v := range file_list{
            <-control_chan
            wg.Add(1)
            go func(file_name string){
                    defer func(){
                        control_chan <- struct{}{}
                        wg.Done()
                    }()
                    fmt.Println(file_name)
                    Downlaod_file(file_name)
              }(v)
    }
    wg.Wait()

}
```

从以上代码就可以看到闭包的便利性了，在开协程时会有变量传入或者复制的情况。如果在线程中要操作某个变量，则用闭包传入会复制一份进去；否则得手动复制。如果直接引用外部变量而不使用闭包，则可能有多线程安全问题，一旦有协程对该变量进行写操作，就可能引发 panic。

闭包函数与普通函数的最大区别就是参数不是值传递，而是引用传递。

2.5.3 defer 语句在 Go 语言中的使用场景

defer 是 Go 语言中的一个关键字，用于在函数执行结束后延迟执行指定的函数调用，它是 Go 语言中特有的。

defer 语句可以完成延迟功能，在当前函数执行完成后再执行 defer 的代码块。通过 defer 语句，可以在代码中优雅地关闭 / 清理代码中所使用的变量。

在 Go 语言中，defer 语句主要用于资源管理、错误处理、简化复杂函数调用。它允许推迟到函数返回前执行某些操作，确保资源被适时释放，从而提高代码的健壮性和清晰度。其中，资源管理是 defer 语句的一个重要使用场景，它可以帮助开发者有效管理文件操作、数据库连接等资源，确保在不再需要这些资源时能够正确地释放它们，避免资源泄露问题。

1. 资源管理

在 Go 语言中处理文件操作时，确保打开的文件最终被关闭是非常重要的。使用 defer 语句可以轻松实现这一点。当用户打开一个文件进行读写操作时，可以立即使用 defer 语句调用文件的 Close 方法。这样，无论函数执行路径如何，文件都将在函数返回前被安全关闭。

例如，当打开一个文件进行读取时：

```
func readFile(filename string) error {
    f, err := os.Open(filename)
    if err != nil {
        return err
}

defer f.Close() // 确保在退出函数前关闭文件
    // 此处对文件进行读取操作
}
```

这种做法简化了错误处理和资源管理逻辑，让用户不必担心会忘记关闭文件，也使代码更加简洁和易于维护。

2. 错误处理

defer 语句在错误处理中同样扮演着重要角色。它可以在函数返回前统一处理错误，特别是在有多个返回的函数中。通过在 defer 语句中使用匿名函数，可以访问函数的返回值，从而在函数执行结束时进行错误记录或其他必要的清理工作。示例代码如下：

```
func executeQuery(query string) (result string, err error) {
   defer func() {
   if r := recover(); r != nil {
      err = fmt.Errorf("query fAIled: %v", r)
   }
}()
// 执行数据库查询操作
}
```

这段代码展示了如何使用 defer 语句来处理可能在执行数据库查询时发生的 panic，确保函数能以错误值正常返回，而不是导致程序崩溃。

3. 简化复杂函数调用

defer 语句还可以用来简化一些复杂的函数调用逻辑，特别是需要成对调用的操作，如加锁和解锁。使用 defer 语句可以保证锁在函数结束时被释放，无论函数是正常结束还是由于错误而提前退出，都能保证资源正确释放。示例代码如下：

```
func secureAccess(resource string) {
    mutex.Lock()
    defer mutex.Unlock()
    // 访问受保护的资源
}
```

在这个例子中，无论 secureAccess 函数的执行路径如何，defer 语句都保证了互斥锁在函数退出时被释放，从而避免了死锁的风险。

综上所述，defer 语句是 Go 语言提供的一种强大工具，通过推迟执行的方式，可以帮助开发者有效管理资源、处理错误并简化复杂的函数调用。恰当地使用 defer 语句不仅可以提升代码的可读性和健壮性，还能避免一些常见的编程错误。掌握 defer 语句的正确使用场景，对于任何 Go 语言开发者来说都是必备的技能。

需要注意的是，defer 语句调用在函数结束时才被执行；不合理地使用 defer 语句会浪费很多资源，甚至造成逻辑错误。

2.5.4　Go 语言中的 error

Go 语言中的错误处理与其他语言不同，它把错误当成一种值来处理，更强调判断错误、处理错误，而不是仅使用 catch 语句捕获异常。

Go 语言提供以下两种错误处理方式：

- 函数返回 error 类型对象判断错误。
- panic 异常。

Go 语言中使用 error 接口来表示错误类型。

```
type error interface {
    Error() string
}
```

error 接口只包含一个方法 Error，这个方法需要返回一个描述错误信息的字符串。

当一个函数或方法需要返回错误时，通常是把错误作为最后一个返回值。例如，下面是标准库 os 中打开文件的函数。

```
func Open(name string) (*File, error) {
   return OpenFile(name, O_RDONLY, 0)
}
```

由于 error 是一个接口类型，默认零值为 nil，因此通常将调用函数返回的错误与 nil 进行比较，以此来判断函数是否返回错误。例如，经常会在 Go 项目代码中看到类似下面的错误判断代码。

```
file, err := os.Open("./test.go")
if err != nil {
     fmt.Println("打开文件失败,err:", err)
     return
}
```

panic 异常是一个严重错误机制，它会导致程序终止。一般来说，panic 适用于重要配置文件不可读、goroutine 产生死锁、程序依赖的数据库连接不上等场景，它会直接导致程序终止。panic 异常可以人为产生，所以 Go 语言官方建议尽量不要使用 panic。

2.5.5 Go 语言单元测试

Go 语言中的测试依赖 go test 命令。编写测试代码和编写普通的 Go 语言代码的过程是类似的，不需要学习新的语法、规则或工具。

go test 命令是一个按照一定约定组织的测试代码。在包目录内，所有以 _test.go 为后缀名的源代码文件都是 go test 命令测试的一部分，不会被 go build 命令编译到最终的可执行文件中。

go test 命令会先遍历所有的 *_test.go 文件中符合上述命名规则的函数，然后生成一个临时的 main 包用于调用相应的测试函数，再构建并运行、报告测试结果，最后清理测试中生成的临时文件。

程序员对单元测试都不陌生。单元测试一般用来测试代码逻辑及其是否按照期望运行，以保证代码质量。

大多数的单元测试都是对某一个函数方法进行测试，以尽可能地保证没有问题或者问题可被预知。为了达到这个目的，可以使用各种手段、逻辑，模拟不同的场景进行测试。

其中单元测试涉及的要点如下：

（1）文件名必须以 _test.go 结尾，Go 语言测试工具只认符合这个规则的文件。

（2）方法必须以 Test[^a-z] 开头，这表明是可以导出的公共函数。

（3）测试函数的签名必须接收一个指向 testing.T 类型的指针，并且不能返回任何值。

（4）正常编译操作（go build）会忽略测试文件。

这里举个简单的例子，代码如下：

```
func Add(a,b int) int{
    return a+b
}

func TestAdd(t *testing.T) {
    sum := Add(1,2)
    if sum == 3{
            t.Log("the result is ok")
    } else {
            t.Fatal("the result is wrong")
    }
}
```

运行测试命令：

```
go test .
```

结果显示如下：

```
ok      woking/test     0.628s
```

这里再举个表组测试（以覆盖更多情况）的例子，代码如下：

```
package main
```

```
import "testing"
func Add(a,b int) int{
	return a+b
}

func TestAdd(t *testing.T) {
	//sum := Add(1,2)
	//if sum == 3{
	//	t.Log("the result is ok")
	//} else {
	//	t.Fatal("the result is wrong")
	//}
	var tests = []struct{
		x int
		y int
		exp int
	}{
		{1,2,3},
		{3,4,7},
		{7,6,13},
	}
	for _, tt := range tests{
		act := Add(tt.x,tt.y)
		if act != tt.exp{
			t.Fatal("the result is wrong")
		}
	}
}
```

运行测试命令：

```
go test .
```

如果是做基准测试，其示例代码如下：

```
package main

import (
	"testing"
)

func add(x,y int) int {
	return x+y
}
```

```
func BenchmarkAdd(b *testing.B) {
     for i:=0;i<b.N;i++{
          _ = add(1,2)
}
}
```

运行命令：

```
go test -bench . -run=none
```

输出结果如下：

```
goos: darwin
goarch: arm64
pkg: woking/test
BenchmarkAdd-8          1000000000                  0.3150 ns/op
PASS
ok      woking/test     0.453s
```

说明：运行基准测试也要使用 go test 命令，不过要加上“-bench=”标记，它接收一个表达式作为参数，匹配基准测试的函数，“.”表示运行所有基准测试。

因为默认情况下 go test 命令会运行单元测试，为了防止单元测试的输出影响查看基准测试的结果，可以使用“-run=”匹配一个不存在的单元测试方法，以过滤掉单元测试的输出，这里使用 none，因为基本上不会创建这个名字的单元测试方法。

下面着重解释以上输出结果。其中，函数后面的 -8 表示运行时对应的 GOMAXPROCS 的值；1000000000 表示运行 for 循环的次数，也就是调用被测试代码的次数；0.3150 ns/op 表示每次需要花费 0.3150 纳秒。

以上测试时间默认是 1 秒，也就是 1 秒的时间，调用一千万次，每次调用花费 0.3150 纳秒。如果想让测试运行的时间更长，可以通过“-benchtime”指定，如 3 秒。

2.6 Go 语言的流程控制

流程控制是编程语言中必不可少的一部分，是整个编程基础的重要一环，Go 语言的流程控制与其他语言类似，这里进行简单的介绍。

2.6.1 条件语句 if

1. 最简单的 if

条件语句需要开发者指定一个或多个条件，然后通过测试条件是否为 true 来决定是否执行指定语句，并在条件为 false 的情况下执行另外的语句。其语法如下（布尔表达式可以省略）：

```
  if 布尔表达式 {
    // 执行语句
    }
```

2. 更多选择的 if

可以在 if 语句后使用可选的 else 语句，else 语句中的表达式在布尔表达式为 false 时执行。其语法如下：

```
if 布尔表达式 {
   // 在布尔表达式为 true 时执行
} else {
  // 在布尔表达式为 false 时执行
}
```

3. if 嵌套语句

if 嵌套语句可以在 if 或 else if 语句中嵌入一个或多个 if 或 else if 语句。其语法如下：

```
if 布尔表达式 1 {
   // 在布尔表达式 1 为 true 时执行
   else if 布尔表达式 2 {
      // 在布尔表达式 2 为 true 时执行
   } else {
      // 不符合以上情况的执行语句
   }
}
```

这里举例说明，代码如下：

```
package main
import "fmt"

func main() {
     a := 14
     // a: = 79
     if a > 20{
            fmt.Println("a 大于 20\n")
     }else if a < 10 {
            fmt.Println("a 小于 10\n")
     } else {
            fmt.Println("a 是介于 10 与 20 之间的值 ")
     }
     fmt.Printf("a 的值是 :%d",a)
}
```

输出结果如下：

```
a 是介于 10 与 20 之间的值
a 的值是 :14
```

2.6.2 条件语句 switch

switch 语句用于基于不同条件执行不同动作，每一个 case 分支都是唯一的，从上至下逐一测试，直到匹配为止。switch 分支表达式可以是任意类型，不限于常量，可省略 break，默认自动终止。其语法如下：

```
switch var1 {
    case val1:
        // 执行语句
    case val2:
        // 执行语句
    default:
        // 执行语句
}
```

这里举例说明，代码如下：

```
package main
import "fmt"

func main() {
     // 定义全局变量
     var grade string = "B"
     var marks int = 90

     switch marks{
     case 90: grade = "A"
     case 80: grade = "B"
     case 50,60,70 : grade = "C"
     default: grade = "D"
     }

     switch{
     case grade == "A" :
             fmt.Printf(" 优秀 !\n" )
     case grade == "B", grade == "C" :
             fmt.Printf(" 良好 \n" )
     case grade == "D" :
             fmt.Printf(" 及格 \n" )
```

```
    case grade == "F":
            fmt.Printf("不及格\n" )
    default:
            fmt.Printf("差\n" )
    }
```

输出结果如下：

```
优秀！
```

2.6.3 循环语句 for

for 语句后面的三个子语句一般称为：初始化语句（init）、条件语句（condition）和赋值表达式（post），这三者不能颠倒顺序。其中，条件语句是必需的，条件语句会返回一个布尔值，该布尔值是 true 则执行代码块，是 false 则跳出循环。代码如下：

```
for init; condition; post {
// 执行语句
}
```

这里举例说明。代码如下：

```
package main
import "fmt"

func main() {
for a := 0; a < 10; a++ {
    fmt.Printf("a 的值为： %d", a)
    // 输出 a 的值为： 0
    a 的值为： 1
    a 的值为： 2
    a 的值为： 3
    a 的值为： 4
    a 的值为： 5
    a 的值为： 6
    a 的值为： 7
    a 的值为： 8
    a 的值为： 9
    }
}
```

如果是 for {} 这样的语句，则表明是一个无限循环（直到按快捷键 Ctrl+C 终止为止），会一直执行下去。例如：

```
package main
import "fmt"

func main() {
    for  {
        fmt.Printf("这是无限循环。\n"); // 一直执行，直到按快捷键Ctrl+C终止
    }
}
```

2.6.4 循环语句 for range

for range 语句可以对 slice、map、array、string，还有 struct 等进行迭代循环。这里以循环 struct 为例，说明 for range 语句的用法。代码如下：

```
package main
import "fmt"

func main() {
     d := [...]struct {
            name string
            age  uint8
     }{
            {"user1", 10},          // 可省略元素类型
            {"user2", 20},          // 别忘了最后一行的逗号
     }

     for _,v := range d{
            fmt.Println("结构体d的各值依次为：", v)
     }
}
```

输出结果如下：

```
结构体d的各值依次为： {user1 10}
结构体d的各值依次为： {user2 20}
```

for 和 for range 语句都可以循环多种数据结构类型，它们的主要区别是使用场景不同。

for 的用途如下：

- 遍历 array 和 slice。
- 遍历 key 为整型递增的 map。
- 遍历 string。

除了具有上述 for 语句的用途，for range 语句还有以下用途：

- 遍历 key 为 string 类型的 map 并同时获取 key 和 value。
- 遍历 channel。

2.6.5 跳出循环

在 Go 语言中执行跳出循环，一般用 break 和 continue 语句，其作用如下：

- break：用于 switch、for、select 语句，终止整个语句块执行。
- continue：仅用于 for 循环，终止后续逻辑，立即进入下一轮循环。

下面举例说明跳出循环语句的用法，代码如下：

```
import "fmt"

func main() {
	for i:= 0;i < 10; i++{
		if i%2 == 0{
			continue
			fmt.Println("跳出当前循环") // 不会执行后面的语句
		}

		if i > 5{
			break
			fmt.Println("终止当前循环") // 不会执行后面的语句
		}
		println(i)
	}
}
```

流程控制语句是一切编程语言的基础，合理使用流程控制可以使程序执行效率更高，希望读者在后续的 DevOps 开发工作中能熟练掌握。

2.7 Go 语言的方法

在结构体中定义的函数一般称为方法。方法与函数的最大区别就是，前者有接收器变量，而后者没有。

其语法如下：

```
func(接收器变量 接收器类型) 方法名(参数列表) (返回参数) {
	// 函数体
}
```

下面对以上语法的各部分进行说明：

（1）接收器变量：在为接收器中的参数变量命名时，官方建议使用接收器类型名的第一个小写字母，而不是 self、this 之类的命名。例如，Socket 类型的接收器变量应该命名为 s；Connector 类型的接收器变量应该命名为 c 等。

（2）接收器类型：接收器类型和参数类似，可以是指针类型和非指针类型。

（3）方法名、参数列表、返回参数：格式与函数定义一致。

（4）指针类型的接收器由一个结构体的指针组成，更接近于面向对象中的 this 或者 self。

这里举例说明方法的具体用法，代码如下：

```
package main
import "fmt"
// 定义属性结构
type Property struct {
    value int  // 属性值
}
// 设置属性值
func (p *Property) SetValue(v int) {
    // 修改 p 的成员变量
    p.value = v
}
// 取属性值
func (p *Property) Value() int {
    return p.value
}
func main() {
    // 实例化属性
    p := new(Property)
    // 设置值
    p.SetValue(100)
    // 打印值
    fmt.Println(p.Value()) // 100
}
```

在实际开发工作中，使用方法时应注意以下问题（这里的 T 是指 Type）：

- 要修改实例状态，用 *T。
- 对于无须修改状态的小对象或固定值，建议用 T。
- 对于大对象建议用 *T，以减少复制成本。
- 对于引用类型、字符串等指针包装对象，直接用 T。
- 对其他无法确定的情况，都用 *T。

2.8　Go 语言的 interface

简单地说，interface（接口）是一组方法签名的组合，通过 interface 来定义对象的一组行为。

interface 就是一组抽象方法的集合，它必须由其他非 interface 类型实现，而不能自我实现，Go 语言通过 interface 实现了 duck-typing，即“如果一只鸟走起来像鸭子、游泳像鸭子、叫起来也像鸭子，那么这只鸟可以被称为鸭子”。其语法定义如下：

```
type 接口类型名 interface{
    方法名 1(参数列表 1) 返回值列表 1
    方法名 2(参数列表 2) 返回值列表 2
    …
}
```

接口具备下列特点：

- 接口只有方法声明，没有实现，没有数据字段。
- 接口可以匿名嵌入其他接口，或嵌入结构中。
- 对象赋值给接口时，会发生复制，而接口内部存储的是指向这个复制品的指针，既无法修改复制品的状态，也无法获取指针。
- 只有当接口存储的类型和对象都为 nil 时，接口才等于 nil。
- 接口调用不会做 receiver 的自动转换。
- 接口同样支持匿名字段方法。
- 接口也可实现类似面向对象中的多态。
- 空接口可以作为任何类型数据的容器。
- 一个类型可实现多个接口。
- 接口命名习惯以 er 结尾。

Go 语言中为什么要使用接口？下面举例说明，代码如下：

```
type Cat struct{}
func (c Cat) Say() string { return "喵喵喵" }

type Dog struct{}
func (d Dog) Say() string { return "汪汪汪" }

func main() {
    c := Cat{}
    fmt.Println("猫:", c.Say())
    d := Dog{}
    fmt.Println("狗:", d.Say())
}
```

上面的代码中定义了猫和狗，它们都会叫，这时会发现 main 函数中有重复的代码。如果后续再加上猪、青蛙等动物，则代码还会一直重复下去。因此，是否能把它们当成“能叫的动物”来处理呢？

类似的例子在编程过程中会经常遇到。例如，三角形、四边形、圆形都能计算周长和面积，能否把它们当成“图形”来处理呢？又如，销售、行政、程序员都能计算月薪，能不能把他们当成“员工”来处理呢？

Go 语言为了解决类似上面的问题，就设计了接口这个概念。接口区别于之前所有的具体类型，是一种抽象的类型。当看到一个接口类型的值时，其实不知道它具体是什么，唯一知道的是通过它的方法能做什么。

空接口 interface{} 没有任何方法签名，也就是任何类型都实现了空接口。其作用类似面向对象语言中的根对象 Object。

这里可以举个简单的例子进行说明。代码如下：

```
package main
import "fmt"
func Print(v interface{}) {
    fmt.Printf("%T: %v\n", v, v)
}

func main() {
   Print(1) // int: 1
   Print("Hello, World!")     // string: Hello, World!
   Print(7.53333)             // float64: 7.53333
}
```

空接口 interface{} 可以存储任意类型的数据，当需要判断数据到底是何种类型时，可以使用以下方法。

（1）直接断言。例如，收到一个类型为 interface{} 的变量 unknown，可以通过如下代码直接断言该变量是否为 string 类型。

```
val, ok := unknown.(string)
```

如果返回 ok 为 true，则变量 unknown 为 string 类型，同时返回一个 val 存储 string 类型的值。如果确定 unknown 为 string 类型，也可以不返回 ok 变量，直接强转获取其值。例如：

```
val := unknown.(string)
```

但是使用这种方法有一定的风险，如果变量不是 string 类型，则会发生 panic 错误：

```
panic: interface conversion: interface {} is int, not string
```

（2）反射。反射位于 reflect 包，使用 reflect.TypeOf 函数获取类型，具体使用方法如下：

```
retType = reflect.TypeOf(unknown)
```

（3）type 关键字。该方法适用于 switch... case 结构，通过不同的 case 进行不同的处理。

```
switch unknown.(type){
   case string:       //string 类型
   case int:          //int 类型
   …
}
```

下面以具体的例子说明上面的各种方法。代码如下：

```
package main

import (
   "fmt"
   "reflect"
)

func main(){
   var str interface{} = "abc"

   retType,val := interfaceAssert1(str)
   fmt.Printf("type:%v, value:%v\n", retType, val)

   retType2,val2 := interfaceAssert2(str)
   fmt.Printf("type:%v, value:%v\n", retType2, val2)

   retType3 := interfaceAssert3(str)
   fmt.Printf("type:%v\n", retType3)

}

//直接断言
func interfaceAssert1(unknown interface{})(retType string, val interface{}){
   val, ok := unknown.(string)

   if ok{
       return "string", val
   }else{
       return "not string", nil
   }

}

//反射
```

```
    func interfaceAssert2(unknown interface{})(retType reflect.Type, val reflect.
Value){
        retType = reflect.TypeOf(unknown)
        val = reflect.ValueOf(unknown)
        return retType,val
    }

    //type 关键字
    func interfaceAssert3(unknown interface{})(retType string){
        switch unknown.(type){
        case string:
            return "string"
        case int:
            return "int"
        default:
            return "other type"
        }
    }
```

运行结果如下：

```
    type:string, value:yhc
    type:string, value:yhc
    type:string
```

在 Go 语言中，map[string]interface{} 是一种数据类型，表示一个字符串到任意值的映射（也称为字典或哈希表）。其中，map 是 Go 语言内置的映射类型；string 和 interface{} 分别表示键类型和值类型。例如，可以定义一个名为 user 的变量，类型为 map[string]interface{}，用来保存用户的基本信息。代码如下：

```
    user := map[string]interface{}{
        "name": "John Smith",
        "age": 30,
        "email": "john@example.com",
    }
```

在这个例子中，将用户的名字、年龄和电子邮件地址分别保存在字符串键 name、age 和 email 对应的值中。由于 interface{}，类型可以表示任意类型的值，因此可以将不同类型的数据存储在同一个映射中。

理解了 map[string]interface{} 之后，map[interface{}]interface{} 就比较好理解了。在 Go 语言中，map[interface{}]interface{} 是一种特殊的映射类型，其中键和值的类型都定义为 interface{}。Go 语言里的 interface{} 是一个空接口，任何类型的对象都可以实现它。因此，当定义一个 map[interface{}]interface{} 时，表示该 map 的键和值可以是任意类型。在使用这种 map 时，可以将任何类型的值作为键和值传入其中，因为所有类型都实现了空接口 interface{}。下面介绍具体实现代码：

```
package main
import (
    "fmt"
)

func main() {
    mapInterface := make(map[interface{}]interface{})
    mapString := make(map[string]string)

    mapInterface["k1"] = 1
    mapInterface[3] = "hello"
    mapInterface["world"] = 1.05
    mapInterface["rt"] = true
    mapInterface["mmap"] = mapString


    for key,value := range mapInterface{
        strKey := fmt.Sprintf("%v",key)
        strValue := fmt.Sprintf("%v",value)

        mapString[strKey] = strValue
    }

    //fmt.Printf("%#v",mapString)
    fmt.Printf("%#v",mapInterface) // map[string]string{"3":"hello",
                                   //"k1":"1", "rt":"true", "world":"1.05"}
}
```

输出结果如下：

```
map[interface {}]interface {}{3:"hello", "k1":1, "mmap":map[string]
string{"3":"hello", "k1":"1", "mmap":"map[3:hello k1:1 rt:true world:1.05]",
"rt":"true", "world":"1.05"}, "rt":true, "world":1.05}
```

结合具体工作中的接口，主要用于多态的实现、隐藏函数的具体实现和中间层，以解耦上下层依赖。

这里介绍接口作为中间层的用法：解耦上下层的依赖。下面使用一个用户权限校验的功能实例进行说明，代码如下：

```
package main

// 假设这是Redis客户端
type Redis struct {
```

```
}

func (r Redis) GetValue(key string) string {
    panic("not implement")
}

// 不通过接口实现：检查用户是否有权限的功能
func AuthExpire(token string, rds Redis) bool {   // 这里要修改
    res := rds.GetValue(token)                    // 这里可能也要修改
    if res == ""{
        return false
    } else {
        // 正常处理
        return true
    }
}
// 这里如果有其他函数引用了 rds Redis，那么肯定也全部要改
// …
func main() {
    token := "test"
    rds := Redis{} // 这里要修改
    AuthExpire(token, rds) // rds 这个名字可能要修改
}
```

如果使用接口的方式来实现，代码如下：

```
package main

type Cache interface {
    GetValue(key string) string
}

// 假设这是 Redis 客户端
type Redis struct {
}

func (r Redis) GetValue(key string) string {
    panic("not implement")
}

// 假设这是一个自定义的缓存器
```

```
type MemoryCache struct {
}

func (m MemoryCache) GetValue(key string) string {
     panic("not implement")
}

// 通过接口实现：检查用户是否有权限的功能
func AuthExpire(token string, cache Cache) bool {
     res := cache.GetValue(token)
     if res == ""{
             return false
     } else {
             // 正常处理
             return true
     }
}

func main() {
     token := "test"

     cache := Redis{}
    // Cache := MemoryCache{}，修改这一句即可
     AuthExpire(token, cache)
}
```

使用接口作为中间层和直接使用接口两种方法的对比如下：

- 两者都可以实现功能，甚至不用接口的代码量更小。
- 如果缓存（存储用户的组件）从Redis换成了MemoryCache，则都需要先编写MemoryCache以实现原有Redis的功能。
- 如果不用接口，则需要更改所有用到此缓存的函数。也就是说，需要更改所有引用Redis这个结构体的代码。示例中可能只是一个AuthExpire，但是实际上可能还有很多类似的函数引用Redis做缓存。
- 如果用了接口，那么需要修改的地方都需要实现具体的功能。

在实际开发工作中会发现，Go语言的接口有很多神奇的地方，它是实现面向对象的主要方式。它的设计与Java、C++语言的实现完全不一样，Java、C++语言是将数据和方法封装在一起，而Go语言是将逻辑进行抽象和组合，这一点在很多成熟且流行的大型Go工程中已证实是成功的设计。

2.9 Go 语言的正则表达式

正则表达式是一个特殊的字符序列，它定义了用于匹配特定文本的搜索模式。在 Go 语言中有一个内置的正则表达式包：regexp，其中包含所有操作列表，如过滤、修改、替换、验证或提取。

模式匹配是能根据正则表达式查询出特定字符集的技术。这种匹配模式允许从字符串中提取所需的数据，并以需要的方式对其进行操作。事实上，文件的处理在 DevOps 业务开发中也是非常重要的需求，而理解和使用正则表达式是处理文本的关键。

Go 语言的正则表达式的元字符见表 2.2。

表2.2 Go语言的正则表达式的元字符

regex 元字符	作 用
.	匹配任意单一字符
?	匹配前面的元素一次或 0 次
+	匹配前面的元素一次或多次
*	匹配前一个元素 0 次或多次
^	匹配字符串中的起始位置
$	匹配字符串中的结束位置
|	交替运算符
[abc]	匹配 a、b 或 c
[a-c]	a~c 的范围，匹配 a、b 或 c
[^abc]	否定，匹配除 a、b、c 之外的一切
\s	匹配空格字符
\w	匹配一个单词字符，等同于 [a-zA-Z_0-9]

Go 语言的 regexp 包中有几个典型的函数，分别为 MatchString、Compile、FindString、FindAllString、FindStringIndex、FindAllStringIndex、FindStringSubmatch、Split、ReplaceAllString 和 ReplaceAllStringFunc 函数。

下面举例介绍常用的正则相关函数，以说明其具体用法。

1. Compile 函数

Compile 函数用于解析正则表达式，如果解析成功，则返回可用于匹配文本的 Regexp 对象。编译的正则表达式能够生成执行速度更快的代码。假如正则表达式非法，则 Compile 函数会返回 error；而 MustCompile 函数在编译非法正则表达式时不会返回 error，而是直接引发 panic。举例如下：

```
package main

import (
	"fmt"
	"log"
	"regexp"
)

func main() {
	words := [...]string{"Seven", "even", "Maven", "Amen", "eleven"}
	re, err := regexp.Compile(".even")

	if err != nil{
		log.Fatal(err)
	}
	for _, word := range words{
		found := re.MatchString(word)
		if found{
			fmt.Printf("%s matches\n", word)
		} else {
			fmt.Printf("%s does not match\n", word)
		}
	}
}
```

返回结果如下：

```
Seven matches
even does not match
Maven does not match
Amen does not match
eleven matches
Even does not match
```

2. MustCompile 函数

下面介绍使用 MustCompile 函数的例子，该函数编译正则表达式并在无法解析表达式时发生 panic 异常。

```
package main

import (
	"fmt"
	"regexp"
```

```
)

func main() {
     words := [...]string{"Seven", "even", "Maven", "Amen", "eleven","Even"}
     re := regexp.MustCompile(".even")
     for _, word := range words{
             found := re.MatchString(word)
             if found{
                  fmt.Printf("%s matches\n", word)
             } else {
                  fmt.Printf("%s does notmatch\n", word)
             }
     }
}
```

输出结果如下：

```
Seven matches
even does not match
Maven does not match
Amen does not match
eleven matches
Even does not match
```

3. FindString 函数

FindString 函数用于返回第一个与正则表达式匹配的字符串结果。如果没有找到匹配的字符串，则会返回一个空字符串。如果正则表达式只是要匹配空字符串，那么 FindString 也会返回空字符串。可以用 FindStringSubmatch 函数区分这两种情况。

FindStringSubmatch 函数是先提取出匹配的字符串，然后通过 []string 返回。第一个匹配到的是这个字符串本身，从第二个开始，才是想要的字符串；如果没有匹配项，则返回 nil 值。这里可以写个简单的例子对比这两者的区别，代码如下：

```
package main

import (
    "fmt"
    "regexp"
)

func main() {
    str := "Golang expressions example"
    regexp,_ := regexp.Compile("Gola([a-z]+)g")
```

```
    fmt.Println(regexp.FindString(str))
    fmt.Println(regexp.FindSubString(str)[1])

}
```

4. FindAllString 函数

FindAllString 函数用于返回正则表达式的所有连续匹配的切片。如果返回 nil 值，代表没有匹配的字符串。

```
package main

import (
    "fmt"
    "regexp"
)

func main() {
    //nodelist := "vcn02/0*4"
    nodelist := "vcn01/0*4+vcn02/0*4"
    node := regexp.MustCompile(`[a-z]{1,3}\d{2,4}\/`)
    nodeslice := node.FindAllString(nodelist, -1)

    fmt.Println(len(nodeslice)) // 2
    fmt.Println(nodeslice[0],nodeslice[1]) // vcn01/ vcn02/
}
```

5. ReplaceAllString 函数

ReplaceAllString 函数的主要功能是替换字符串，该函数返回修改后的字符串。例如：

```
package main

import (
    "fmt"
    "regexp"
)

func main() {
    str := "Golang regular expressions example"
    regexp, _ := regexp.Compile(`examp([a-z]+)e`)
    match := regexp.ReplaceAllString(str, "tutorial")
    //fmt.Println("Match:", match, " Error: ", err)
    fmt.Println(match)
}
```

正则表达式是很重要的，因为在很多业务开发工作中，字符串的相关处理工作都与业务息息相关。下面介绍实际业务开发中的例子，代码如下：

```
package main

import (
	"fmt"
	"regexp"
)

// 计算真正的用户名
func getCompUser(s string) string{
	//用户名不仅可以用::分隔，还可以用:分隔
	sep := "[: ::]"
	tmpuser := regexp.MustCompile(sep).Split(s, -1)
	fmt.Println("tmpuser",tmpuser)
	fmt.Println("len",len(tmpuser))

	// :分隔符的切片长度为 2; ::分隔符的切片长度为 3，目前只有这两种情况
	if len(tmpuser) == 2{
		compuser := regexp.MustCompile(sep).Split(s, -1)[1]
		return compuser
	} else {
		compuser :=regexp.MustCompile(sep).Split(s, -1)[2]
		return compuser
	}
}


func main(){
	tests := "CFD::yuhongchun" // 这里分两种情况分别进行测试
	// tests := "CFD:yuhongchun"
	test := getCompUser(tests)
	fmt.Println(test)
}
```

这样的例子在实际工作中还有很多，特别是在实际的 DevOps 开发过程中，应该还会涉及正则表达式，希望读者能够在工作中熟练掌握。

2.10 Go 语言中的并发编程

Go 语言被称为互联网的 C 语言，这也得益于 Go 语言中的 goroutine 和 channel。在 Java/C++

语言中实现并发编程时，通常需要用户维护一个线程池，并且需要包装一个又一个的任务，同时还需要调度线程执行任务并维护上下文切换，这一切通常会耗费程序员大量的时间和心力。在早期，处理程序的开发问题是较复杂的事情，因为无论是单进程还是多进程 / 线程，程序的上下文切换、内存消耗、各种锁都是不可避免的问题，如图 2.1 所示。

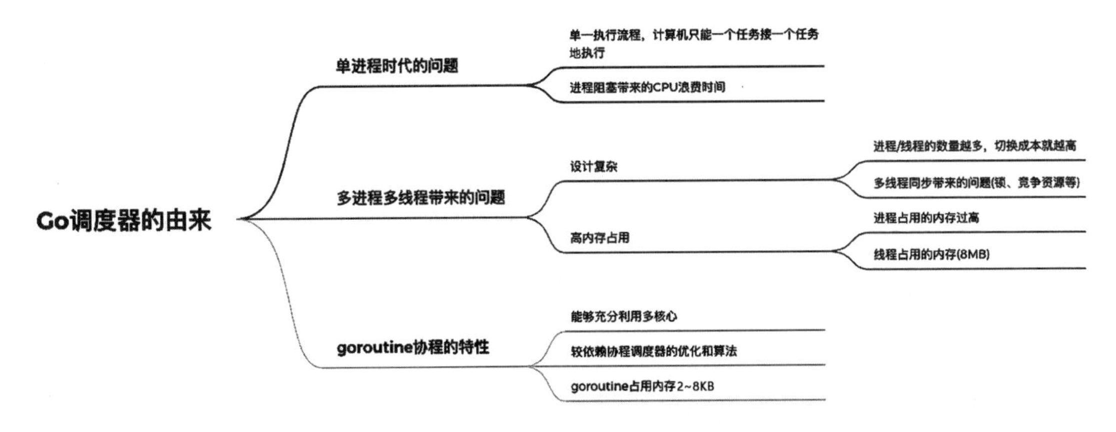

图 2.1　单进程 / 多进程 / 多线程在并发处理时遇到的问题

事实上，在 Go 语言中，基本上所有的一切都以并发方式运行，如系统监控、网络通信、垃圾回收，还有业务开发任务等。所有这一切都需要一个高效的调度器来指挥协调，所以 GMP 模型应运而生。GMP 模型能够充分利用多核心，让 Go 语言能够轻松实现并发编程。

GMP 模型是 Go 调度器的核心组成部分。它由以下三个主要的组件组成：

（1）G：goroutine 是 Go 语言中的轻量级线程，它代表一个并发任务。每个 goroutine 都有自己的栈和相关的上下文信息。goroutine 的创建和销毁由 GMP 模型负责管理，也可以说它是基于用户态的线程。

（2）M：Machine 是 Go 调度器中的执行线程，它负责将 goroutine 映射到操作系统线程上。每个 M 都有自己的调用栈和寄存器状态。Go 调度器会根据需要创建或销毁 M，以适应并发任务的数量和系统负载。

（3）P：P 是 M 的上下文，一般会在当前 goroutine 运行的上下文环境（函数指针，堆栈地址及地址边界）中，维护一组可运行的 goroutine 队列。每个 P 都与一个 M 关联，并负责将 goroutine 调度到 M 上执行。当一个 goroutine 被调度执行时，它会占用所在的 P，并在执行完成后释放。

> **注**：M与P的数量没有绝对关系，一个M阻塞，P就会创建或者切换另一个M，所以，即使P的默认数量是1，也有可能会创建很多个M。

Go 语言的 GMP 模型工作流如图 2.2 所示。

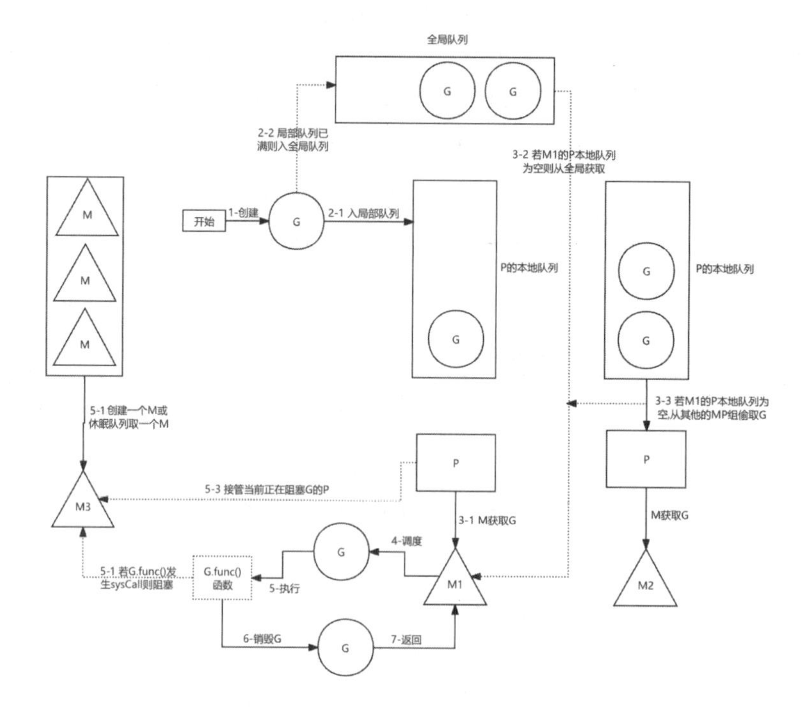

图 2.2　Go 语言的 GMP 模型工作流

GMP 模型工作的完整步骤如下：

（1）通过 go func() 创建一个 goroutine。

（2）有两个存储 G 的队列，一个是局部调度器 P 的本地队列，另一个是全局 G 队列。新创建的 G 会先保存在 P 的本地队列中，如果 P 的本地队列已经满了，就会保存在全局队列中。

（3）G 只能运行在 M 中，一个 M 必须持有一个 P，M 与 P 是 1 ∶ 1 的关系。M 会从 P 的本地队列弹出一个可执行状态的 G 来执行，如果 P 的本地队列为空，就会向其他的 MP 组合窃取一个可执行的 G 来执行（不可能让 CPU 空闲）。

（4）M 调度 G 的执行过程是一个循环机制。

（5）当 M 执行某一个 G 时，如果发生了 syscall 或者其他阻塞操作，M 会阻塞，如果当前有一些 G 在执行，runtime 会把这个线程 M 从 P 中摘除（detach），然后创建一个新的操作系统的线程（如果有空闲的线程可用，就复用空闲线程）来服务于这个 P。

（6）当 M 系统调用结束时，它对应的 G 会尝试获取一个空闲的 P 执行，并放入这个 P 的本地队列。如果获取不到 P，那么这个线程 M 会变成休眠状态，并加入空闲线程中，然后它对应的 G 会被放入全局队列中。

2.10.1　goroutine

Go 语言中的 goroutine 就是一种机制，它来自协程的概念，可以让一组可复用的函数运行在一组线程之上。即使有协程阻塞，该线程的其他协程也可以被 runtime 调度，从而将协程转移到其他可运行的线程上。Go 程序会智能地将 goroutine 中的任务合理地分配给每个 CPU。goroutine 非常轻量，一个只占几 KB，并且这几 KB 就足够 goroutine 运行完，因此可以在有限的内存空间支持大量 goroutine，从而支持更多的并发。Go 语言之所以被称为现代化的编程语言，就是因为它在语言层面内置了调度和上下文切换的机制。

在 Go 语言编程中不需要用户写进程、线程、协程，只需记住 goroutine，当需要让某个任务并发执行时，只需把这个任务包装成一个函数，开启一个 goroutine 去执行这个函数就可以了。

1. 启动单个 goroutine

启动 goroutine 的方式非常简单，只需在调用的函数（普通函数和匿名函数）前面加上一个 go 关键字。

例如：

```
func hello() {
    fmt.Println("Hello Goroutine!")
}
func main() {
    hello()
    fmt.Println("main goroutine done!")
}
```

执行结果如下：

```
Hello Goroutine!
main goroutine done!
```

以上示例中的 hello 函数和下面的语句是串行执行的，执行结果是先打印“Hello Goroutine！”后打印“main goroutine done！”。

接下来，在调用的 hello 函数前面加上关键字 go，也就是启动一个 goroutine 去执行 hello 函数。

```
func hello() {
    fmt.Println("Hello Goroutine!")
}

func main() {
    go hello()
    fmt.Println("main goroutine done!")
}
```

这次的执行结果只打印了“main goroutine done!”，并没有打印“Hello Goroutine!”。因为在程序启动时，Go 程序就会为 main 函数创建一个默认的 goroutine。当 main 函数返回时，该 goroutine 就结束了，所有在 main 函数中启动的 goroutine 会一同结束，main 函数相当于主进程，只要它一结束，程序就会结束。所以要想办法让 main 函数等一等 hello 函数，最简单的方式就是 time.Sleep。代码如下：

```
func hello() {
    fmt.Println("Hello Goroutine!")
}
func main() {
    go hello()          // 启动另外一个 goroutine 去执行 hello 函数
    fmt.Println("main goroutine done!")
    time.Sleep(time.Second)
}
```

执行结果如下：

```
main goroutine done!
Hello Goroutine!
```

从执行结果可以发现，这次先打印“main goroutine done!”，然后打印“Hello Goroutine!”。

2. 利用 sync.WaitGroup 启动多个 goroutine

上面是启动单个 goroutine，如果要启动多个 goroutine，可以利用 sync.WaitGroup 实现。例如：

```
var wg sync.WaitGroup

func hello(i int) {
    defer wg.Done()    // goroutine 结束就登记 -1
    fmt.Println("Hello Goroutine!", i)
}
func main() {
    for i := 0; i < 10; i++ {
        wg.Add(1)               // 启动一个 goroutine 就登记 +1
        go hello(i)
    }
    wg.Wait()                   // 等待所有登记的 goroutine 都结束
}
```

多次执行上面的代码，会发现每次打印的数字的顺序都不同。这是因为 10 个 goroutine 是并发执行的，而 goroutine 的调度是随机的。

3. WaitGroup 工作机制

WaitGroup 用于等待一组 goroutine 结束，用法很简单。它有以下三个方法：

```
func (wg *WaitGroup) Add(delta int)
func (wg *WaitGroup) Done()
func (wg *WaitGroup) Wait()
```

WaitGroup 对象内部有一个计数器，最初从 0 开始，用来控制计数器的数量。其中，Add(n) 把计数器设置为 n；Done() 每次把计数器的数量减 1; Wait() 会阻塞代码的运行，直到计数器的值减为 0。

WaitGroup 对象不是一个引用类型，在通过函数传值时需要使用地址。下面可以看一个例子。

```
func main() {
    wg := sync.WaitGroup{}
    wg.Add(100)
    for i := 0; i < 100; i++ {
        go f(i, &wg)
    }
    wg.Wait()
}

// 一定要通过指针传值，不然进程会进入死锁状态
// 程序运行后会出现 fatal error: all goroutines are asleep - deadlock!
func f(i int, wg *sync.WaitGroup) {
    fmt.Println(i)
    wg.Done()
}
```

2.10.2 channel

channel（通道）是一个用来传递数据的数据结构。channel 可以在两个 goroutine 之间通过传递一个指定类型的值实现同步运行和通信。操作符 <- 用于指定通道的方向，即发送或接收；如果未指定通道的方向，则视为双向通道。

channel 还可以作为函数参数传递，默认是双向通道。channel 用于 goroutine 之间的通信，让它们之间可以进行数据交换，就像管道一样，一个 goroutine 向 channel_A 中放数据，另一个 goroutine 从 channel_A 中取数据。

一般 channel 的声明形式如下：

```
var chanName chan ElementType
```

例如，声明一个传递类型为 int 的 channel：

```
var ch chan int
```

其语法如下：

```
ch <- v              // 把变量 v 发送到通道 ch
v := <-ch            // 从 ch 接收数据，并把值赋给变量 v
v,ok = <-ch          // 从 ch 读取一个值，判断是否读取成功，如果成功，则保存到变量 v 中
```

声明一个通道很简单，使用 chan 关键字即可，通道在使用前必须先创建。声明形式如下：

```
ch := make(chan int)
```

默认情况下，通道是不带缓冲区的。使用无缓冲通道装入数据时，装入方将会被阻塞，直到通道在另外一个 goroutine 中被取出数据。同样，如果通道中没有装入任何数据，接收方试图从通道中获取数据时，同样也会被阻塞。发送和接收的操作是同步完成的。发送端发送数据，同时必

须有接收端接收相应的数据。下面用 Go 语言来简化生产者 - 消费者模型。代码如下：

```
package main

import (
	"fmt"
	"time"
)

// chan 作为函数参数
func produce(ch chan<- int) {
	for i := 0; i < 10; i++{
		ch <- i
		fmt.Println("Send:", i)
	}
}
func consumer(ch <-chan int) {
	for i := 0; i < 10; i++{
		v := <-ch
		fmt.Println("Receive:", v)
	}
}
/* 因为 channel 没有缓冲区，所以当生产者给 channel 赋值后，生产者线程会阻塞，直到消费者线程将数据从 channel 中取出。消费者第一次将数据取出后，进行下一次循环时，消费者的线程也会阻塞，因为生产者还没有将数据存入，这时程序会去执行生产者的线程。程序就这样在消费者和生产者两个线程之间不断切换，直到循环结束
*/
func main() {
	ch := make(chan int)
	go produce(ch)
	go consumer(ch)
	time.Sleep(1 * time.Second)
}
```

输出结果如下：

```
Receive: 0
Send: 0
Send: 1
Receive: 1
Receive: 2
Send: 2
Send: 3
```

```
Receive: 3
Receive: 4
Send: 4
Send: 5
Receive: 5
Receive: 6
Send: 6
Send: 7
Receive: 7
Receive: 8
Send: 8
Send: 9
Receive: 9
```

再举个例子，代码如下：

```
package main

import (
    "fmt"
    "time"
)

func main() {
    ch := make(chan string)
    go sender(ch)
    go recver(ch)
    time.Sleep(time.Second)
}

func sender(ch chan string) {
    ch <- "malongshuai"
    ch <- "gaoxiaofang"
    ch <- "wugui"
    ch <- "tuner"
}

func recver(ch chan string) {
    var recv string
    // 读取 channel 的操作放在了无限 for 循环中，表示 recver goroutine 将一直阻塞
        for{
        recv = <-ch
```

```
        fmt.Println(recv)
        }
}
```

输出结果如下：

```
malongshuai
gaoxiaofang
wugui
tuner
```

以上代码激活了两个goroutine，其中一个用于执行sender函数，该函数每次向channel ch中发送一个字符串；另一个用于执行recver函数，该函数每次从channel ch中读取一个字符串。

注意以上代码的recv = <-ch，当channel中没有数据可读时，recver goroutine将会在此行阻塞。由于recver中读取channel的操作放在了无限for循环中，表示recver goroutine将一直阻塞，直到从channel ch中读取到数据，读取到数据后进入下一轮循环又被阻塞在recv = <-ch上。直到main函数中的time.Sleep()指定的时间到了，main函数终止，所有的goroutine将全部被强制终止。因为recver函数要不断地从channel中读取可能存在的数据，所以recver函数一般都使用一个无限循环来读取channel，以避免sender函数发送的数据被丢弃。

channel还支持close操作，用于关闭channel，随后对基于该channel的任何发送操作都将导致panic异常。对一个被close过的channel进行接收操作，依然可以接收到之前成功发送的数据；如果channel中没有数据，将产生一个零值的数据。

使用内置的close函数就可以关闭channel，代码如下：

```
close(ch)
```

关于关闭通道需要注意的是，只有在通知接收方goroutine所有的数据都发送完毕时才需要关闭通道。通道可以被垃圾回收机制回收，它和关闭文件不一样，在结束操作之后关闭文件是必须要做的，但关闭通道不是必须的。

关闭通道之后需要注意以下事项：

（1）对一个关闭的通道再发送数据就会导致panic异常。

（2）对一个关闭的通道进行接收会一直获取数据，直到通道为空。

（3）对一个关闭且没有数据的通道执行接收操作会得到对应类型的零值。

（4）关闭一个已经关闭的通道会导致panic异常。

每个channel都有3种操作：send、receive和close。详细说明如下：

（1）send：表示sender端的goroutine向channel中投放数据。

（2）receive：表示receiver端的goroutine从channel中读取数据。

（3）close：表示关闭channel。

使用channel的注意事项：

（1）channel close并非强制需要使用close(ch)来关闭channel，在某些时候channel可以自动关闭。

（2）如果使用close，建议条件允许的情况下加上defer语句。

（3）只需在 sender 端显式使用 close 关闭 channel。因为关闭通道意味着没有数据再需要发送。

（4）注意死锁的问题。

1. channel 死锁的几种情况以及例子

channel 死锁是指两个或两个以上的协程在执行过程中，由于竞争资源或彼此通信而造成的一种阻塞的现象，若无外力作用，它们将无法推进下去。以下是在实际工作中总结出来的几种 channel 死锁情况（以下示例均会产生“fatal error: all goroutines are asleep - deadlock!”报错）。

（1）一个通道在一个 Go 主程里同时进行读和写。示例如下：

```
func main() {
// 死锁 1
    ch := make(chan int)
    ch <- 100
    num := <-ch
    fmt.Println("num=", num)
}
```

（2）Go 主程开启之前使用通道。示例如下：

```
func main()  {
    ch := make(chan int)
    ch <- 100  // 此处死锁，优于 Go 主程之前使用通道
    go func() {
       num := <-ch
       fmt.Println("num=", num)
     }()
     //ch <- 100  此处不死锁
     time.Sleep(time.Second*3)
     fmt.Println("finish")
}
```

（3）读取空 channel 会产生死锁。示例如下：

```
func main() {
     // 死锁 1
     ch := make(chan int)
     //close(ch) 向关闭的 channel 中读取数据，默认是初始值 nil
     num := <-ch
     fmt.Println("num=", num)
```

（4）超过 channel 缓存继续写入数据会导致死锁。示例如下：

```
func main() {
     ch := make(chan int, 2)
     ch <- 1
     ch <- 2
```

```
        ch <- 3
        num := <-ch
        fmt.Println("num=", num)
}
```

（5）向已关闭的 channel 中写入数据不会导致死锁，但是会触发 panic 异常。示例如下：

```
func main() {
ch := make(chan int, 2)
    close(ch)
    ch <- 1
    num := <-ch
    fmt.Println("num=", num)
}
```

2. 用 channel 控制并发数

前面章节已经介绍过，可以采取 sync.WaitGroup+struct{}{} 的方式控制 goroutine 并发数。下面看一下只使用 channel 的方式。代码如下：

```
package main

import (
      "fmt"
      "time"
)
// chan 作为参数入参
func worker(id int, jobs <-chan int, results chan<- int) {
      for j := range jobs{
                fmt.Printf("worker %d processing job %d\n", id, j)
                time.Sleep(2*time.Second)
                // 此处有 <-chan，表示阻塞，直到 chan 能读到数据
               results <- j * 2
      }
}

func main() {
      jobs := make(chan int, 10)
      results := make(chan int, 10)
      for w := 1; w <= 3; w++{
           go worker(w, jobs, results)
      }
      for j := 1; j <= 9; j++{
                jobs <- j
```

```
        }
        close(jobs)
        for a := 1; a <= 9; a++{
            v := <-results
            fmt.Println("results 的值为 :",v)
        }
}
```

channel 的高级用法如下：

- 用 channel 控制并发数。
- 用 channel 实现异步任务。

3. 异步执行长时间任务

在 Go 语言中可以使用 goroutine 和 channel 实现异步执行任务的功能。

具体步骤如下：

（1）在 Web 请求处理函数中开启一个 goroutine 去执行长时间任务。

（2）在 goroutine 中执行任务，并将任务结果通过 channel 传递给主线程。

（3）在主线程中等待任务结果的到达，并将结果返回给 Web 请求。

其逻辑代码如下：

```
func handleRequest(w http.ResponseWriter, r *http.Request) {
    // 开启 goroutine 执行任务
    resultChan := make(chan string)
    go longTimeTask(resultChan)

    // 等待任务结果并返回
    result := <-resultChan
    w.Write([]byte(result))
}

func longTimeTask(resultChan chan<- string) {
    // 执行长时间任务
    time.Sleep(10 * time.Second)
    resultChan <- " 任务完成 "
}
```

在上面的代码中，首先，定义了一个 handleRequest 函数来处理 Web 请求。在该函数中，开启了一个 goroutine 去执行 longTimeTask 函数中的长时间任务。然后，在 longTimeTask 函数中使用 time 包中的 Sleep 函数模拟耗时的任务，并在任务结束后将结果通过 resultChan 这个 channel 传递给主线程。最后，主线程等待任务结果的到达，并将结果返回给 Web 请求。

2.10.3 Go 并发中的锁

在 Go 代码中可能会有多个 goroutine 同时操作一个资源（临界区），这种情况会发生竞态问题（数据竞态）。类比现实生活中的例子，如十字路口被来自各个方向的汽车竞争，导致有碰撞发生。

这里举例说明：

```
package main

import (
	"fmt"
	"sync"
)

var x int
var wg sync.WaitGroup

func add() {
	for i := 0; i < 5000; i++{
		x = x + 1
	}
	wg.Done()
}
func main() {
	wg.Add(2)
	go add()
	go add()
	wg.Wait()
	fmt.Println(x)
}
```

在上面的代码中，开启了两个 goroutine 累加变量 x 的值，这两个 goroutine 在访问和修改变量 x 时就会存在数据竞争，导致最后的结果与期待的不符。

那么如何解决上面的问题呢？这里可以采用 sync 包的 Mutex 互斥锁。它用于主动控制 Mutex 类型的变量，或者将 Mutex 类型作为 struct 的元素的变量，在同一时间只被一个 go routine 访问，这个 Mutex 有两个方法：Lock 和 Unlock。互斥锁不区分读和写，即无论是读操作还是写操作都是互斥的。

```
package main

import (
	"fmt"
```

```
    "sync"
)

var x int
var wg sync.WaitGroup
var lock sync.Mutex

func add() {
    for i := 0; i < 5000; i++{
        lock.Lock()                    // 加锁
        x = x + 1
        lock.Unlock()                  // 解锁
    }
    wg.Done()
}
func main() {
    wg.Add(2)
    go add()
    go add()
    wg.Wait()
    fmt.Println(x)              // 10000
}
```

使用互斥锁能够保证同一时间有且只有一个 goroutine 进入临界区，其他的 goroutine 则在等待锁；当互斥锁释放后，等待的 goroutine 才可以获取锁进入临界区。前面提到 map 的线程安全时已经用到了 Mutex 锁，相信用户对此不陌生。

下面以实际工作中的业务代码进行说明。

```
func (g *GitlabReleaseInfo) GetDockerInfo(ctx context.Context, params
*gitmodel.PipelineEvents) *model.DockerInfo {
        projectId := params.Project.ID
        jobs := params.Builds
        lock := sync.Mutex{}
        wg := sync.WaitGroup{}
          // 并发读取 DockerInfo 信息
        wg.Add(len(jobs))
        var dockerInfo *model.DockerInfo
        for _, job := range jobs {
            jobid := job.ID
            go func(){
                    d, err := getDockerUrlFromJobLog(ctx, projectId, jobid)
                    if err != nil{
```

```
                        nlog.WithContext(ctx).Infof("id:%d %v", jobid, err)
                }
                if d != nil{
                        lock.Lock()         // 加互斥锁
                        dockerInfo = d
                        lock.Unlock()       // 解互斥锁
                }
                wg.Done()
            }()
    }
    wg.Wait()
    // if dockerInfo!=nil
&&len(dockerInfo.Repository)!=0&&len(dockerInfo.Tag)!=0{
    //}
    return dockerInfo
}
```

生活中有这么一种场景，当用户去银行存钱或取钱时，对账户余额的修改是需要加锁的。如果对账户金额的修改不加锁，此时若有人汇款到这个账户，则很可能导致最后的金额发生错误。读取账户余额也需要等待修改操作结束，才能读取到正确的余额。大部分情况下，读取账户余额的操作会很频繁，如果能保证读取账户余额的操作并发执行，程序效率会得到很大提高。如果要保证读操作的安全，则只需保证并发读时没有写操作即可。在这种场景下需要一种特殊类型的锁，其允许多个只读操作并行执行，但写操作会完全互斥。

这种锁称为多读单写锁（multiple readers，single writer lock），简称读写锁。读写锁分为读锁和写锁，读锁是允许同时执行的，但写锁是互斥的。一般来说，有以下几种情况：

- 读锁之间是不互斥的，在没有写锁的情况下，读锁是无阻塞的，多个协程可以同时获得读锁。
- 写锁之间是互斥的，如果存在某个写锁，则其他写锁阻塞。
- 写锁与读锁是互斥的。如果存在读锁，则写锁阻塞；如果存在写锁，则读锁阻塞。

Go 标准库中提供了 sync.RWMutex 互斥锁类型及其 4 个方法：

- Lock 用于加写锁。
- Unlock 用于释放写锁。
- RLock 用于加读锁。
- RUnlock 用于释放读锁。

读写锁的存在是为了解决读多写少时的性能问题，在读场景较多时，读写锁可有效地减少锁阻塞的时间。下面的例子可以很好地说明读锁和写锁之间的互斥，代码如下：

```
package main

import (
```

```
    "log"
    "runtime"
    "sync"
    "time"
)

var rwMutex sync.RWMutex

func runReadLock() {
    log.Println("来到读锁方法")
    rwMutex.RLock() // 与写锁构成互斥，在读的时候不允许写
    defer rwMutex.RUnlock()
    log.Println("运行的是读锁方法，并不一定要读数据，这里休眠10秒")
    time.Sleep(time.Second * 10)
}

func runWriteLock() {
    log.Println("来到写锁方法")
    rwMutex.Lock()
    defer rwMutex.Unlock()
    log.Println("运行的是写锁方法，并不一定要写数据，这里休眠10秒")
    time.Sleep(time.Second * 10)
}

func main() {
    runtime.GOMAXPROCS(10)
    for q := 0; q < 4; q++ {
        go runWriteLock()
    }
    for k := 0; k < 3; k++ {
        go runReadLock()
    }
    time.Sleep(time.Second * 100)
}
```

运行结果如下：

```
2024/03/23 21:41:16 来到写锁方法
2024/03/23 21:41:16 运行的是写锁方法，并不一定要写数据，这里休眠10秒
2024/03/23 21:41:16 来到读锁方法
2024/03/23 21:41:16 来到读锁方法
2024/03/23 21:41:16 来到写锁方法
2024/03/23 21:41:16 来到写锁方法
2024/03/23 21:41:16 来到读锁方法
```

```
2024/03/23 21:41:16 来到写锁方法
2024/03/23 21:41:26 运行的是读锁方法，并不一定要读数据，这里休眠10秒
2024/03/23 21:41:26 运行的是读锁方法，并不一定要读数据，这里休眠10秒
2024/03/23 21:41:26 运行的是读锁方法，并不一定要读数据，这里休眠10秒
2024/03/23 21:41:36 运行的是写锁方法，并不一定要写数据，这里休眠10秒
2024/03/23 21:41:46 运行的是写锁方法，并不一定要写数据，这里休眠10秒
2024/03/23 21:41:56 运行的是写锁方法，并不一定要写数据，这里休眠10秒
```

既然有了 channel，为什么还要引入互斥锁呢？这是由于两者的使用场景不同，channel 倾向于解决逻辑层次的并发处理流程，而互斥锁则用于保护局部范围内的数据安全。

2.11 小结

本章主要介绍了 Go 语言的基础语法，包括其基础数据结构及特征、函数（方法）定义、struct 和 interface 接口以及 Go 正则等，也介绍了与 Go 语言的并发编程相关的知识体系，包括 goroutine 和 channel。事实上，以作者自身的经验来说，DevOps 开发工作中涉及的就是 Go 语言基础，很多时候开发过程中产生的 Bug 乃至疑惑，根源就在于对 Go 语言的基础知识掌握不够牢固，所以深入理解 Go 语言基础的知识点，对于 DevOps 的开发工作是有帮助的。

第3章　用脚本语言开发云原生 CI SDK

3.1　CI SDK 基本流程

为什么有了 Jenkins 作为复杂项目的 CI 流水线，还需要一个 CI SDK 工具呢？例如，对于公司的核心产品 F，由于历史原因，其本身就包含几十个 artifact 和 image 产出，每次打包构建都长达 3 小时，CI 过程效率太慢了。下面总结 CI 痛点：

（1）全量构建时，模块与模块之间依赖太多。

（2）构建日志分散，查看历史日志很麻烦。

（3）构建行为的共同需求和细分需求实现，需要各个项目自己维护。

（4）公司核心项目库 Bitbucket 迁移到 gitlab 成本较大，所以暂时也没有办法用 gitlab CI。

（5）团队早期各小组对于 CI 系统的使用能力参差不齐。

基于以上痛点，在团队早期，很难通过开放 Jenkins CI 或者 GitLab CI 流水线的能力让各个小组自己定义持续构建作业，所以 DevOps 团队决定开发统一的 CI SDK 工具来解决以上痛点。

这里为了提升 CI 执行作业的并发效率，Jenkins 以集群的方式来工作，每台 Jenkins 物理机器都运行最新的稳定版本分支的 CI SDK，并且最终会分别产出 artifact。例如，像 jar 包和 tar 包等 artifact 产出会上传至 RAM 仓库，image 产出会上传至 Docker 仓库。云原生 CI SDK 的工作流如图 3.1 所示。

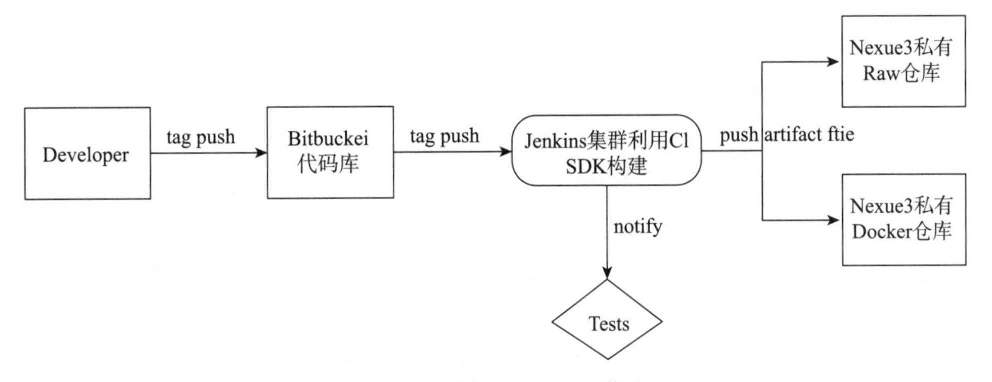

图 3.1　云原生 CI SDK 工作流

CI SDK 的最终交付产物为 artifact 和 image，最终功能实现如下：

（1）目前支撑内部近百个构建 / 发布作业，日均运行 500 多次构建 / 发布任务。

（2）构建 / 发布策略支持 webhook、on demand、nightly 等。

（3）自建的私有仓库，支持 Docker、maven、pypi、npm、raw 等格式的构建产物管理。

（4）自建的内部缓存镜像源，可极大提高构建的稳定性。

（5）基于 Mesos/Kubernetes 的 DevOps 系统和微服务的弹性伸缩，提供多用户支持。

（6）统一构建日志收集机制，提高 CI Debug 效率。

这里首先说明 CI SDK 的目录文件明细，此工程是基于 Python 2.7 版本（另加上 Shell 脚本）开发的，除了几个外部库，不需要任何第三方依赖，主要是考虑到后续 Jenkins 出现性能瓶颈时，能在 Kubernetes 平台下以 Pod 的方式弹性扩展，所以开发环境和依赖包越简单越好。其源码结构如下：

```
├── Readme.md                // CI SDK 帮助说明文件
├── changelog.md             // 整个工程的功能更新说明文件
├── getsdkenv.py             // 导入外部环境变量
├── requirements.txt         // CI SDK Python 依赖库文件
├── sdkbuildArtifacts.sh     // 构建 artifact 功能文件
├── sdkbuildImage.sh         // 构建 Docker 功能文件
├── sdkbuildreport.py        // 生成 report.json 报告文件
├── sdkmain.py               // CI SDK 主体函数文件
├── sdkpublish.sh            // 构建 artifact 和 Docker 成功以后的上传文件
├── sdktransprocess.py       // CI SDK 解析 struct.settings.json 文件
└── tools                    // 各种测试脚本及依赖文件
    ├── debian
    │   └── sources.list
    ├── deploy_sendmail.py
    ├── queue.png
    └── testing
        ├── apt
        │   └── sources.list
        ├── pypi
        │   ├── pip.conf
        │   └── requirements.txt
        ├── testcienv.sh    //测试 Nexus3 缓存服务脚本
        └── yum
            └── example.repo
```

Readme.md 是 CI SDK 帮助说明文件，它指明了程序运行时需要代入的参数。例如：

```
    Options supported by program execution:
    sdkmain.py  [-r REGISTRY_DOMAIN] [-a ARTIFACT_HOST] [-l ARTIFACT_HOST_
PULL] [-f | -i | -o] [-h help]
            -r: REGISTRY_DOMAIN, if not exists, will not push
            -a: ARTIFACT_HOST
```

```
            -l: ARTIFACT_HOST_PULL
            -f: Full build
            -i: Incremental build
            -o: Just generate delta files for CD tasks
            -h: help"
    Program execution method:
    python ${BDOS_SDK_DIR}/sdkbuild/sdkmain.py -r ${DOCKER_HOSTED_DEV} -a
${RAW_HOSTED_DEV} -l ${RAW_GROUP_DEV} -i
```

CI SDK 是串行处理逻辑（多任务并行时，Jenkins 会分布式处理执行，不用担心并发问题），整个过程较简单，其函数处理逻辑如图 3.2 所示。

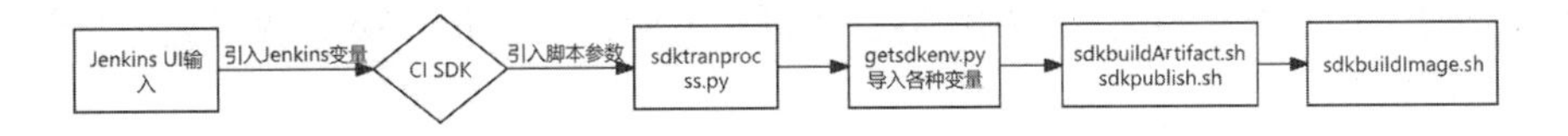

图 3.2 CI SDK 函数处理逻辑

CI SDK 的主要核心文件为 sdkmain.py、sdkbuildArtifact.sh 和 sdkbuildImage.sh 等，后面会分功能点依次介绍此工程的实际功能。

3.2 利用 Nexus 3 配置 CI 工作流的私有仓库

从图 3.1 中可以发现，私有仓库在整个 CI 工作流中的比重还是很大的。它主要用于存放 Artifact File Server 在 CI 过程中产生的 Artifact 文件，如 jar、tar、rpm 或 Docker 等文件。除此之外，还要满足以下需求：

（1）有 Docker 仓库，这是基础需求。

（2）有 Raw 仓库，可以存放 tar 和 rpm 等任何对象的文件，Nexus 可以像管理文件一样管理它们。

（3）有 pypi、yun 及 apt、Go/Java 的 Proxy 缓存代理。

（4）性能很好，不能受多并发的 CI 任务影响。

（5）有强大的授权 / 安全功能，可以做到特定用户对资源的可读或可写控制，建议不要分配线上环境的相关仓库权限。

此外，引用外部镜像源加速时，都存在一个问题：本地的测试环境（包括阿里云 JKS 环境）因为网络中断及源仓库更新了缓存，导致 CI 构建失败。

随着构建任务涉及的操作系统版本、开发语言版本越来越多，基于公网镜像源构建的不稳定性会越来越高；同时，基于单个公网镜像源会有单点失效的情况。

综合以上需求考虑，最终选用 Nexus 3 作为整个 CI 工作流的私有仓库和缓存代理服务器；另外，由于测试环境和研发环境的 Jenkins CI + CI SDK 都是放在本地内部机房，可以免去外部阿里云主机成本及流量成本费用，也算是在某种意义上起到了“降本”的作用。

选用 Nexus3 来搭建公司内部的镜像源，优势在于：

（1）覆盖公司所有开发语言、操作系统源，所有源码项目不需要自己解决镜像加速。

（2）代理国内、外多个镜像站，大大降低了单个公网镜像源失效后导致的 CI 构建失败的概率。

（3）有效缓存了开发语言、操作系统开源社区已停止维护版本的软件包，可保证“新”“老”项目均能正常构建。

（4）本身可以形成一套产品外部项目实施的解决方案，解决了客户在 air-gapped 环境中 CI 构建难的问题。

3.2.1 Nexus 3 的权限控制

Nexus 3 的权限控制有以下几种：

（1）browse。允许查看相关仓库的内容。不同于 read，拥有 browse 的权限类型只能在页面上查看和管理仓库内容。

（2）create。允许在仓库管理器中创建适用的配置。因为需要用 read 查看配置内容，所以 read 关联了绝大多数的现有 create 权限。

（3）delete。允许删除仓库管理器的配置、仓库内容、脚本。因为删除前需要先用 read 查看配置内容，所以一般用 read 关联 delete action。

（4）edit。允许修改关联的脚本、仓库内容、仓库管理配置内容。

（5）read。允许查看各种配置列表和脚本。如果没有 read，任何关联的 action 都会允许查看这些列表，但不允许查看其内容。read 也允许利用“能够从命令行查看内容”的工具。

（6）update。允许更新仓库管理器配置。绝大多数现有的拥有 update 的权限也拥有 read 权限。因此，如果创建自定义的权限也拥有 update 的功能，创建者应该考虑同时添加 read 到该权限，以便查看仓库管理器配置的更新。

3.2.2 Nexus 3 的仓库类型

Nexus 3 的仓库类型主要分为 hosted、proxy、group 三种，其具体作用见表 3.1。

表 3.1 Nexus 3 的仓库类型及其作用

仓库类型	具体作用
hosted	本地存储，像官方仓库一样提供本地私库功能
proxy	提供代理其他仓库的类型，作用类似于镜像功能
group	组类型，能够组合多个仓库为一个地址提供服务，可以理解为 group = hosted + proxy

下面介绍 Nexus 3 的安装方式。为了方便安装及后面 Nexus 3 的数据迁移，这里选用 Docker 安装方式。命令如下：

```
docker pull sonatype/nexus3
```

显示结果如下：

```
Using default tag: latest
```

```
latest: Pulling from sonatype/nexus3
865dc90c13b3: Pull complete
886bc343b9fd: Pull complete
37044071693a: Pull complete
3fd6bb466e42: Pull complete
Digest: sha256:8926032ab7eb9389351df78e68f21d1452dd57200152f424a57bcb26094e50c4
Status: Downloaded newer image for sonatype/nexus3:latest
```

接下来，以 Docker 的方式运行 Nexus 3，命令如下：

```
mkdir -p /data/nexus3/data
chown -R 200 /data/nexus3/data
docker run -d --name nexus3 --restart=always -v
/data/nexus3/data:/nexus-data -p 8081-8095:8081-8095 \
-e SONATYPE_DIR=/opt/sonatype    -e NEXUS_HOME=/opt/sonatype/nexus -e
NEXUS_DATA=/nexus-data  \
-e SONATYPE_WORK=/opt/sonatype/sonatype-work   sonatype/nexus3
```

安装成功以后，以 admin 的身份登录 Nexus 3 的工作界面，其地址为 http://10.1.0.201:8081，密码在 /data/nexus/admin.password 文件中。首次登录以后，系统会强制更新密码，这里改为 yhc@654321；8082 为 group 端口，8084 则为 hosted 端口（proxy 端口为 8083）；这里如果执行 docker pull 命令，则使用 group 仓库；如果执行 docker push 命令，则使用 hosted 仓库。宿主机 IP 地址为 10.1.0.201。

Nexus 3 配置 Docker 仓库的流程比较简单，此处略过。

这里用另一台主机来进行相关测试。

首先修改其中的 /etc/docker/daemon.json 文件，内容如下：

```
{
"insecure-registries":["10.1.0.201:8084","10.1.0.201:8082","10.1.0.201:8083"],
  "exec-opts": [
    "native.cgroupdriver=systemd"
  ]
}
```

然后，用 admin:yhc@6754321 进行 docker login 动作，这一步提前做好。

```
docker login 10.1.0.201:8082
```

如果正确登录，则有下面的显示：

```
Login Succeeded
```

如果没有出现预期的结果，则需要检查相应的端口设置和密码；接下来测试与 Docker 相关的 pull 及 push 动作，即尝试先从 group 仓库中拉取镜像使用 docker push，然后再将镜像推送到 hosted 仓库中使用 docker push，以此来检查流程是否顺畅。

（1）测试拉取 group 仓库的镜像。其步骤如下：

1）登录认证，其命令如下：

```
docker login 10.1.0.201:8082
```

2）输入密码，显示结果如下：

```
Username: admin
Password:
WARNING! Your password will be stored unencrypted in /root/.docker/config.json.
Configure a credential helper to remove this warning. See
https://docs.docker.com/engine/reference/commandline/login/#credentials-store

Login Succeeded
```

3）验证 docker pull 的动作，命令如下：

```
docker pull 10.1.0.201:8082/grafana:latest
```

4）输出结果如下：

```
latest: Pulling from grafana
Digest: sha256:33be3c2d3f457192a284777e24ae6bd264598896451c01ec2c7f329da4707af2
Status: Downloaded newer image for 10.1.0.201:8082/grafana:latest
10.1.0.201:8082/grafana:latest
```

从输出结果可以看出步骤（1）是没问题的，接下来执行步骤（2）。

（2）测试推送到 hosted 仓库的动作。其步骤如下：

1）登录认证，其命令如下：

```
docker login 10.1.0.201:8083
```

2）输出结果如下：

```
Authenticating with existing credentials...
WARNING! Your password will be stored unencrypted in /root/.docker/config.json.
Configure a credential helper to remove this warning. See
https://docs.docker.com/engine/reference/commandline/login/#credentials-store

Login Succeeded
```

3）验证 docker push 的动作，判断能否正常地推送到 hosted 仓库中。其步骤如下：

```
docker tag 10.1.0.201:8082/grafana:latest 10.1.0.201:8083/grafana:latest
docker push 10.1.0.201:8083/grafana:latest
```

输出结果如下：

```
The push refers to repository [10.1.0.201:8083/grafana]
7cb09c9a4372: Layer already exists
f0be059d864b: Layer already exists
ef88b45a119f: Layer already exists
3b2b7c569a4c: Layer already exists
```

```
    839b0b40ab56: Layer already exists
    adff21d3bcca: Layer already exists
    1e2c29677398: Layer already exists
    8675ede87d30: Layer already exists
    34d5ebaa5410: Layer already exists
    latest: digest:
sha256:33be3c2d3f457192a284777e24ae6bd264598896451c01ec2c7f329da4707af2 size:
2203
```

综上所述，说明此流程也没有问题。

下面可以测试 Nexus 3 的 raw 文件私库功能，利用 Nexus 3 建立一个高性能的文件器是一件非常简单的事情。事实上，这种需求在工作中是很常见的，有很多 CI 过程中产生的 artifact 组件，如 tar/jar/war 包，都可以放在 raw-hosted 私库中。

这里选择 Repositories → raw-hosted，设置如图 3.3 所示。

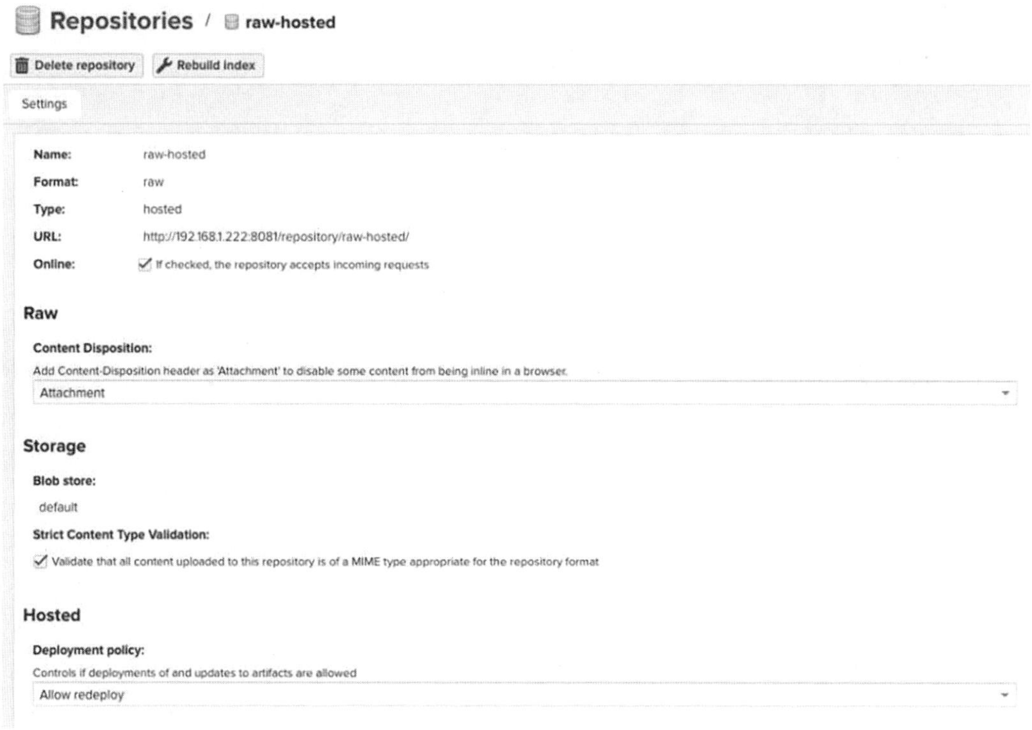

图 3.3 建立 Nexus 3 的 raw-hosted 仓库

建立成功后，可以用 curl 命令来上传文件（注意账号的权限及保密性），以确定 raw-hosted 仓库的上传流程是否成功。命令如下：

```
    curl -v --user 'admin:admin@yhc' --upload-file ./14844_rev1.json
http://10.1.0.201:8081/repository/raw-hosted/14844_rev1.json
```

命令显示结果如下：

```
* Trying 10.1.0.201...
* TCP_NODELAY set
* Connected to 10.1.0.201 (10.1.0.201) port 8081 (#0)
* Server auth using Basic with user 'admin'
> PUT /repository/raw-hosted/14844_rev1.json HTTP/1.1
> Host: 10.1.0.201:8081
> Authorization: Basic YWRtaW46YWRtaW5AeWhj
> User-Agent: curl/7.64.1
> Accept: */*
> Content-Length: 32812
> Expect: 100-continue
>
< HTTP/1.1 100 Continue
* We are completely uploaded and fine
< HTTP/1.1 201 Created
< Date: Sun, 21 Jan 2024 12:01:27 GMT
< Server: Nexus/3.50.0-01 (OSS)
< X-Content-Type-Options: nosniff
< Content-Security-Policy: sandbox allow-forms allow-modals allow-popups
allow-presentation allow-scripts allow-top-navigation
< X-XSS-Protection: 1; mode=block
< Content-Length: 0
<
* Connection #0 to host 10.1.0.201 left intact
* Closing connection 0
```

从以上输出结果可以证明，文件已经上传成功，说明 raw-hosted 文件仓库的上传流程是成功的。

这时需要注意的是，笔者在工作中使用 Nexus 3 时发现，如果 Nexus 3 的磁盘空间低于 4GB，数据库会变成只读模式，所有任务都不能执行。因此，在使用 Nexus 3 时要尽量多预留空间，不要把磁盘用到这种临界阈值。

Nexus 3 作为 APT、YUM 缓存服务器的过程较简单，这里不再赘述。

3.3 CI SDK 功能介绍

前面章节介绍过，早期团队在持续集成（CI）方面遇到了很多痛点问题，为了解决这些问题开发了 CI SDK。下面介绍其具体的功能。

- 支持全量构建。

- 支持增量构建。
- 按需构建，自助定义构建模块。
- 日志和构建元数据归档。
- Auto Tagging，镜像自动打上 Git SHA。
- 自定义 pre/post 脚本。
- SonarQube 流程集成。

3.3.1　struct.settings.json 文件介绍

每个源码项目只要在项目根目录下定义一个项目配置文件，取名为 struct.settings.json 文件（类似 .gitlab-ci.yml），即可通过 CI SDK 进行 CI 构建集成，主要内容包括：

- 构建阶段，Artifact 和 Docker 任选其一，或者两者兼而有之。
- 构建上下文路径。
- Git 探针路径，方便实现增量构建功能。
- 构建产物信息，包括名称、版本号、类型，可以以日志的方式归档。
- 自定义 Preinstall/Postinstall 脚本，方便做功能扩展。
- Docker 构建参数。
- 其他参数。

下面以某个产品线源码项目中的 struct.settings.json 文件为例来说明，其内容如下：

```
    {
    /* 首先需要定义 artifact_build 或 image_build，整个 sdkbuild 分成两个逻辑流程来进行，先是 artifact_build，后是 image_build*/
      "artifact_build": [
        {
          "path": [
            "executor/instance/labeling-slave-spark"
    /* 这里定义 labeling-slave-sparky 应用的源码路径，主要是考虑到 sdkbuild 的增量构建，这里可以把 path 理解成起文件路径探针的作用，此处是一个列表，如果是多级的，后面就并列写，以逗号分隔 */
          ],
          "build": {
            "app_name": "labeling-slave-spark", // 定义 artifact 的 app 名字
            "version": "1.0", // 定义 artifact 的版本名字
            "type": "artifact", // 定义 artifact 的类型，此处填写 artifact 即可
            "params": {
              /* 定义 labeling-slave-sparky 应用的上下文环境，用户自定义的 preset.sh 脚本真正操作的目录地址 */
              "ctx":"executor/instance/labeling-slave-spark",
```

```
                "file_type": "jar", /* 定义 artifact 的打包类型，可以是 tar，也可以是 jar
或 war，如果是 jar 或 war，要保证 ctx 路径中存在 install.sh 脚本 */
                "artifact_command": "preset.sh",  /* 用户自定义的 preset.sh 脚本，这
里是项目用户在打 tar 包或 jar 包之前的预置动作；如果没有，此处就填写默认值 ""*/
                "is_binary_action": "" /* 用户定义的加密动作，如 setup.py 脚本；如果没有，
此处就填写 ""*/
              }
            }
          },
          {
            "path": [
              "executor/instance/checkHivePartition",
              "executor/common"
              /* 这里定义 checkHivePartion 应用的源码路径，主要是考虑到 sdkbuild 的增量构建，
可以把它们理解成起文件路径探针的作用，此处是一个列表；如果是多级的，后面就并列着写，以逗号分
隔 */
            ],
            "build": {
              "app_name": "checkHivePartion",
              "version": "1.0",
              "type": "artifact",
              "params": {
                "ctx":"executor/instance/checkHivePartition",
                "file_type": "tgz",
                "artifact_command": "preset.sh",
                "is_binary_action": "setup.py"   // 用户定义的加密动作，此处是存在的，
                                                 // 这里就写 setup.py
              }
            }
          }
        ],
        /* 这里需要定义 artifact_build 或 image_build，整个 sdkbuild 分成两个逻辑流程来进行，
先是 artifact_build，后是 image_build*/
        "docker_build": [
          {
            "path": [
              "docker/master",
              "master"
            ],
              /* 这里定义 etl-master 应用的源码路径，主要是考虑到要支持增量构建，这里可以把它
```

```
们理解成起文件路径探针的作用，此处是一个列表；如果是多级的，后面就并列写，以逗号分隔 */
          "build": {
            "app_name": "etl-master", // 定义 etl-master 的 app 名字
            "version": "1.0", // 定义 etl-master 的版本
              "type": "image", /* 定义 artifact 的类型，此处填写 image 即可，即表示
Docker 构建 */
            "preinstall_command": "preinstall.sh",
            /* 用户自定义的 preinstall.sh 脚本，这里是项目用户在打包 image 之前的预置动作；
如果没有，此处就填写默认值 ""*/
            "params": {
              /* 定义 etl-master docker build 的上下文环境，用户自定义的 preset.sh 脚
本真正操作的目录地址 */
              "ctx": "docker/master",
              "args": {
                "target": ""
              /* 在 Dockerfile 中定义的打包编译阶段。如果是 dev，就填 dev；如果是 release，
就填 release；如果没有，就填写 ""*/
              }
            }
          }
        },
        {
          "path": [
            "docker/basic-etl-worker",
            "worker"
          ],
          "build": {
            "app_name": "basic-etl-worker",
            "version": "1.0",
            "type": "image",
            "preinstall_command": "preinstall.sh",
            "params": {
              "ctx": "docker/basic-etl-worker",
              "args": {
                "target": ""
              }
            }
          }
        }
      ]
    }
```

struct.settings.json 文件中的各参数说明如下：

- path：表示具备一系列操作动作的目录列表，这里可以把它们理解成起文件路径探针的作用。
- build：表示构建的动作，这里会分得较细。
- app_name：表示对应 app 的名字。
- version：表示 app 的对应版本，这里无论是 artifact 版本还是 image 版本，建议保持一致。
- type：表示具体构建的到底是 artifact 还是 image。
- params：表示 artifact 或 image 的构建明细参数。
- file_type：表示 artifact 具体的打包类型。
- install_sh：表示打包动作前执行项目的自定义文件，一般是 install_sh；如果为空，就填一个 ""。
- target：表示后面就是具体的 image 了，params 的 target 对应的是分步骤构建动作，这里主要是 dev 或 release，具体写在项目的 Dockerfile 中，如果没有，就填 ""。
- ctx：表示打包镜像的上下文环境，这里一般填写当前的路径即可。

下面是需要注意的事项：

- struct.settings.json 文件需要提前放在项目根目录下，否则执行 CI SDK 时会报错，程序会退出。
- struct.settings.json 文件是定义的项目的明细项目路径，这个信息要保证准确，不然后面执行 sdkbuild 流程时会出现问题。
- 有些源码项目比较特殊，特别是单项目单 Dockerfile 的场景，当需要构建上下文时，这里填写 "." 即可。

3.3.2 基于 CI SDK 处理 struct.settings.json 文件的流程

前面章节提到过，CI SDK 对开发环境和依赖越简单越好，所以处理 struct.settings.json 文件的流程较简单。数据以文件的方式保持持久化，这个过程没有引入任何中间件或数据库，如 redis 或 MySQL，其源码由 sdktransprocess.py 文件实现，主要是读取 struct.settings.json 文件，然后处理生成项目的 process_artifact_list.txt 和 process_image_list.txt 文件。其源码实现如下：

```
#!/usr/bin/env python
#-*- coding:utf-8 -*-
import json
import os

'''
读取 struct.settings.json 文件，然后处理生成项目的 process_artifact_list.txt 和
process_image_list.txt 文件
清理上一次构建过程中产生的 process_docker_list.txt 和 process_artifact_list.txt 文件
'''
def get_struct_settings():
    print "SDK remove duplicate files..."
    if os.path.exists('process_artifact_list.txt'):
        os.remove('process_artifact_list.txt')
```

```
    if os.path.exists('process_image_list.txt'):
        os.remove('process_image_list.txt')

    with open("struct.settings.json",'r') as f:
        settings = json.load(f)
        #print settings
    return settings


def get_artifact_var(settings):
    artifact_list = []
    global_version = None
    if "global_version" in settings.keys():
        global_version = settings['global_version']
    for data in list(settings['artifact_build']):
        artifact_path_list=data["path"]
        app_name=data["build"]['app_name']
        app_version = None
        if 'version' in data['build']:
            app_version=data['build']['version']
        app_type=data['build']['type']
        app_file_type=data['build']['params']['file_type']
        app_artifact_mode=data['build']['params']['artifact_command']
        is_binary_code = data['build']['params']['is_binary_action']
        app_ctx=data['build']['params']['ctx']
        tagFlag = os.getenv("TAGFLAG")

        if not app_version and global_version:
            app_version = global_version
        try:
            preinstall_code = data['build']['params']['preinstall_command']
        except KeyError:
            preinstall_code = ""

        app_post_action = ""
        if 'post_command' in data['build']['params']:
            app_post_action = data['build']['params']['post_command']

        for artifact_path in artifact_path_list:
            s = ""
```

```
                for i in
artifact_path,app_name,app_version,app_type,app_file_type,app_artifact_
mode,is_binary_code,app_ctx,preinstall_code,tagFlag,app_post_action:
                    s += '{},'.format(i)
                    seq=s[:-1]
                f = open('process_artifact_list.txt','a')
                f.write( seq + '\n')
                f.close()
                artifact_list.append(seq)

        return artifact_list

    def get_image_var(settings):
        image_list = []
        global_version = None
        if "global_version" in settings.keys():
            global_version = settings['global_version']
        for data in list(settings['docker_build']):
            #print data
            docker_path_list=data['path']
            app_name = data['build']['app_name']
            app_version = None
            if 'version' in data['build']:
                app_version = data['build']['version']
            app_type = data['build']['type']
            app_action = data['build']['preinstall_command']
            app_target = data['build']['params']['args']['target']
            app_ctx = data['build']['params']['ctx']
            app_dockerfile_path = ""
            if 'app_dockerfile_path' in data['build']['params']:
                app_dockerfile_path = data['build']['params']['app_dockerfile_path']

            sourceslist = "sources.list"
            if 'sourceslist' in data['build']['params']['args']:
                sourceslist = data['build']['params']['args']['sourceslist']

            app_post_action = ""
            if 'post_command' in data['build']:
                app_post_action = data['build']['post_command']

            if not app_version and global_version:
```

```
                app_version = global_version
            # 增加 docker_version 的 Tag 功能
            # TAGFLAG 由 Jenkins UI 传递进来
            tagFlag = os.getenv("TAGFLAG")
            for docker_path in docker_path_list:
                s = ""
                for i in
docker_path,app_name,app_version,app_type,app_action,app_target,app_
ctx,tagFlag,app_dockerfile_path,sourceslist,app_post_action:
                    s += '{},'.format(i)
                    seq=s[:-1]
                f = open('process_image_list.txt','a')
                f.write( seq + '\n')
                f.close()
                image_list.append(seq)

        return image_list


    def process():
        artifact_list = []
        image_list = []
        settings = get_struct_settings()
        if "artifact_build" in settings.keys():
            artifact_list = get_artifact_var(settings)
        if "docker_build" in settings.keys():
            image_list = get_image_var(settings)

        return artifact_list,image_list

    if __name__ == "__main__":
        process()
```

3.2.3 增量构建的实现

增量构建的具体实现逻辑如下：

CI SDK 会比对最后一次的 git diff，然后将生成的文件目录与项目根目录下的 struct.settings.json 文件进行比较，从中摘取 struct.settings.json 文件中提供的 PATH 目录对应的 artifact/image 资源，然后进行下一步的构建动作。

artifact 的文件会取前三级目录，然后汇总和去重；这里如果涉及 common 库，那么所有与其关联的 artifact 需要重新构建一次。

Docker 文件的路径处理是同样的流程。

其具体源码实现如下：

```
# -*- coding: utf-8 -*-
import argparse
import os
import sys
import shutil
import filecmp
import git
import subprocess
import pystache
import json
import time
import traceback
from datetime import datetime
from getsdkenv import *
import sdktransprocess

class Sdk(object):
  …
    def gen_git_updated_dir_list(self):
        git_diff_file_list = []
        gen_git_update_dir_list = []
        gen_git_update_dir_list_tmp = []
        gen_git_update_dir_list_txt_tmp = []
        repo = git.Repo(self.bdos_src_dir)
               #其执行效果等于git diff --name-only HEAD~ HEAD
        hc = repo.head.commit
        df = hc.diff("HEAD~")
        for d in df:
            path = d.a_rawpath  # b_rawpath
            git_diff_file_list.append(path)
            path_list = path.split("/")
            if len(path_list) < 4:
                gen_git_update_dir_list.append(os.path.dirname(path))
            else:

gen_git_update_dir_list.append(os.path.join(path_list[0],path_list[1],path_
list[2]))

        gen_git_update_dir_list_tmp = gen_git_update_dir_list
        gen_git_update_dir_list_tmp = list(set(gen_git_update_dir_list_tmp))
```

```
        for d2 in gen_git_update_dir_list_tmp:
            p2 = d2.split("/")
            nums = len(p2)
            if nums == 1:
                gen_git_update_dir_list_txt_tmp.append(d2)
            elif nums == 2:
                path = os.path.dirname(d2)
                gen_git_update_dir_list_txt_tmp.append(d2)
                gen_git_update_dir_list_txt_tmp.append(path)
            else:
                path = os.path.dirname(d2)
                par_path = os.path.dirname(path)
                gen_git_update_dir_list_txt_tmp.append(d2)
                gen_git_update_dir_list_txt_tmp.append(path)
                gen_git_update_dir_list_txt_tmp.append(par_path)

        gen_git_update_dir_list_txt = []
        for d3 in set(gen_git_update_dir_list_txt_tmp):
            if "." not in d3:
                gen_git_update_dir_list_txt.append(d3)
        self.gen_git_update_dir_list_txt = gen_git_update_dir_list_txt
        self.git_diff_file_list = git_diff_file_list

        gen_git_update_dir_list_file =
"{}/{}".format(self.JENKINS_SDK_LOGS,"gen_git_update_dir_list")
        write_list_in_file(gen_git_update_dir_list,
gen_git_update_dir_list_file)

        git_diff_file_list_file = "{}/{}".format(self.JENKINS_SDK_LOGS, "git_
diff_file_list")
        write_list_in_file(git_diff_file_list, git_diff_file_list_file)

        gen_git_update_dir_list_txt_file =
"{}/{}".format(self.JENKINS_SDK_LOGS, "gen_git_update_dir_list.txt")
        write_list_in_file(gen_git_update_dir_list_txt,
gen_git_update_dir_list_txt_file)

        gen_git_update_dir_list_txt_tmp_file =
"{}/{}".format(self.JENKINS_SDK_LOGS, "gen_git_update_dir_list.txt.tmp")
        write_list_in_file(gen_git_update_dir_list_txt_tmp, gen_git_update_
```

```
dir_list_txt_tmp_file)
    ...
```

3.2.4 报告当前构建的信息

CI 在当前面临的一个问题是，有些产品线有很多私有化定制项目的开发分支并且处于快速迭代和 bugfix 的过程；除了研发团队在频繁使用 CI 外，项目交付团队也在使用，所以都希望每次构建完成就立即能看到详细的报表输出。这个具体是由 sdkbuildreport.py 实现的。其源码内容如下：

```
#!/usr/bin/env python
#-*- coding:utf-8 -*-
import json
import os
import re
import collections


temp_dict = {}
result = []
# 直接在项目根目录下生成 jenkins_build_report.json 文件，可以在 Jenkins workspace
# 工作目录中查看
submit = './jenkins_build_report.json'


temp_dict['gitbranch'] = os.getenv('GITBRANCH')
temp_dict['gitrepo'] = os.getenv('JOBNAME')
temp_dict['memo'] = os.getenv('GITLOG')
temp_dict['releaseEngineer'] = os.getenv('COMMITUSER')
temp_dict['dockerurl'] = os.getenv('DOCKER_URL')
result.append(temp_dict)


# 这里重新写 gitbranch 的逻辑，不符合命令规则的分支默认为 internal 项目
# 项目分支有独立的命名规则，会以项目名作为开头，所以可以利用正则抓取
gitbranch = os.getenv('GITBRANCH')
try:
    # 利用正则获取项目名
    projectName = re.findall(r'#(.*?)#', gitbranch)[0]
except:
    projectName = ""
if projectName == "":
    # 表示是内部运营项目，非外部项目
```

```
    PROJECTNAME = "internal"
else:
    PROJECTNAME = projectName

f = open('process_image_list.txt')
# 注意 Python 中列表添加字典元素的问题
lis = []
lines = f.readlines()
for line in lines:
    #dic = {}
    dic = collections.OrderedDict()
    dic['appName']=re.findall(r',(.*?),', line)[0]
    dic['dockerVersion']=re.findall(r',.*?,(.*?),.*?', line)[0]
    #lis.append(sorted(dic.items(), key=lambda d: d[0]))
    lis.append(dic)

temp_dict['project_name'] = PROJECTNAME
temp_dict['apps'] = lis
temp_dict['artifact'] = []
temp_dict['git_diff_file_list'] = []
temp_dict['process_artifact_list'] = []
temp_dict['process_image_list'] = []
temp_dict['gittag'] = ""
temp_dict['skipCheck'] = "false"

# 写个循环加入字典
with open(submit, 'w') as f:
    json.dump(result, f)
```

3.2.5　构建 Docker 日志明细归档

这个功能也是为了解决之前的一个痛点问题：Docker 构建某应用镜像时，无法确定是哪个步骤慢了或有没有经过 Docker Cache 缓存。另外，项目交付团队也希望能够保留 15 天左右的应用镜像构建详细日志，以便定位和调试镜像构建过程中出现的各种问题。其具体实现在 sdkbuildImage.sh 文件中，相关源码如下：

```
#!/usr/bin/env bash
basedir=${BDOS_SDKBUILD_HOME}
timestr=$(date +%Y%m%d%H%M)
commitID=`echo ${GIT_COMMIT} | cut -c1-8`

# 如果是 wget Artifact 文件，则强制 docker build 运行 docker --no-cache 逻辑
```

```
    export MISS_CACHE="--build-arg CACHEBUST=$(date +%s)"

    # 生成build_artifact_result文件，方便Jenkins Console输出
    export BUILD_IMAGE_RESULT=$basedir/${BUILD_IMAGE_LOGS}/build_image_result
    if [  ! -f ${BUILD_IMAGE_RESULT} ]; then
      touch ${BUILD_IMAGE_RESULT}
    fi

    AUTH=""
    if [[ "$ARTIFACT_DOWNLOAD_USER" != "" && "$ARTIFACT_DOWNLOAD_PWD" != "" ]];
then
        AUTH="--user=${ARTIFACT_DOWNLOAD_USER}
--password=${ARTIFACT_DOWNLOAD_PWD}"
    fi
    upload_user=
    if [[ "$ARTIFACT_DOWNLOAD_USER" != "" && "$ARTIFACT_DOWNLOAD_PWD" != "" ]];
then
        upload_user="--user ${ARTIFACT_DOWNLOAD_USER}:${ARTIFACT_DOWNLOAD_PWD}"
    fi

    echo -e "######      start to build $imageName:$VERSION image in
$PACKAGE_INSTANCE_DIR/"
    echo ".git"     > .dockerignore
    echo "*/.git"  >> .dockerignore
    echo "*~"   >> .dockerignore
    echo "*.tgz" >> .dockerignore
    echo "#*" >> .dockerignore

    set +e
    echo -e "[notice]You can tail -f ${build_docker_result} to follow up log
info ..."
    # 定义build_image函数
    function build_image()
    {
    if [[ "$app_dockerfile_path" != "" ]]; then
        app_dockerfile_path=" -f ${BDOS_SDKBUILD_HOME}/$app_dockerfile_path "
    fi

    echo "func_appfile:"${app_dockerfile_path}
    # 进入镜像的上下文环境目录中
```

```
    cd ${SDKBUILD_HOME}/$imagectx
    docker build ${MISS_CACHE} --build-arg
DOCKER_HOSTED_DEV=${DOCKER_HOSTED_DEV} --build-arg AUTH="${AUTH}" --build-arg
ARTIFACT_HOST=${ARTIFACT_HOST_PULL} --build-arg SOURCES_LIST=${SOURCES_LIST}
-t $imageName:$VERSION ${app_dockerfile_path} . >${build_app_docker_result}
2>&1
    if [[ $? -ne 0 ]]; then
         FLAG="failed"
         echo -e "######      build images $imageName:$VERSION ${FLAG} ..."
         # 如果构建失败，则直接在 Jenkins Console 界面显示最后 100 行的日志，方便查看
         tail -n 100 ${build_app_docker_result}
         echo "$imageName:$VERSION  Build image failed" >> ${BUILD_IMAGE_RESULT}
         exit 1
    else
         FLAG="success"
         echo -e "######      build images $imageName:$VERSION ${FLAG} ..."
         echo "$imageName:$VERSION  Build image success" >> ${BUILD_IMAGE_RESULT}
         cp -f ${BUILD_IMAGE_RESULT} $basedir
    fi
    # 兼容旧的 repo 代码结构
    cd ${BDOS_SDKBUILD_HOME}

    if [[ "$REGISTRY_DOMAIN" != "" ]]; then
        docker tag $imageName:$VERSION ${REGISTRY_DOMAIN}/${imageName}:$VERSION
        if [[ $? -ne 0 ]]; then
            echo -e "######      docker tag $imageName:$VERSION to
${REGISTRY_DOMAIN}/${imageName}:$VERSION failed ..."
            exit 1
        fi
        docker push ${REGISTRY_DOMAIN}/${imageName}:$VERSION
        if [[ $? -ne 0 ]]; then
            echo -e "######      docker push
${REGISTRY_DOMAIN}/${imageName}:$VERSION failed ..."
            exit 1
        fi
    else
        echo -e "REGISTRY_DOMAIN is not set. Don't push to registry."
    fi
    }
```

3.2.6 自动给镜像打 Tag 版本

自动给镜像打 Tag 版本功能在 CI 流水线体系中较为重要。有了此功能，由于 Docker 镜像 Tag 在整个 CI/CD 体系中有唯一标识性，这样就不会出现后面构建的应用镜像覆盖前面已完成的应用镜像的问题了。这个功能在开发密集阶段及多项目分支并行开发时非常有用。

```
#!/usr/bin/env bash
basedir=${BDOS_SDKBUILD_HOME}
timestr=$(date +%Y%m%d%H%M)
# 镜像小版本号取 commitId 前 8 位数字，保证唯一性
commitID=`echo ${GIT_COMMIT} | cut -c1-8`

# 循环执行 process_list.txt 的文件列表，并进行 docker build 和 docker push 动作
mapfile myarr < process_image_list.txt
for(( i=0;i<${#myarr[@]};i++))
do
    export  target=`echo ${myarr[i]} | awk -F',' '{print $6}'`
    export  imageName=`echo ${myarr[i]}  | awk -F',' '{print $2}'`
    # export IMAGE_NAME environment variable for preinstall.sh file
    export IMAGE_NAME=$imageName
    export appversion=`echo ${myarr[i]}  | awk -F',' '{print $3}'`
    export imagectx=`echo ${myarr[i]}  | awk -F',' '{print $7}'`
    export preinstall=`echo ${myarr[i]}  | awk -F',' '{print $5}'`
    export tagFlag=`echo ${myarr[i]}  | awk -F',' '{print $8}'`
    export app_dockerfile_path=`echo ${myarr[i]}  | awk -F',' '{print $9}'`
    export SOURCES_LIST=`echo ${myarr[i]}  | awk -F',' '{print $10}'`
    app_post_action=`echo ${myarr[i]}  | awk -F',' '{print $11}'`
    # tagFlag 参数由 Jenkins UI 传进来，分成 False 和 True 两种情况处理
    if [ "$tagFlag"x == "False"x ];then
        export VERSION=$appversion
    else
        export VERSION=$appversion-$commitID
    fi
    export
build_app_docker_result="$basedir/${BUILD_IMAGE_LOGS}/${imageName}-build-
docker-result.log"
    set -e
    if [[  "${preinstall}" != "" ]]; then
        echo "preinstall:"${preinstall}

        if [[  "${preinstall}" == "preinstall.sh" ]]; then
```

```
            bash ${BDOS_SDKBUILD_HOME}/$preinstall
        else
            /bin/sh -c "${preinstall}"
        fi
    fi
    set +e
    # 执行buid_image函数
    build_image
    if [[  "${app_post_action}" != "" ]]; then
        echo "app_post_action:"${app_post_action}
        /bin/sh -c "${app_post_action}"
    fi
done
...
# 进入镜像的上下文环境目录中
cd ${SDKBUILD_HOME}/$imagectx
# 注意docker build中的VERSION变量
docker build ${MISS_CACHE} --build-arg
DOCKER_HOSTED_DEV=${DOCKER_HOSTED_DEV} --build-arg AUTH="${AUTH}" --build-arg
ARTIFACT_HOST=${ARTIFACT_HOST_PULL} --build-arg SOURCES_LIST=${SOURCES_LIST}
-t $imageName:$VERSION ${app_dockerfile_path} . >${build_app_docker_result}
2>&1
if [[ $? -ne 0 ]]; then
     FLAG="failed"
     echo -e "######      build images $imageName:$VERSION ${FLAG} ..."
     # 如果构建失败，则直接在Jenkins Console界面显示最后100行日志，方便查看
     tail -n 100 ${build_app_docker_result}
     echo "$imageName:$VERSION  Build image failed" >> ${BUILD_IMAGE_RESULT}
     exit 1
else
     FLAG="success"
     echo -e "######      build images $imageName:$VERSION ${FLAG} ..."
     echo "$imageName:$VERSION  Build image success" >> ${BUILD_IMAGE_RESULT}
     cp -f ${BUILD_IMAGE_RESULT} $basedir
fi
```

3.2.7 自动测试Nexus 3的缓存代理功能

自动测试Nexus 3的缓存代理功能主要通过运行testcienv.sh文件实现。它会测试内部的Nexus 3缓存功能，每天自动在工作日凌晨运行，如果失败，则会在研发企微群里发送失败通知（若有错误提前处理，免得影响CI SDK构建镜像的功能）。其具体代码如下：

```
    #!/bin/bash
    set -e
    Set -u
    Set -o pipefail
    # 定义 Docker 镜像的名字
    container_name_apt=testDebian9
    container_name_yum=testCentos7
    container_name_pip=testpip
    container_name_apttwo=testDebian9-2

    # 如果存在 Docker 镜像则清理
    dockerName=`docker ps | awk '{print $10}'`
    echo $dockerName | egrep
    "${container_name_apt}|${container_name_yum}|${container_name_pip}|$container_
name_apttwo"
    if [[ $? -eq 0 ]];then
        docker rm -f ${container_name_apt} ${container_name_yum}
${container_name_pip} $container_name_apttwo || true
    fi

    docker run -ti -d --name ${container_name_apt} openjdk:8-jdk-stretch /bin/bash
    docker cp ./apt/sources.list ${container_name_apt}:/etc/apt/sources.list
    docker exec ${container_name_apt} /bin/bash -c 'apt-get update --fix-missing
&& DEBIAN_FRONTEND=noninteractive apt-get install --fix-missing -y apt-
transport-https bash curl gcc krb5-user libkrb5-dev libsasl2-dev locales jq
mysql-client openssh-server python-dev python-kerberos python-pip vim psmisc
sasl2-bin libsasl2-2 libsasl2-modules libsasl2-modules-gssapi-mit alien
libaio1 sudo curl unzip zip git libpam-krb5'

    if [[ $? -eq 0 ]];then
      echo "Test Apt Proxy Pass"
      docker rm -f $container_name_apt
    else
      echo "Test Apt Proxy Failed"
      docker rm -f $container_name_apt
      exit 1
    fi

    docker run -ti -d --name ${container_name_yum} centos:7 /bin/bash
    docker exec ${container_name_yum} /bin/bash -c ' rm -rf /etc/yum.repos.d/*'
    docker cp ./yum/internal.repo testCentos7:/etc/yum.repos.d/
```

```
    docker exec ${container_name_yum} /bin/bash -c 'yum clean all && yum makecache'
    docker exec ${container_name_yum} /bin/bash -c 'yum -y update && yum groupinstall -y "Development tools" && yum remove -y git && yum install -y bzip2 git222 gzip jq krb5-workstation mysql openldap openldap-devel openssh openssh-clients openssh-server openssl openssl-devel python-devel python-pip  sudo unzip wget nodejs && yum install -y --setopt=obsoletes=0 docker-ce-17.12.1.ce'

    if [[ $? -eq 0 ]];then
      echo "Test YUM Proxy Pass"
      docker rm -f $container_name_yum
    else
      echo "Test YUM Proxy Failed"
      docker rm -f $container_name_yum
      exit 1
    fi


    docker run -ti -d --name ${container_name_pip} python:2.7.13 /bin/bash

    docker exec ${container_name_pip} /bin/bash -c 'mkdir -p /root/.pip '
    docker cp ./pypi/pip.conf ${container_name_pip}:/root/.pip/pip.conf
    docker cp ./pypi/requirements.txt ${container_name_pip}:/root/.pip/requirements.txt
    docker exec ${container_name_pip} /bin/bash -c 'pip install --upgrade pip && pip uninstall -y paramiko && pip uninstall -y ipaddress'
    docker exec ${container_name_pip} /bin/bash -c 'pip download --destination-directory /root/pypi pip && pip download --destination-directory /root/pypi -r /root/.pip/requirements.txt'


    if [[ $? -eq 0 ]];then
      echo "Test pip Proxy Pass"
      docker rm -f $container_name_pip
    else
      echo "Test pip Proxy Failed"
      docker rm -f $container_name_pip
      exit 1
    fi
```

```
    docker run -ti -d --name ${container_name_apttwo} debian:stretch /bin/bash
    docker cp ./apt/sources.list ${container_name_apttwo}:/etc/apt/sources.list
    docker exec ${container_name_apttwo} /bin/bash -c 'apt-get update --fix-missing && DEBIAN_FRONTEND=noninteractive apt-get -qq  install openssh-server wget vim curl apt-transport-https locales krb5-user python-kerberos python-pip python-dev libsasl2-dev gcc python-kerberos libkrb5-dev'

    if [[ $? -eq 0 ]];then
      echo "Test Apttwo Proxy Pass"
      docker rm -f $container_name_apttwo
    else
      echo "Test Apttwo Proxy Failed"
      docker rm -f $container_name_apttwo
      exit 1
    fi
```

3.4 小　　结

CI SDK 目前应用于两套 CI 环境，即前面提到的本地环境（对应测研环境）和阿里云环境（对应预发布和线上环境），可以支撑内部近百个构建 / 发布作业，日均运行 100 多次构建 / 发布任务，解决了以前的 CI 痛点问题，缩短了每次 CI 构建的时间，极大地提升了研发效率。

CI SDK 是 DevOps 效率型工具，是为产品 / 项目在产研阶段而做的解决方案；它充分利用了 Jenkins/Nexus3/Docker 等开源组件，利用脚本语言实现了基于 Jenkins CI 的增量 / 按需 / 全量 / webhook 等构建策略，缩短了增量构建的时间，极大地提高了构建的稳定性；站在工程的角度看，它在设计上并不完美，但确实能在某个阶段解决 CI 的痛点问题，从而提升整个 CI 的效率，进一步提高产品的研发效率。如果在产品研发的过程中遇到类似的痛点问题，可以参考 CI SDK 的解决思路。

第 4 章　用 Go 语言开发 CD 自动化发版工具

4.1　项目开发概述

项目背景：SaaS 产品线全面采用腾讯云无托管版 Kubernetes 集群与容器微服务架构进行应用的发布和运行，按照研发环境、测试环境、预发布环境、线上环境及项目 POC 环境，主要部署 Go 应用（但每个环境的配置是不一样的），5 种环境的容器总量达到了 8000+，所以这里采取开发 CD（Continuous Deployment，以下简称 CD）自动化发版工具，并且也做了研发 / 测试环境自动化发版，预发布 / 线上环境手动发版的效果。

开发目的：减少 CD 自动化发版过程中的重复性工作，遵循云原生 DevOps 流程，能够支持快速迭代和频繁发版，实现自动化 CD。

CD 自动化发版工具流程如图 4.1 所示。

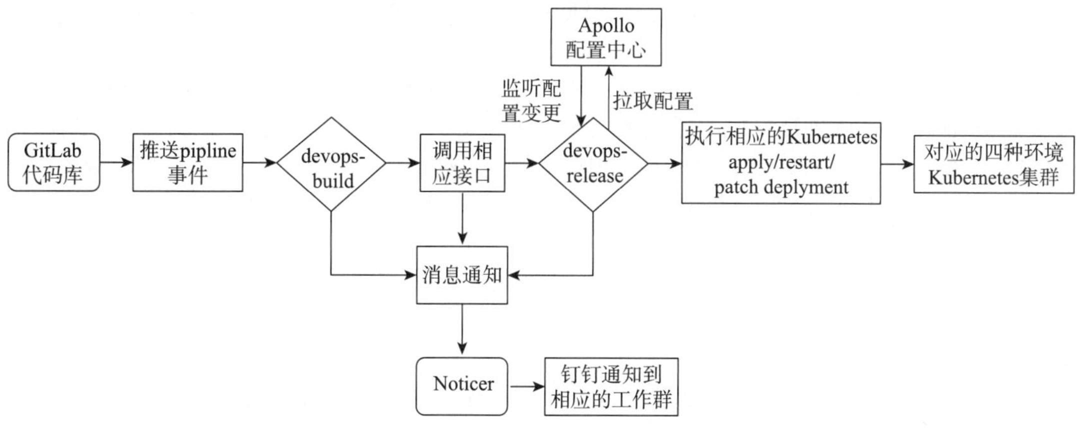

图 4.1　CD 自动化发版工具流程

CD 工具目前的版本（Version）：1.0。

开发语言：Golang，版本 1.17。

主要开源组件：GitLab + PostgreSQL 数据库 + Apollo。

主要开发组件：devops - build + devops - release。

使用的 Go 第三方库有以下部分：gin/gin-swagger/client-go/mapstruct/viper/cobra/logrus。

CD 发版工具的主要功能如下：

（1）支付多集群 / 多租户环境（namespace）部署。

（2）利用 Go 语言打通了 Kubernetes/Postgres/Apollo/GitLab。

（3）devops_build 负责对外提供 API 和 PostgresSQL 操作，devops_release 负责对 Kubernetes 集群进行操作，并且开发了 apollo_sdk 作为常驻程序监听 Apollo。

（4）Kubernetes 集群采取 AES+ 自定义私匙加密的方式，提升了安全性。

（5）支持从 0 到 1 发布一个完整的项目，即 Deployment+Service，操作动作为 build。

（6）支持版本更新，如从 1.0 到 1.1，操作动作为 patch_image。

（7）Apollo 功能正常，能通过 Golang 程序自动生成 configMap 并挂载到应用上，并且相应配置发生更新时，能够自动应用到 Deployment 且重启应用。

（8）程序已经兼容了 CD 发版流程上单 ConfigMap 对应多 Deployment 的方式，但实际操作略复杂，建议还是采用单 ConfigMap 对应单 Deployment 的方式。

（9）支持 route 路径识别发版。

（10）除了支持 build/restart/patch_image 动作以外，还支持 patch Deployment 的动作。

总结了 CD 发版工具的主要功能以后，下面介绍 CD 发版工具的局限性。

（1）整个工程代码（加上单元测试）接近 30000 万行，后期需要考虑精简流程，以减少整体代码量。

（2）重度依赖 GitLab 和 GitLab CI，目前只能适用于 GitLab 代码库环境，其他代码管理版本不支持。

自动化发版工具的很多操作还是要用前端界面执行，后续需要前端开发人员介入。

目前市场上有很多基于 Kubernetes 的成熟云原生发版工具，如 ArgoCD。企业可以基于自身的开发 / 运维现状合理挑选工具和方案。在研发人员较少且不愿意在这方面投入人力的情况下，可以使用 ArgoCD。区别在于，像当前这种情况，如果使用 ArgoCD，那么配置中心的引入将是一个比较麻烦的事情，各种环境的配置只能以 YAML 的方式放在 GitLab 源码中（包括很多客户项目上的敏感信息）。相比之下，Apollo 配置中心则具备权限管理功能，能严格控制查看和修改权限，从而降低安全风险。因此，对于云原生发版工具的选择，还是得结合自身的实际情况来判断，以做出最优选择。

在 CD 自动化发版工具中，比较重要的环节是 Apollo 和 GitLab CI。下面介绍这两者的相关应知识点。

4.2 GitLab CI 和 Runner 简介

4.2.1 GitLab CI 的基本概念

GitLab CI 是 GitLab 提供的持续集成（Continuous Integration）服务，可以帮助自动化构建、测试和部署。下面介绍 GitLab CI 的基本概念和使用方法。

（1）Runner：Runner 是 GitLab CI 的执行代理程序，可以在不同的操作系统中运行。GitLab CI 通过 Runner 来执行 CI/CD 任务。

（2）pipeline：pipeline 代表 GitLab CI 中的一次完整的 CI/CD 流程。它由多个阶段（Stage）组成，每个阶段又包含多个作业（Job）。pipeline 会按照定义好的顺序依次执行 Stage 和 Job。

（3）Job：Job 是 GitLab CI 中最基础的执行单元。每个 Job 都必须归属于某个 Stage，并且要在 Runner 上执行。

（4）Stage：Stage 是一个逻辑分组，用于将相关联的 Jobs 归在一组。只有当前 stage 前面的所有 stage 都执行成功后，该 Stage 才能开始执行。

（5）Artifacts：Artifacts 是 Job 执行结果的输出物，可以是文件、归档等。这些 Artifacts 可供后续的 Job 使用。

GitLab CI 的配置文件是 .gitlab-ci.yml，该文件必须位于项目的根目录下，采用 YAML 格式，用缩进表示层次关系。下面是一个配置文件的简单示例：

```
stages:
  - build
  - test
  - deploy
build_job:
  stage: build
  script:
    - echo "Hello, GitLab CI!"
test_job:
  stage: test
  script:
    - echo "Testing..."
deploy_job:
  stage: deploy
  script:
    - echo "Deploying..."
```

以上配置文件定义了三个 Stage：build、test 和 deploy。每个 Stage 包含一个 Job，分别是 build_job、test_job 和 deploy_job。每个 Job 都有一个 script 属性，表示 Job 要执行的命令。

在 GitLab 中启用 CI/CD 功能需要管理员权限。在项目设置中选择 CI/CD，然后选择“自动部署”，接着按照提示进行配置即可。具体步骤如下。

（1）使用默认的 Runner。GitLab 提供了一些 Runner 可以直接使用，只需将该项目的 gitlab-ci.yml 文件提交到 GitLab 即可。GitLab 会自动检测到该文件并开始构建项目。

（2）自定义 Runner。如果需要更多的控制权，可以自己搭建 Runner。GitLab 提供了 Docker 镜像，可以很方便地搭建 Runner。测试过很多类型的 Runner，如物理机（虚拟机）和 Docker/Kubernetes 类型，最稳定的就是使用物理机（虚拟机）来安装 Runner。下面会重点介绍。

4.2.2 使用物理机（虚拟机）安装 Runner

下面是在一台 CentOS 7.9 x86_64 机器上安装 Runner 的步骤：

（1）登录 GitLab 服务器，通过地址 http://10.1.0.200:30808/admin/runners 拿到 Token，如图 4.2 所示。

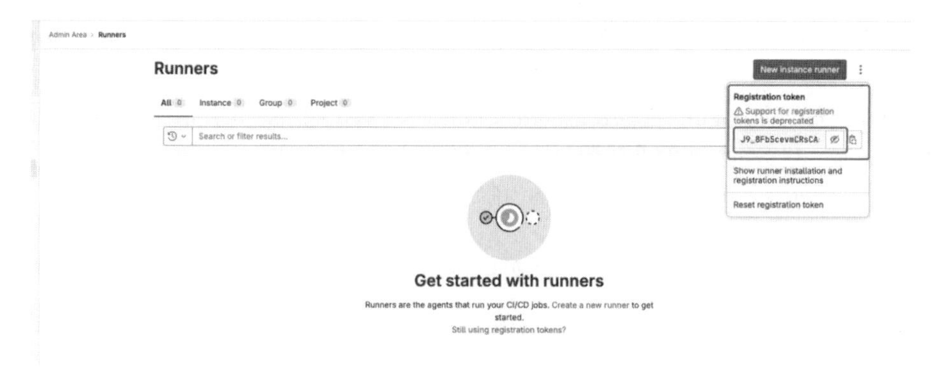

图 4.2　通过 GitLab 获取 Token

（2）安装及启动 gitlab-runner 服务。

安装 gitlab-runner 服务，其命令如下：

```
    curl -L
https://packages.gitlab.com/install/repositories/runner/gitlab-runner/script.
rpm.sh | sudo bash
    yum install gitlab-runner  -y
```

启动 gitlab-runner 服务，其命令如下：

```
    systemctl  daemon-reload                    # 重新加载配置
    systemctl start gitlab-runner               # 启动服务
    systemctl enable gitlab-runner              # 设置开机启动
    systemctl restart gitlab-runner             # 重启服务
```

（3）通过 gitlab-runner register 命令注册 Runner：

```
    Runtime platform arch=amd64 os=linux pid=6245 revision=ac8e767a
version=12.6.0
    Running in system-mode.

    # 输入公司的 gitlab 公网地址
    Please enter the gitlab-ci coordinator URL (e.g. https://gitlab.com/):
    http://10.1.0.200:30808

    # 输入 gitlab 的 token，找到第三步 token，然后选择第四步，复制 token
    Please enter the gitlab-ci token for this runner:
    8sjydnrsPuSRhKiQTQky

    # 输入 runner 名称
    Please enter the gitlab-ci description for this runner:
    [k8s-node02]: my-runner

    # 输入 runner 的标签
```

```
Please enter the gitlab-ci tags for this runner (comma separated):
my-tag,another-tag
Registering runner... succeeded runner=8sjydnrs

# 输入 runner 执行器的环境
Please enter the executor: custom, docker-ssh, parallels, kubernetes, docker-ssh+machine, docker, shell, ssh, virtualbox, docker+machine:
shell
Runner registered successfully. Feel free to start it, but if it's running already the config should be automatically reloaded!
```

成功安装 gitlab-runner 后的界面，如图 4.3 所示。

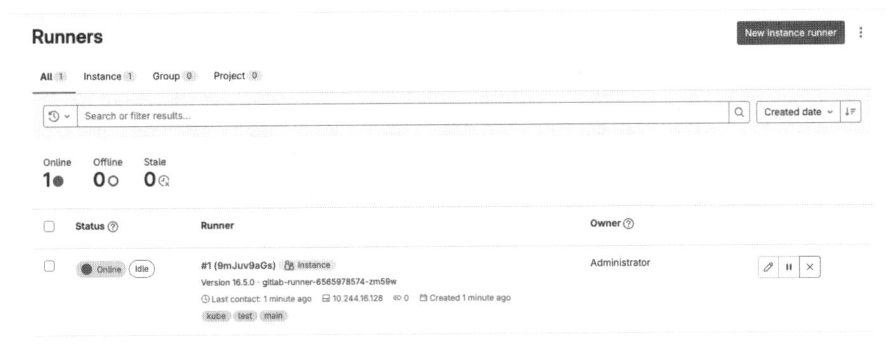

图 4.3　成功安装 gitlab-runner 后的界面

（4）使用 gitlab-runner 遇到的问题如图 4.4 所示。

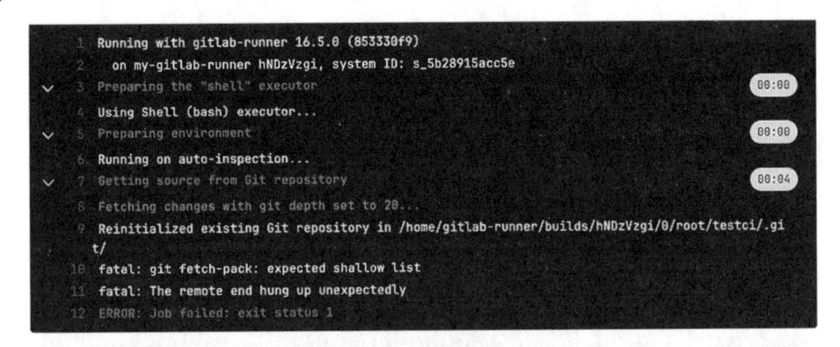

图 4.4　使用 gitlab-runner 遇到的问题

下面用修复上面遇到的问题的方法升级 git 版本，其过程如下：

```
git --version
```

显示结果如下：

```
git version 1.8.3.1
yum install http://opensource.wandisco.com/centos/7/git/x86_64/wandisco-git-release-7-2.noarch.rpm
yum install git
```

```
yum update git
git --version
```

显示结果如下：

```
git version 2.39.1
```

另外，gitlab-runner 用户会报 Docker 权限使用不足的错误，其处理命令如下：

```
sudo  -u gitlab-runner -H docker info
```

输出结果如下：

```
Client:
 Context:    default
 Debug Mode: false
 Plugins:
  buildx: Docker Buildx (Docker Inc.)
    Version:  v0.10.4
    Path:     /usr/libexec/docker/cli-plugins/docker-buildx
  compose: Docker Compose (Docker Inc.)
    Version:  v2.17.3
    Path:     /usr/libexec/docker/cli-plugins/docker-compose

Server:
ERROR: permission denied while trying to connect to the Docker daemon socket at unix:///var/run/docker.sock: Get "http://%2Fvar%2Frun%2Fdocker.sock/v1.24/info": dial unix /var/run/docker.sock: connect: permission denied
errors pretty printing info
```

将 gitlab-runner 用户添加到 docker 组，其操作命令如下：

```
sudo usermod -aG docker gitlab-runner
```

最后，用一个简单的 YAML 来测试，输出结果如图 4.5 所示。

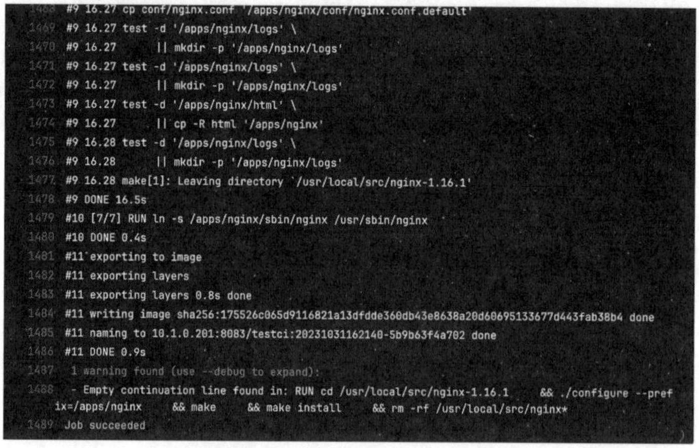

```
1468 #9 16.27 cp conf/nginx.conf '/apps/nginx/conf/nginx.conf.default'
1469 #9 16.27 test -d '/apps/nginx/logs' \
1470 #9 16.27       || mkdir -p '/apps/nginx/logs'
1471 #9 16.27 test -d '/apps/nginx/logs' \
1472 #9 16.27       || mkdir -p '/apps/nginx/logs'
1473 #9 16.27 test -d '/apps/nginx/html' \
1474 #9 16.27       || cp -R html '/apps/nginx'
1475 #9 16.28 test -d '/apps/nginx/logs' \
1476 #9 16.28       || mkdir -p '/apps/nginx/logs'
1477 #9 16.28 make[1]: Leaving directory `/usr/local/src/nginx-1.16.1'
1478 #9 DONE 16.5s
1479 #10 [7/7] RUN ln -s /apps/nginx/sbin/nginx /usr/sbin/nginx
1480 #10 DONE 0.4s
1481 #11 exporting to image
1482 #11 exporting layers
1483 #11 exporting layers 0.8s done
1484 #11 writing image sha256:175526c065d9116821a13dfdde360db43e8638a20d60695133677d443fab38b4 done
1485 #11 naming to 10.1.0.201:8083/testci:20231031162140-5b9b63f4a702 done
1486 #11 DONE 0.9s
1487  1 warning found (use --debug to expand):
1488  - Empty continuation line found in: RUN cd /usr/local/src/nginx-1.16.1     && ./configure --pref
   ix=/apps/nginx     && make     && make install     && rm -rf /usr/local/src/nginx*
1489 Job succeeded
```

图 4.5　显示 gitlab-runner 安装成功

输出结果为“Job succeeded”，表示利用 gitlab-runner 执行 CI 任务是成功的。

4.3　Apollo 的主要功能和设计应用

随着程序功能的日益复杂，程序的配置日益增多，如功能开关、参数设置、服务器地址等。开发人员对程序配置的要求越来越高，如配置修改后实时生效，支持灰度发布，分环境、分集群管理配置，完善的权限、审核机制等。在这样的大环境下，传统的配置文件、数据库等方式已经无法满足开发人员对配置管理的需求。基于这种情况，Apollo（阿波罗）配置中心应运而生。

Apollo 自开发之初采用的就是开源模式，服务端基于 Spring Cloud 和 Spring Boot 框架用 Java 开发。客户端目前提供 Java 和 .Net 两种实现。

Apollo 是携程框架部门研发的开源配置管理中心，能够集中化管理使用不同环境、不同集群的配置，配置修改后能够实时推送到应用端，并且具备规范的权限、流程治理等特性。

Apollo 支持从 4 个维度管理 Key-Value 格式的配置，分别为应用（application）、环境（environment）、集群（cluster）和命名空间（namespace）。

4.3.1　Apollo 的主要功能

Apollo 的主界面如图 4.6 所示。

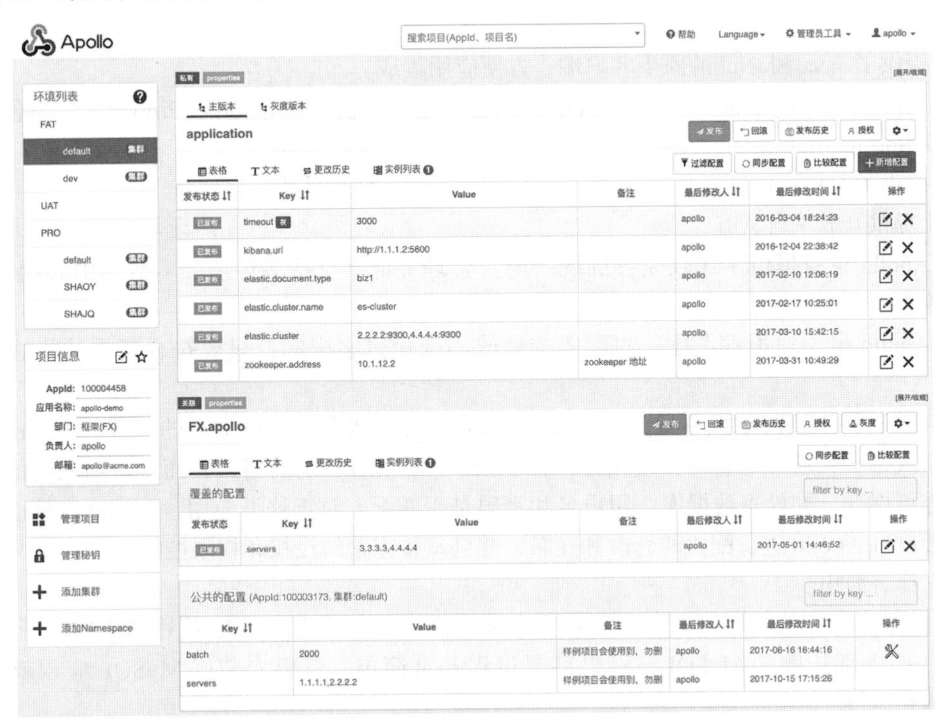

图 4.6　Apollo 的主界面

Apollo 的主要功能如下：

（1）统一管理不同环境、不同集群的配置。

1）Apollo 提供了一个统一界面，可集中式管理不同环境、不同集群、不同命名空间的配置。

2）同一份代码部署在不同的集群，可以有不同的配置，如 zookeeper 地址等。

3）通过命名空间可以很方便地支持多个不同应用共享同一份配置，同时还允许对共享的配置进行覆盖。

4）配置界面支持多种语言，如中文、英文等。

（2）配置修改实时生效（热发布）。用户在 Apollo 修改配置并发布后，客户端能实时（1 秒）接收到最新的配置，并通知到应用程序。

（3）版本发布管理。所有的配置发布都有版本记录，便于配置的回滚和管理。

（4）灰度发布。支持配置的灰度发布。例如，发布配置后，只对部分应用实例生效，等观察一段时间没问题后再推给所有应用实例。

（5）权限管理、发布审核、操作审计。

1）应用和配置的管理都有完善的权限管理机制，对配置的管理分为编辑和发布两个环节，可减少人为的错误。

2）所有的操作都有审计日志，可以方便追踪问题。

（6）客户端配置信息监控。可以方便地看到配置在被哪些实例使用。

（7）提供 Java 和 .Net 原生客户端。

1）提供了 Java 和 .Net 的原生客户端，方便应用集成。

2）支持 Spring Placeholder、Annotation 和 Spring Boot 的 ConfigurationProperties，方便应用使用（需要 Spring 3.1.1+）。

3）提供了 HTTP 接口，非 Java 和 .Net 应用也可以方便地使用。

（8）提供开放平台 API。

1）Apollo 自身提供了比较完善的统一配置管理界面，支持多环境、多数据中心配置管理、权限、流程治理等特性。

2）Apollo 出于通用性考虑，对配置的修改不会做过多限制，只要符合基本的格式就能够保存。

3）在调研中发现，对于有些使用方，它们的配置可能会有比较复杂的格式，如 xml、json，需要对格式进行校验。还有一些使用方（如 DAL），不仅有特定的格式，而且对输入的值也需要校验后方可保存，如检查数据库、用户名和密码是否匹配。对于这类应用，Apollo 支持应用方通过开放接口对 Apollo 进行配置的修改和发布，并且具备完善的授权和权限控制。

（9）部署简单。

1）配置中心作为基础服务，可用性要求非常高，这就要求 Apollo 对外部依赖尽可能地少。目前唯一的外部依赖是 MySQL，所以部署过程非常简单，只要安装好 MySQL 应用就可以让 Apollo 顺利运行。

2）在 Kubernetes 集群中，利用 Helm 3 和 Yaml 安装非常方便，安装流程在后面的章节中详细介绍。

4.3.2 Apollo 的主要设计说明

Apollo 的主架构模块如图 4.7 所示。

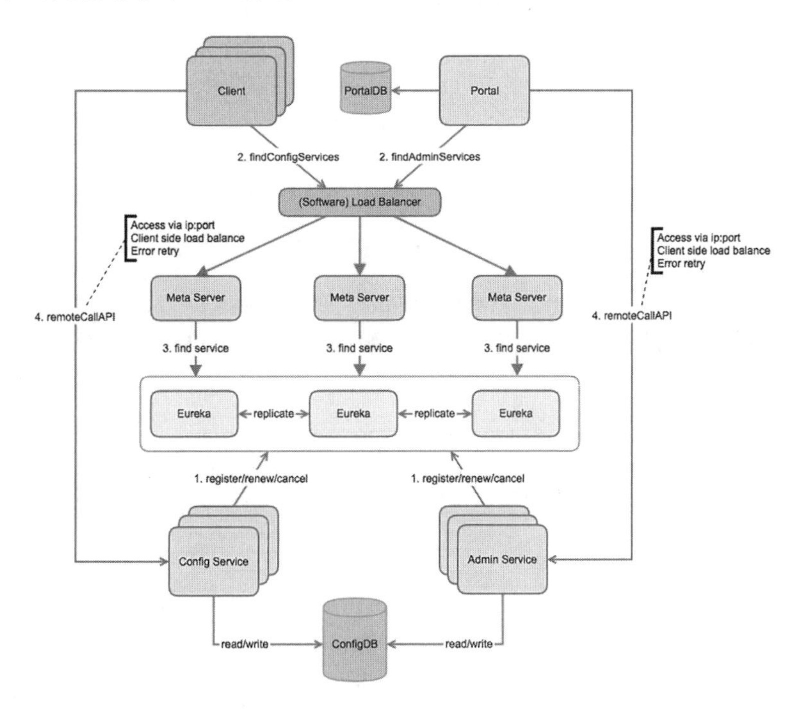

图 4.7　Apollo 的主架构模块

Apollo 用户端工作流程如图 4.8 所示。

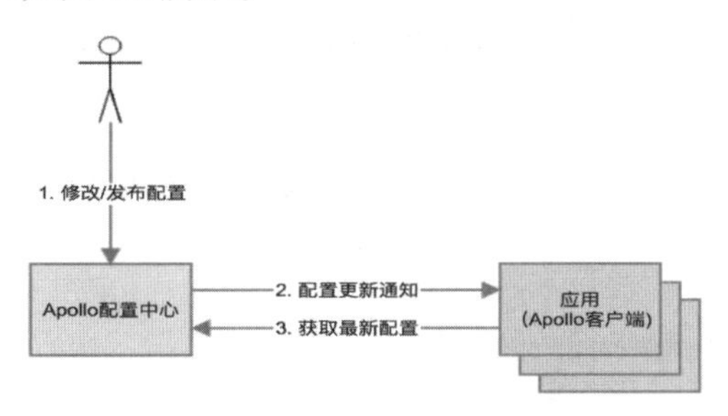

图 4.8　Apollo 用户端工作流程

Apollo 用户端的主要工作流程：

（1）用户在配置中心对配置进行修改并发布。

（2）配置中心通知 Apollo 客户端有配置更新。

（3）Apollo 客户端从配置中心获取最新的配置、更新本地配置并通知到应用。

可以将图 4.7 中 Apollo 的主架构模块细分，了解各组件的工作流程。Config Service 提供配置的读取、推送等功能，服务对象是 Apollo 客户端，其工作流如图 4.9 所示。

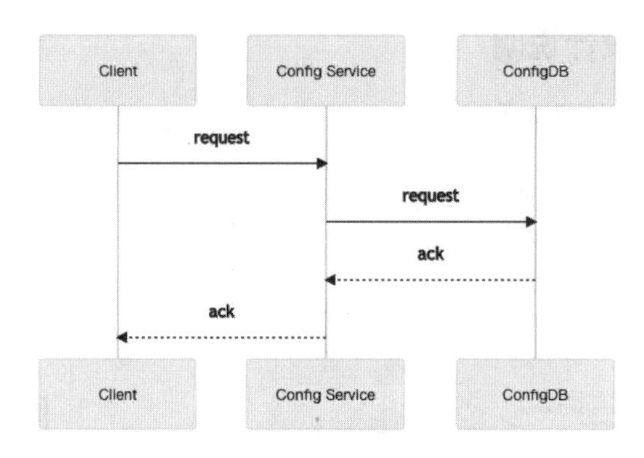

图 4.9 Apollo Config Service 工作流

Admin Service 提供配置的修改、发布等功能，服务对象是 Apollo Portal（管理界面），其工作流如图 4.10 所示。

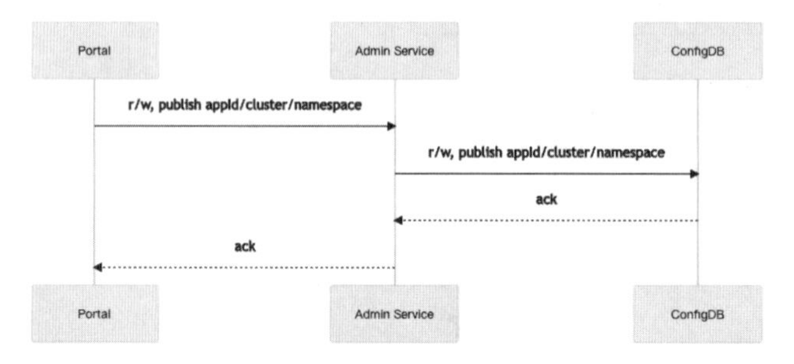

图 4.10 Apollo Admin Service 工作流

Portal 通过域名访问 Meta Server 获取 Admin Service 服务列表（IP+Port），然后直接通过 IP+Port 访问服务，同时在 Portal 端会做 load balance、错误重试，其工作流如图 4.11 所示。

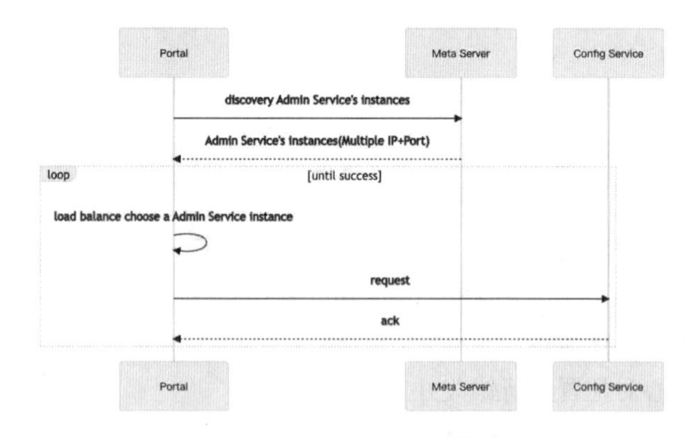

图 4.11 Apollo Portal 工作流

Client 端通过域名访问 Meta Server 获取 Config Service 服务列表（IP+Port），然后直接通过 IP+Port 访问服务，同时在 Client 端会做 load balance、错误重试，其工作流如图 4.12 所示。

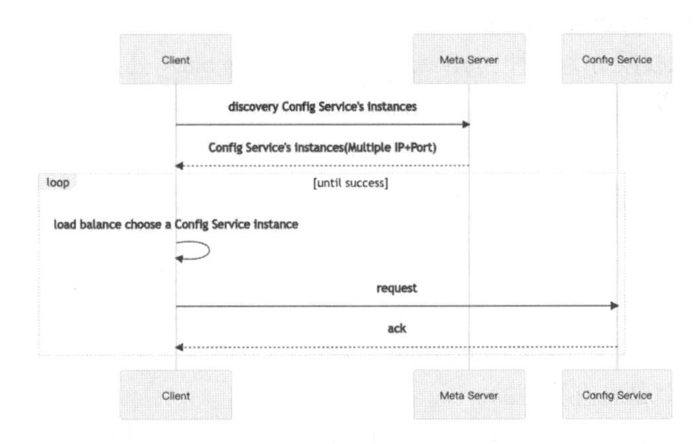

图 4.12　Apollo Client 工作流

Meta Server 在 Apollo 中是一个重要角色，也是一个逻辑角色，在部署时与 Config Server 在同一个 JVM 进程中，所以 IP 和端口与 Config Server 保持一致，它的作用有如下几点：

（1）Portal 通过域名访问 Meta Server，以获取 Admin Service 服务列表（IP+Port）。

（2）Client 端通过域名访问 Meta Server，以获取 Config Service 服务列表（IP+Port）。

（3）Meta Server 从 Eureka 获取 Config Service 和 Admin Service 的服务信息，相当于一个 Eureka Client。

Apollo 设计增设一个 Meta Server 角色，主要是为了封装服务发现的细节。对于 Portal 和 Client 而言，永远通过一个 HTTP 接口来获取 Admin Service 和 Config Service 的服务信息，而无须关心其背后的实际服务注册和发现组件。

4.3.3　Apollo 的核心概念 namespace

Apollo 中也有 namespace，注意与 Kubernetes 中的 namespace 进行区分。

1. Apollo 的 namespace

namespace 是 Apollo 配置项的集合，类似于一个配置文件。什么是 application 的 namespace？

Apollo 在创建项目时，都会默认创建一个 application 的 namespace。顾名思义，application 是给应用自身使用的，熟悉 Spring Boot 的读者都知道，Spring Boot 项目都有一个默认的配置文件 application.yml。在这里，application.yml 等同于 application 的 namespace。对于 90% 的应用来说，application 的 namespace 已经能够满足日常配置使用场景了。

在客户端获取 application 的 namespace 的代码如下：

```
Config config = ConfigService.getAppConfig();
```

在客户端获取非 application 的 namespace 的代码如下：

```
Config config = ConfigService.getConfig(namespaceName);
```

namespace 配置文件有多种格式，如 properties、xml、yml、yaml、json 等。在 Portal UI 中可以看到，application 的 namespace 上有一个 properties 标签，表示 application 是 properties 格式的。

2. Apollo 的 namespace 权限分类

namespace 的获取权限分为两种：private（私有的）和 public（公共的）。这里的获取权限是相对于 Apollo 客户端来说的。

（1）private 权限。具有 private 权限的 namespace，只能被其所属的应用获取。若一个应用尝试获取其他应用 private 的 namespace，Apollo 配置中心会返回“404”异常。

（2）public 权限。具有 public 权限的 namespace，能被任何应用获取。

namespace 的类型有三种，分别是私有类型、公共类型和关联类型（有时也称为继承类型）。下面详细介绍这三种类型。

（1）私有类型。私有类型的 namespace 具有 private 权限。例如，每个应用都有的默认 application namespace 就是私有类型的。

（2）公共类型。公共类型的 namespace 具有 public 权限。公共类型的 namespace 相当于游离于应用之外的配置，并且通过唯一的 namespace 名称标识，所以公共的 namespace 的名称必须全局唯一。

推荐使用的业务场景有：部门级别共享的配置、小组级别共享的配置、几个项目之间共享的配置和中间件客户端的配置。

（3）关联类型又可称为继承类型，关联类型具有 private 权限。关联类型的 namespace 继承于公共类型的 namespace，用于覆盖公共 namespace 的某些配置。

> **注**：在实际的CD自动化发版工具设计中，为了简化操作和流程，暂时只用到了私有类型。

4.3.4 使用 Apollo 创建应用

登录 Apollo Portal 界面之前，应仔细查看 Apollo 的相关 Pod 日志是否有报错的现象，如果有报错则应该修复相应问题。详细安装过程见 4.3.5 小节。

Apollo 的创建应用界面如图 4.13 所示。

图 4.13　Apollo 的创建应用界面

Apollo 的创建应用界面中的参数说明如下。

- 部门：选择应用所在的部门。部门数据来自 ApolloPortalDB 库中 ServerConfig 表的 Key =

organizations 对应的记录，这里选择默认值：样例部门 1（TEST1）。

- AppId：用来标识应用身份的唯一 id，格式为 string，需要和客户端 app.properties 中配置的 app.id 对应。
- 应用名称：应用名，仅用于界面展示。
- 应用负责人：默认具有项目管理员权限。
- 应用管理员：可以创建 namespace、集群和分配用户权限。

4.3.5 在 Kubernetes 集群上安装 Apollo

在 Kubernetes 集群上安装 Apollo 的步骤较为简单，主要利用 YAML 文件进行配置和安装。

首先，需要安装 MySQL，其安装文件 mysql-singel.yaml 的内容如下：

```
apiVersion: apps/v1
kind: Deployment
metadata:
  labels:
    app: mysql
  name: mysql-deployment
  namespace: devops
spec:
  replicas: 1
  selector:
    matchLabels:
      app: mysql
  template:
    metadata:
      labels:
        app: mysql
    spec:
      containers:
        - name: mysql
          image: 10.1.0.201:8083/mysql:5.7
          imagePullPolicy: IfNotPresent
          env:
            - name: MYSQL_ROOT_PASSWORD
              value: "123456"
          volumeMounts:
            - name: mysql-persistent-storage
              mountPath: /var/lib/mysql
          livenessProbe:
            exec:
```

```
            command: ["mysqladmin", "-ppassword", "ping"]
          initialDelaySeconds: 30
          periodSeconds: 10
          timeoutSeconds: 5
        readinessProbe:
          exec:
            # Check we can execute queries over TCP (skip-networking is off).
            command: [ "mysql", "-p123456", "-h", "127.0.0.1", "-e", "SELECT 1" ]
          initialDelaySeconds: 5
          periodSeconds: 2
          timeoutSeconds: 1
    imagePullSecrets:
      - name: odregistry
    volumes:
      - name: mysql-persistent-storage
        hostPath:
          path: /data/mysql/mysql-data
---
apiVersion: v1
kind: Service
metadata:
  labels:
    app: mysql
  name: mysql-svc
  namespace: devops
spec:
  type: ClusterIP
  ports:
    - port: 3306
      protocol: TCP
      targetPort: 3306
  selector:
    app: mysql
```

然后，需要导入两条初始化 SQL 语句来创建 apolloconfigdb 及 apolloportaldb 数据库，下载地址为 https://github.com/apolloconfig/apollo-quick-start/blob/master/sql/apolloconfigdb.sql。

数据库需要提前创建，在 MySQL 容器中执行以下命令：

```
source /tmp/apolloconfigdb.sql;
source /tmp/apolloportaldb.sql;
```

再依次安装 apollo-config、apollo-admin 及 apollo-portal 组件。apollo-config 组件对应的 YAML 文件内容如下：

```yaml
    apiVersion: apps/v1
    kind: Deployment
    metadata:
      annotations:
        deployment.kubernetes.io/revision: "2"
      creationTimestamp: "2023-04-19T08:38:44Z"
      generation: 2
      labels:
        appName: apollo
      name: apollo-config
      namespace: devops
    spec:
      progressDeadlineSeconds: 600
      replicas: 1
      revisionHistoryLimit: 10
      selector:
        matchLabels:
          name: apollo-config
      strategy:
        rollingUpdate:
          maxSurge: 25%
          maxUnavailable: 25%
        type: RollingUpdate
      template:
        metadata:
          creationTimestamp: null
          labels:
            app: apollo
            ccse_app_name: apollo
            name: apollo-config
            source: CCSE
        spec:
          containers:
          - env:
            - name: SPRING_DATASOURCE_URL
              value:
jdbc:mysql://mysql-svc:3306/ApolloConfigDB?characterEncoding=utf8   #IP 为已创建
                                                                    # 好的数据库 IP
            - name: SPRING_DATASOURCE_USERNAME
              value: "root"
            - name: SPRING_DATASOURCE_PASSWORD
              value: "123456"
```

```
        image: apolloconfig/apollo-configservice:latest
        imagePullPolicy: Always
        name: apollo-config
        resources:
          limits:
            cpu: "4"
            memory: 4Gi
          requests:
            cpu: 100m
            memory: 128Mi
        terminationMessagePath: /dev/termination-log
        terminationMessagePolicy: File
        volumeMounts:
        - mountPath: /etc/localtime
          name: localtime
          readOnly: true
      dnsPolicy: ClusterFirst
      imagePullSecrets:
      - name: odregistry
      restartPolicy: Always
      schedulerName: default-scheduler
      securityContext: {}
      terminationGracePeriodSeconds: 30
      volumes:
      - hostPath:
          path: /etc/localtime
          type: ""
        name: localtime
---
apiVersion: v1
kind: Service
metadata:
  creationTimestamp: "2023-04-19T08:41:39Z"
  labels:
    appName: apollo
    workloadKind: Deployment
    workloadName: apollo-config
  name: apollo-config
  namespace: devops
spec:
  clusterIP: 10.96.69.115
  externalTrafficPolicy: Cluster
```

```
  ports:
  - name: "8080"
    nodePort: 32109
    port: 32109
    protocol: TCP
    targetPort: 8080
  selector:
    name: apollo-config
  sessionAffinity: None
  type: NodePort
```

apollo-admin 组件对应的 YAML 文件内容如下：

```
apiVersion: apps/v1
kind: Deployment
metadata:
  annotations:
    deployment.kubernetes.io/revision: "1"
  creationTimestamp: "2023-04-19T08:53:03Z"
  generation: 1
  labels:
    appName: apollo
  name: apollo-admin
  namespace: devops
spec:
  progressDeadlineSeconds: 600
  replicas: 1
  revisionHistoryLimit: 10
  selector:
    matchLabels:
      name: apollo-admin
  strategy:
    rollingUpdate:
      maxSurge: 25%
      maxUnavailable: 25%
    type: RollingUpdate
  template:
    metadata:
      creationTimestamp: null
      labels:
        app: apollo
        ccse_app_name: apollo
        name: apollo-admin
        source: CCSE
```

```
      spec:
        containers:
        - env:
          - name: SPRING_DATASOURCE_URL
            value:
jdbc:mysql://mysql-svc:3306/ApolloConfigDB?characterEncoding=utf8
          - name: SPRING_DATASOURCE_USERNAME
            value: "root"
          - name: SPRING_DATASOURCE_PASSWORD
            value: "123456"
          image: apolloconfig/apollo-adminservice:latest
          imagePullPolicy: Always
          name: apollo-admin
          resources:
            limits:
              cpu: "4"
              memory: 4Gi
            requests:
              cpu: 100m
              memory: 128Mi
          terminationMessagePath: /dev/termination-log
          terminationMessagePolicy: File
          volumeMounts:
          - mountPath: /etc/localtime
            name: localtime
            readOnly: true
        dnsPolicy: ClusterFirst
        imagePullSecrets:
        - name: odregistry
        restartPolicy: Always
        schedulerName: default-scheduler
        securityContext: {}
        terminationGracePeriodSeconds: 30
        volumes:
        - hostPath:
            path: /etc/localtime
            type: ""
          name: localtime
---
apiVersion: v1
kind: Service
metadata:
```

```yaml
  creationTimestamp: "2023-04-19T08:56:52Z"
  labels:
    appName: apollo
    workloadKind: Deployment
    workloadName: apollo-admin
  name: apollo-admin
  namespace: devops
spec:
  clusterIP: 10.96.178.0
  externalTrafficPolicy: Cluster
  ports:
  - name: "8090"
    nodePort: 32108
    port: 32108
    protocol: TCP
    targetPort: 8090
  selector:
    name: apollo-admin
  sessionAffinity: None
  type: NodePort
```

apollo-portal 组件对应的 YAML 文件内容如下：

```yaml
apiVersion: apps/v1
kind: Deployment
metadata:
  annotations:
    deployment.kubernetes.io/revision: "1"
  creationTimestamp: "2023-04-19T08:43:15Z"
  generation: 1
  labels:
    appName: apollo
  name: apollo-portal
  namespace: devops
spec:
  progressDeadlineSeconds: 600
  replicas: 1
  revisionHistoryLimit: 10
  selector:
    matchLabels:
      name: apollo-portal
  strategy:
    rollingUpdate:
      maxSurge: 25%
```

```
        maxUnavailable: 25%
      type: RollingUpdate
    template:
      metadata:
        creationTimestamp: null
        labels:
          app: apollo
          ccse_app_name: apollo
          name: apollo-portal
          source: CCSE
      spec:
        containers:
        - env:
          - name: SPRING_DATASOURCE_URL
            value:
jdbc:mysql://mysql-svc:3306/ApolloPortalDB?characterEncoding=utf8    #数据库地址
          - name: SPRING_DATASOURCE_USERNAME
            value: "root"
          - name: SPRING_DATASOURCE_PASSWORD
            value: "123456"    #数据库密码
          - name: APOLLO_PORTAL_ENVS
            value: dev
          - name: DEV_META
            value: http://10.96.69.115:32109    #apollo-config的ServiceIP+NodePort端口
          image: apolloconfig/apollo-portal:latest
          imagePullPolicy: Always
          name: apollo-portal
          resources:
            limits:
              cpu: "4"
              memory: 4Gi
            requests:
              cpu: 100m
              memory: 128Mi
          terminationMessagePath: /dev/termination-log
          terminationMessagePolicy: File
          volumeMounts:
          - mountPath: /etc/localtime
            name: localtime
            readOnly: true
        dnsPolicy: ClusterFirst
        imagePullSecrets:
```

```
      - name: odregistry
      restartPolicy: Always
      schedulerName: default-scheduler
      securityContext: {}
      terminationGracePeriodSeconds: 30
      volumes:
      - hostPath:
          path: /etc/localtime
          type: ""
        name: localtime
---
apiVersion: v1
kind: Service
metadata:
  creationTimestamp: "2023-04-19T08:55:49Z"
  labels:
    appName: apollo
    workloadKind: Deployment
    workloadName: apollo-portal
  name: apollo-portal
  namespace: devops
spec:
  clusterIP: 10.96.210.42
  externalTrafficPolicy: Cluster
  ports:
  - name: "8070"
    nodePort: 32110
    port: 32115
    protocol: TCP
    targetPort: 8070
  selector:
    name: apollo-portal
  sessionAffinity: None
  type: NodePort
```

组件安装完成后，可以访问apollo-portal组件对应的nodePort地址，其访问地址为http://10.1.0.200:32110/，初始账号为apollo，密码为admin。

4.4　使用CD发版工具的流程

使用CD发版工具的流程如下：

（1）触发CI：当GitLab分支上有代码推送或打tag操作时，会触发CI事件（这些事件可以根据需求自定义，包括格式化检查、构建镜像、推送镜像等）。CI 事件完成后，GitLab 会通过 webhook 将结果推送到 devops-build 服务。

（2）事件接收与过滤：devops-build 服务接收这些事件信息，并将 CI 的结果通知到钉钉或企业微信（企微）群聊。接着，它会根据数据库中存储的正则表达式（ref_rep 可配置）对事件进行过滤，匹配的内容通常是分支名或 tag 名（由 enabled_branches 和 ref_rep 配置）。只有满足正则匹配条件的分支才会进入下一步流程。

（3）获取发版信息：devops-build 会获取此分支对应的发版 channel（类似于 dev、test、stage 和 prod 的环境概念）。同时，它会从 CI 事件的工作日志中提取 docker 仓库与版本信息（如果日志中没有这些信息，则默认使用 deployment 配置中指定的 docker 仓库加上 tag 名）。之后，将包含 channel、project_name、docker_tag 等信息的发版信息发送到 devops-release 服务。

（4）处理发版：devops-release 根据接收到的发版信息，确定指定项目发版的目的地。例如，对于项目名为 we-work-server 且环境为 dev 的情况，它会从数据库中获取 we-work-server 在 dev 环境的发版路径（包括集群和命名空间）。

（5）获取部署资源：devops-release 会从数据库中获取集群的必要信息（如 token、CA 证书和 URL），从 Apollo 配置中心获取服务的配置并拼接为 ConfigMap（获取配置的关键字包括项目名、环境和集群命名空间）。同时，它还会从数据库中获取项目的 Deployment 和 Service 配置内容。

（6）部署：最后，通过获取 k8sclient（Kubernetes 客户端），将准备好的 ConfigMap、Deployment 和 Service 应用到 Kubernetes 集群中。整个 CD 发版工具的流程相对复杂，主要由 DevOps 组内成员使用。由于 dev/test/stage 环境已经实现了全自动流程，因此目前暂时不需要前端开发人员参与。

4.4.1 devops-build 组件服务

devops-build 组件服务的主要作用是提供资源管理、规则匹配过滤，并统一对外提供 API 服务。同时，它与 devops-release 组件进行交互，而 devops-release 则只与 devops-build 交互。其工作流如图 4.14 所示。

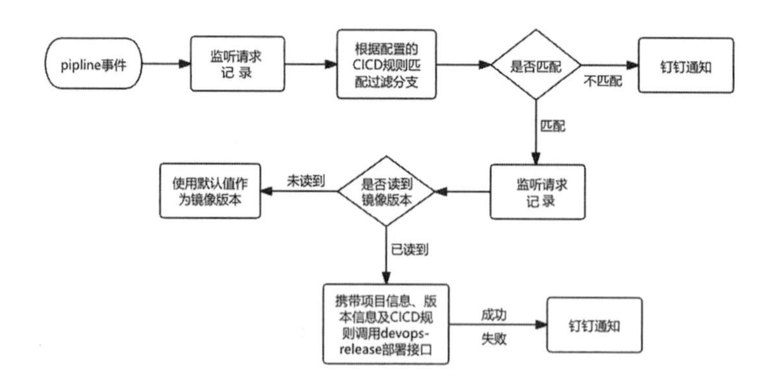

图 4.14　devops-build 组件工作流

devops-build组件的目录结构可以用下面的命令查看：

```
tree /workspace/cloudnative/devops_build
```

显示结果如下：

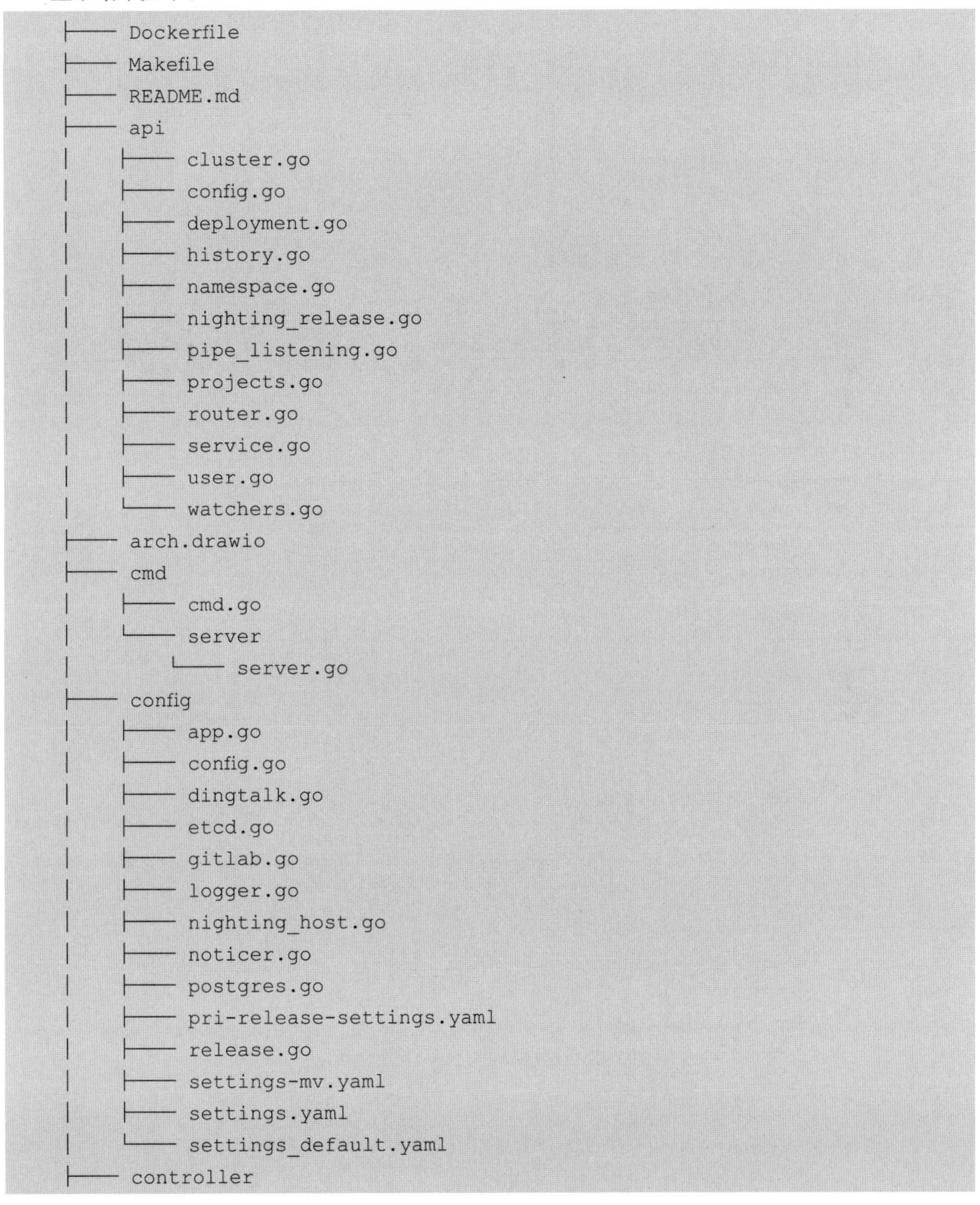

```
├── Dockerfile
├── Makefile
├── README.md
├── api
│   ├── cluster.go
│   ├── config.go
│   ├── deployment.go
│   ├── history.go
│   ├── namespace.go
│   ├── nighting_release.go
│   ├── pipe_listening.go
│   ├── projects.go
│   ├── router.go
│   ├── service.go
│   ├── user.go
│   └── watchers.go
├── arch.drawio
├── cmd
│   ├── cmd.go
│   └── server
│       └── server.go
├── config
│   ├── app.go
│   ├── config.go
│   ├── dingtalk.go
│   ├── etcd.go
│   ├── gitlab.go
│   ├── logger.go
│   ├── nighting_host.go
│   ├── noticer.go
│   ├── postgres.go
│   ├── pri-release-settings.yaml
│   ├── release.go
│   ├── settings-mv.yaml
│   ├── settings.yaml
│   └── settings_default.yaml
├── controller
```

```
|       └── nighting_release
|           └── nighting_release.go
├── database
|   ├── database.go
|   ├── etcd.go
|   ├── model
|   |   ├── model.go
|   |   └── user.go
|   ├── postgres.go
|   └── relational
|       ├── db_test
|       |   └── db_test.go
|       ├── postgres
|       |   ├── postgres.go
|       |   └── utils.go
|       └── relational.go
├── docs
|   ├── docs.go
|   ├── models
|   |   └── doc_models.go
|   ├── swagger.json
|   └── swagger.yaml
├── go.mod
├── go.sum
├── internal
|   ├── common_service
|   |   └── watcher.go
|   ├── gitlabapi
|   |   ├── gitlabapi.go
|   |   └── gitlabapi_test
|   |       └── gitlabapi_test.go
|   ├── gitlabsdk
|   |   └── gitlabsdk.go
|   ├── middleware
|   |   ├── cors.go
|   |   ├── middleware.go
|   |   ├── requset_history.go
|   |   ├── role.go
|   |   └── session.go
|   ├── model
|   |   └── resp.go
```

```
│   ├── project
│   │   └── project.go
│   ├── release
│   │   ├── gitlab
│   │   │   └── gitlab.go
│   │   ├── model
│   │   │   └── model.go
│   │   └── release.go
│   ├── router
│   │   └── router.go
│   └── send
│       ├── filechange.go
│       └── send.go
├── main
├── main.go
├── manage.sh
├── nlog
│   └── main.go
├── pkg
│   └── env.go
├── rest.http
├── scripts
│   └── main.go
├── util
│   ├── aes_util
│   │   ├── aes_test
│   │   │   └── main.go
│   │   └── aes_util.go
│   ├── dingtalk_robot.go
│   ├── dingtalk_util
│   │   └── dingtalk_util.go
│   ├── http_util
│   │   └── http_util.go
│   ├── noticer
│   │   └── noticer.go
│   ├── parse_ref_mapping.go
│   └── status_code
│       └── devops_err.go
├── v0.1.1-ddl.sql
└── v0.1.2-ddl.sql
```

由于工作目录路径较长，因此后面用 workspace 别名代指 /Users/yuhongchun/repo/workspace/cloudnative/。

devops-build 的主要数据库是 PostgreSQL，Go 语言利用 sqlx 库来做 CURD 工作，其主要核心表为 projects、cmcfg、deployment、service、routes、k8s_cluster 和 k8s_namespace，表结构较简单，这里只对几个关键表结构进行介绍。

projects 表中记录了项目信息，其表结构见表 4.1。

表 4.1　projects 表结构

名　字	类　型	说　明	是否主键
id	int4	唯一标识	是
repo_url	text	仓库地址	否
project_name	text	项目名称	否
repo_type	text	仓库类型	否
tags	text	项目标签	否
enabled	bool	保留字段	否
proejct_token	text	webhook 回调 token	否
enabled_branch	text	保留字段	否
descript	text	项目描述	否
topic	text	gitlab 代码库中的 topic	否
group	text	gitlab 代码库中的 group	否
gitlab_id	int4	gitlab 代码库中的 project_id	否

deployment 表记录了 deployment 名称及其对应内容，其表结构见表 4.2。

表 4.2　deployment 表结构

名　字	类　型	说　明	是否主键
id	int4	唯一标识	是
deployment_name	text	deployment 名称	否
project_id	int4	项目 id	否
channel_name	text	保留字段	否
content	text	deployment 内容	否
enabled	bool	保留字段	否
docker_repo_id	int	保留字段	否
namespace_id	int	集群命名空间 id	否

service 表记录了与服务相关的信息和内容，其表结构见表 4.3。

表 4.3 service 表结构

名 字	类 型	说 明	是否主键
id	int4	唯一标识	是
deployment_id	int4	deployment 对应的 id	否
channel_name	text	保留字段，可不填	否
content	text	service 内容	否

routes 表记录了与发版路径相关的信息和内容，其表结构见表 4.4。

表 4.4 routes 表结构

名 字	类 型	说 明	是否主键
id	int4	唯一标识	是
project_id	int4	项目 id	否
ref_rep	text	分支 /tag 正则表达式	否
docker	text	保留字段	否
create_time	date	创建时间	否
update_time	date	更新时间	否
cluster_id	int4	集群 id	否
enabled	bool	是否生效	否
namespace_id	int4	命名空间 id	否

cmcfg 表与 configMap 资源相关，后面会详细说明，这里暂时略过，其他表暂且略过。首先介绍 devops-build 的核心 API，其相关代码 /workspace/devops_build/internal/router/router.go 如下：

```
package router

import (
	"net/http"
	handle "devops_build/api"
	_ "devops_build/docs"
	"devops_build/internal/middleware"
	"github.com/gin-gonic/gin"
	gs "github.com/swaggo/gin-swagger"
	"github.com/swaggo/gin-swagger/swaggerFiles"
)

func ProjectsRouter(r *gin.RouterGroup) {
	// list projects
	r.GET("/project", handle.ListProjects)
```

```
    r.GET("/projectname", handle.ListProjectsFuzzy)
    r.PUT("/project", handle.AddProjects)
    r.PATCH("/project", handle.PatchProjects)
    r.PUT("/project/resource/default", handle.AddNewProResourceDefault)
    r.PUT("/project/sync_gitlab", handle.SyncProjectFromGitlab)
    r.DELETE("/project", handle.DeleteProject)
    r.GET("/dingtalk_bot", handle.AddDingTalkBot)
    r.DELETE("/dingtalk_bot", handle.DeleteDingTalkBot)
    r.GET("/list_dingtalk_bot", handle.ListDingTalkBot)
    r.PUT("/dingtalk_bot", handle.AddDingTalkBot)
}

func DeploymentRouter(r *gin.RouterGroup) {
    r.GET("/selectdeployment", handle.ListDeploymentsByProIdAndNspIdAndPageSize)
    r.GET("/deployment", handle.ListDeployments)
    r.PUT("/deployment", handle.AddDeployments)
    r.PATCH("/deployment", handle.PatchDeployments)
    r.GET("/deploymentJson", handle.DeploymentJson)
    r.GET("/deploymentPatchData", handle.DeploymentPatchData)
    r.POST("/deploymentPatch", handle.DeploymentPatch)
    r.DELETE("/deployment", handle.DeleteDeployment)
    r.GET("/deployment_fuzzy", handle.ListDeploymentByFuzzy)
}

func ServiceRouter(r *gin.RouterGroup) {
    r.GET("/selectservice", handle.ListServicesByDeploymentId)
    r.GET("/service", handle.ListServices)
    r.PUT("/service", handle.AddServices)
    r.PATCH("/service", handle.PatchServices)
    r.DELETE("/service", handle.DeleteService)
}

func RouteRouter(r *gin.RouterGroup) {
    r.GET("/selectroute", handle.ListRoutesByProIdAndNspIdAndPageSize)
    r.GET("/route", handle.ListRoutes)
    r.PUT("/route", handle.AddRoutes)
    r.PATCH("/route", handle.PatchRoutes)
    r.DELETE("/route", handle.DeleteRoute)
}

func RouteUser(r *gin.RouterGroup) {
```

```
	r.POST("/login", handle.Login)
	r.GET("/currentUser", handle.GetCurrentUser)
	r.POST("/createUser", middleware.VerifyAdmin, handle.CreateUser)
	r.GET("/loginout", handle.LoginOut)
}

func ConfigRouter(r *gin.RouterGroup) {
	r.GET("/list_config_of_pro", handle.ListAllConfigOfProject)
	r.GET("/list_config_of_pro_and_nsp", handle.ListConfigOfProjectAndNamespace)
	r.GET("/list_config_of_deployment", handle.ListConfigsOfDeployment)
	r.PATCH("/config", handle.PatchConfig)
	r.DELETE("/config", handle.DeleteConfig)
	r.PUT("/config", handle.AddConfig)
}

func K8sRouter(r *gin.RouterGroup) {
	r.GET("/cluster", handle.ListCluster)
	r.GET("/namespace", handle.ListNspByClusterId)
	r.GET("/single_namespace", handle.GetNamespaceById)
}

// 与 devops-release 交互的 API
func NightingReleaseRoute(r *gin.RouterGroup) {
	r.GET("/release/imageList", handle.NightingReleaseImageList)

	r.POST("/release/callback", handle.NightingReleaseCallback)

	r.GET("/release/pod_info", handle.PodsInfo)
	r.POST("/release/compare_config", handle.CompareConfig)
	r.POST("/release/convert_yaml_kv", handle.ConvertYamlOrKv)
	r.GET("/release/get_config_content", handle.GetConfigContent)
	r.POST("/release/set_config", handle.SetConfigContent)
	r.POST("/release/pub_config", handle.PubConfig)
}

func InitRouter(r *gin.Engine) *gin.Engine {
	r.POST("/", func(c *gin.Context) { c.JSON(http.StatusTeapot, struct{}{}) })

	r.GET("/ping", func(c *gin.Context) {
		c.Data(http.StatusOK, "text", []byte("pong"))
	})
```

```
    r.POST("/test", func(c *gin.Context) {})
    r.GET("/swagger/*any", gs.WrapHandler(swaggerFiles.Handler))
    r.StaticFile("/swagger_file/swagger.json", "./docs/swagger.json")
    k8sapi := r.Group("/api/nighting-build")
    ProjectsRouter(k8sapi)
    DeploymentRouter(k8sapi)
    ServiceRouter(k8sapi)
    RouteRouter(k8sapi)
    K8sRouter(k8sapi)
    NightingReleaseRoute(k8sapi)
    ConfigRouter(k8sapi)

    k8sapi.POST("/gitlab_callback", handle.PipeListening)
    return r
}
```

/workspace/devops_build/api/pipe_listening.go 文件的作用是在数据库的 projects 中查找项目路径和发版路径，并用此发版路径与 CI 接收的事件信息中的 branch 或 tag 匹配。如果匹配成功，则走下一步流程；否则就拒绝，发送钉钉通知。其源码如下：

```
package api

import (
    "context"
    "regexp"
    "strings"
    dbmodel "devops_build/database/model"
    "devops_build/config"
    "devops_build/database"
    "devops_build/internal/model"
    "devops_build/internal/send"
    "github.com/gin-gonic/gin"
    "github.com/sirupsen/logrus"
    "go.elastic.co/apm"
)

type MyParams struct {
}

func PipeListening(c *gin.Context) {
    ctx := c.Request.Context()
    span, ctx := apm.StartSpan(ctx, "PipeListening", "POST")
```

```
    defer span.End()

    xGitlabEvent := c.Request.Header.Get("X-Gitlab-Event")
    if xGitlabEvent != "Pipeline Hook" {
        logrus.WithContext(ctx).Error("ERROR: XGitlabEvent is not Pipeline Hook")
        return
    }

    params := model.PipelineEvents{}
    if err := c.Bind(&params); err != nil{
        logrus.WithContext(ctx).Errorf("ERROR: Params bind failed, err: %s", err)
        return
    }
    if params.ObjectKind != "pipeline"{
        logrus.WithContext(ctx).Error("ERROR: Object kind not pipeline")
        return
    }

    labels := make(map[string]string)
    labels["project_name"] = params.Project.Name

    if params.ObjectAttributes.Tag{
        labels["project_tag"] = params.ObjectAttributes.Ref
    } else {
        labels["project_branch"] = params.ObjectAttributes.Ref
    }

    status := params.ObjectAttributes.Status
    labels["webhook_status"] = status

    devopsdb := database.GetDevopsDb()
    project, err := devopsdb.GetProjectByGitlabId(ctx, params.Project.ID)
    if err != nil{
        logrus.Error("查询project失败! err:", err)
        return
    }
    //拦截不能发版的分支或者tag
    enabled_flag := false
    if len(config.GitlabConfig.Keywords) == 0{
            namespace_id := 0
            enabled_flag, namespace_id = RefFilter(ctx, *project, params)
```

```
                logrus.Infof("查询到%v的发版路径 namespace:%d", project,
namespace_id)
        } else {
                for _, keyword := range config.GitlabConfig.Keywords{
                    if strings.Contains(params.ObjectAttributes.Ref,
keyword){
                        enabled_flag = true
                    }
                }
        }
        if !enabled_flag{
            logrus.Info("RefRep不匹配，或者发版路径未启用！ ")
            return
        }

        if status != "success"{
                if status == "created" || status == "failed" || status ==
"canceled" || status == "skipped"{
                    logrus.Info("status is not success")
                }
                return
        }

        if err != nil{
            logrus.WithContext(ctx).Error("select project err!err:", err)
            return
        }
        if len(project.ProjectToken) == 0{
            logrus.WithContext(ctx).Errorf("ERROR: Get Web hook token failed, err:
%s", err)
            return
        }
        if string(project.ProjectToken) !=
c.Request.Header.Get("X-Gitlab-Token"){
            logrus.WithContext(ctx).
                    Errorf("ERROR: Validate Web hook token failed, request
token: %s, should be: %s",

        c.Request.Header.Get("X-Gitlab-Token"),project.ProjectToken)
            return
        }
```

```
        err = send.SendRelease(ctx, &params, *project)
        if err != nil{
            logrus.WithContext(ctx).Errorf("Error: send to release failed, err: 
%s", err)
            params.ObjectAttributes.Status = "error"
            params.IsFailed = true
            params.ErrorMsg = err.Error()
            return
        }

    }
    // 查找发版路径并且进行检查，如果没有正则匹配，则拒绝
    func RefFilter(ctx context.Context, project dbmodel.Project, params model.
PipelineEvents) (bool, int) {
        devopsdb := database.GetDevopsDb()
        routes, err := devopsdb.GetRoutesRepNamespaceIdByProId(ctx, 
project.Id)
        if err != nil{
            logrus.Infof("未找到发版路径! project:%v err:%v", project, err)
            return false, 0
        }
        for _, route := range routes{
            logrus.Info(*route)
            r, err := regexp.Compile(route.RefRep)
            if err != nil{
                    logrus.Errorf("正则表达式错误! pro:%s env:%s rep: %s", 
params.Project.Name, route.Channel, route.RefRep)
                    continue
            }
            if r.FindString(params.ObjectAttributes.Ref) != ""{
                    return true, route.NamespaceId
            }
            if !route.Enabled{
                    return false, 0
            }
        }
        return false, 0

    }
```

/workspace/cloudnative/devops_build/api/nighting_release.go 文件实现了 devops-build 与 devops-release 交互的接口。其相关代码如下：

```
package api

import (
      "encoding/json"
      "net/http"
      "strconv"

      "github.com/gin-gonic/gin"
      nightingrelease "devops_build/controller/nighting_release"
      statuscode "devops_build/util/status_code"
      "github.com/sirupsen/logrus"
)

// Callback devops-release 回调接口
// @Summary devops-release 回调接口
// @Description event_type 为 build|restart|patch_image
// @Tags 发版相关接口
// @Accept application/json
// @Produce application/json
// @Param object body nightingrelease.ReleaseInfo true " 主要是获取 event_type
// 字段等 "
// @Success 200 {object} nightingrelease.OpsInfo
// @Router /api/nighting-build/release/callback [post]
func NightingReleaseCallback(c *gin.Context) {
      // TODO: 操作通知，给出操作人，操作对象，操作结果发送通知到钉钉
      releaseInfo := nightingrelease.ReleaseInfo{}
      err := c.BindJSON(&releaseInfo)
      if err != nil{
            c.JSON(http.StatusBadRequest, gin.H{"err_code":
statuscode.PARAMS_BIND_ERR, "err_msg":
statuscode.ErrMsg[statuscode.PARAMS_BIND_ERR]})
      }
      body, err := nightingrelease.CallBack(c, releaseInfo)
      if err != nil{
              c.JSON(http.StatusOK, gin.H{"err_code": 0, "err_msg": err.
Error()})
              return
      }
```

```
        c.JSON(http.StatusOK, body)
    }
```

CallBack 函数相关源码内容如下：

```
    func CallBack(ctx *gin.Context, releaseInfo ReleaseInfo)
(*CallbackReturnData, error) {
        data, err := json.Marshal(releaseInfo)
        if err != nil{
                logrus.Error("json 编码错误! err:", err)
                return nil, err
        }
        if len(releaseInfo.Deployments) == 0{
                return nil, fmt.Errorf("deployments is empty!")
        }
        devopsDb := database.GetDevopsDb()
        changeBeforeDeployments := []model.Deployment{}
        for _, d := range releaseInfo.Deployments{
                deployment, err := devopsDb.GetDeploymentById(ctx, d)
                if err == nil{
                    changeBeforeDeployments =
append(changeBeforeDeployments, *deployment)
                } else {
                        return nil, err
                }
            }
        body, err := httputil.SendHttpRequest("POST",
map[string]string{"Content-type": "application/json"},
config.NightingHostConfig.NightingReleaseHost+"/api/nighting-release/callback",
data)
        if err != nil{
                logrus.Error(" 发送请求错误! err:", err)
                return nil, err
        }
        session := sessions.Default(ctx)
        userId, ok := session.Get("uuid").(int)
        if !ok{
                userId = 0
        }

        for i, d := range releaseInfo.Deployments{
                deployment, err := devopsDb.GetDeploymentById(ctx, d)
```

```
                if err != nil{
                        logrus.Error(err)
                        break
                }
                opsHistory := model.OpsHistory{
                        ResourceType: "deployment",
                        OpsType:      releaseInfo.EventType,
                        ResourceId:   d,
                        UserId:       userId,
                        UpdateTime:   time.Now(),
                        ChangeBefore: changeBeforeDeployments[i].Content,
                        ChangeAfter:  deployment.Content,
                }
                if strings.TrimSpace(opsHistory.ChangeBefore) !=
strings.TrimSpace(opsHistory.ChangeAfter){
                        _, err = devopsDb.InsertIntoOpsHistory(ctx,
opsHistory)
                        if err != nil{
                                logrus.Error(err)
                        }
                }
        }
        returnData := CallbackReturnData{}
        err = json.Unmarshal(body, &returnData)
        if err != nil{
                return nil, err
        }
        return &returnData, nil
    }
```

读取与 devops-build 配置（即启动 devops-build 应用需要的配置）相关的文件 /workspace/cloudnative/devops_build/config/config.go，其源码内容如下：

```
    package config
    import (
        "github.com/spf13/viper"
    )

    func SetUp(path string) {
        settingCfg := viper.New()
        settingCfg.SetConfigFile(path)
        if err := settingCfg.ReadInConfig(); err != nil{
```

```
        panic("读取配置文件失败:" + err.Error())
    }

    cfgLogger := settingCfg.Sub("logger")
    if cfgLogger == nil{
        panic("config not found logger")
    }
    LoggerConfig = InitLogger(cfgLogger)

    // 启动参数
    cfgApplication := settingCfg.Sub("application")
    if cfgApplication == nil{
        panic("config not found application")
    }
    ApplicationConfig = InitApplication(cfgApplication)

    cfgRelease := settingCfg.Sub("release")
    if cfgRelease == nil{
        panic("config not found release")
    }
    ReleaseConfig = InitRelease(cfgRelease)

    // gitlab配置初始化
    cfgGitlab := settingCfg.Sub("gitlab")
    if cfgGitlab == nil{
        panic("config not found gitlab")
    }
    GitlabConfig = InitGitlab(cfgGitlab)
    // 钉钉机器人配置文件初始化
    cfgDingtalkRobot := settingCfg.Sub("dingtalk_robot")
    if cfgDingtalkRobot == nil{
        panic("config not found dingtalk_robot")
    }
    DingtalkRobotConfig = InitDingtalkRobot(cfgDingtalkRobot)

    cfgPostgres := settingCfg.Sub("postgres")
    if cfgPostgres == nil{
        panic("config not found cfgpostgres")
    }
    PostgresConfig = InitPostgresConfig(cfgPostgres)
    cfgNightingHost := settingCfg.Sub("nighting_host")
```

```
    if cfgNightingHost == nil{
            panic("config not found cfgNightingHost")
    }
    NightingHostConfig = InitNightingHost(cfgNightingHost)
}
```

配置文件 settings.yaml 的内容如下：

```
nighting_host:
  nighting_release_host: http://localhost:5000
  nighting_authorization_host:
  nighting_config_host:
gitlab:
  api_host: 10.1.0.200:30808/api/v4
  pri_token: glpat-ysbisTks6UBH6bs9Kbtz
  ci_job_log_docker_key: enter-devops-dockertag
  keywords:
    - main
    - dev
    - test
    - master
noticer:
  host: ''
dingtalk_robot:
  token:
https://oapi.dingtalk.com/robot/send?access_token=d23d372e22a5d8c90e943e6b584
7c843be254a63ebfd4e387e4ef211aef50200
  sercret: ""
  keywords: '['
  member_map:
    余洪春 : +86-16666666666
release:
  enabled_chanel:
    - beta
    - release
    - dev
    - test
    - main
logger:
  loglevel: info
  apm:
    filepath: ./log/nlog.log
```

```
    maxfilesize: 100
    maxbackups: 5a
    maxage: 30
    compress: false
```

此外，根据业务需求，这里分以下两种情况来处理镜像版本：

（1）devops-build 会调用 GitLab 接口 /projects/:id/jobs/:job_id/trace，获取项目指定作业的日志，如果其中有关键字 enter-devops-dockertag，则直接从日志中获取镜像应用名及版本 tag 号。

（2）如果没有关键字 enter-devops-dockertag，则直接从 deployment 表中相应的字段获取镜像应用名及版本 tag 号。

其代码由 /workspace /devops_build/internal/release/gitlab/gitlab.go 文件来实现，内容如下：

```
package gitlab

import (
    "context"
    "devops_build/config"
    gitmodel "devops_build/internal/model"
    "devops_build/internal/release/model"
    "fmt"
    "github.com/sirupsen/logrus"
    "io"
    "net/http"
    "strings"
    "sync"
)

type GitlabReleaseInfo struct {
}

var ReleaseInfo = &GitlabReleaseInfo{}

func (g *GitlabReleaseInfo) GetDockerInfo(ctx context.Context, params
*gitmodel.PipelineEvents) *model.DockerInfo {
    projectId := params.Project.ID
    jobs := params.Builds
    lock := sync.Mutex{}
    wg := sync.WaitGroup{}
    wg.Add(len(jobs))
    var dockerInfo *model.DockerInfo
    for _, job := range jobs{
        jobid := job.ID
```

```
            go func(){
                    d, err := getDockerUrlFromJobLog(ctx, projectId, jobid)
                    if err != nil{
                    logrus.WithContext(ctx).Infof("id:%d %v", jobid, err)
                    }
                    if d != nil{
                            lock.Lock()
                            dockerInfo = d
                            lock.Unlock()
                    }
                    wg.Done()
            }()
        }
        wg.Wait()
        return dockerInfo
    }

    /* 获取项目指定作业的日志或跟踪日志
    GET /projects/:id/jobs/:job_id/trace
    参考 GitLab API 文档 https://docs.gitlab.cn/jh/api/jobs.html
    */
    func getDockerUrlFromJobLog(ctx context.Context, projectId int, jobId int)
(*model.DockerInfo, error) {
        fmt.Println(jobId)
        url := fmt.Sprintf("http://%s/projects/%d/jobs/%d/trace",
config.GitlabConfig.ApiHost, projectId, jobId)
        request, err := http.NewRequest("GET", url, nil)
        if err != nil {
                return nil, err
        }
        request.Header.Add("PRIVATE-TOKEN", config.GitlabConfig.PriToken)
        client := http.Client{}
        resp, err := client.Do(request)
        if err != nil {
                return nil, err
        }
        bodyBytes, err := io.ReadAll(resp.Body)
        if err != nil{
                return nil, err
        }
        logString := string(bodyBytes)
        logStringSplit := strings.Split(logString, "\n")
```

```
        for _, line := range logStringSplit{
                if strings.Contains(line, "echo"){
                        continue
                }
                /* 如果项目指定作业的日志中包含 enter-devops-dockertag 字段，则走下面的逻辑，否则返回 nil 空值（即从 deployment 表中获取数据）*/
                if strings.Contains(line,
strings.TrimSpace(config.GitlabConfig.CIJobLogDockerKey)) {
                        logrus.Info(config.GitlabConfig.CIJobLogDockerKey)
                        line = strings.TrimSpace(line)
                        lineSplit := strings.Split(line,":")
                        if len(lineSplit) != 3{
                                logrus.WithContext(ctx).Errorf("dockerurl 格式错误！ ")
                                return nil, err
                        } else {
                                dockerRepo := strings.TrimSpace(lineSplit[1])
                                dockerTag := strings.TrimSpace(lineSplit[2])

                                return &model.DockerInfo{Repository:
dockerRepo, Tag: dockerTag}, nil
                        }
                }
        }
        return nil, fmt.Errorf(" 未在 job 日志里找到 dockertag！ ")
    }
```

打包编译完成以后会生成 main 二进制命令，可以用下面的命令来启动 devops-build 组件（此时会监听本地的 5001 端口）：

```
ELASTIC_APM_ACTIVE=false ./main server -c config/settings.yaml
```

正常启动会有如下输出：

```
2023/11/26 22:10:19 starting server...
[GIN-debug] [WARNING] Running in "debug" mode. Switch to "release" mode in
production.
 - using env:   export GIN_MODE=release
 - using code:  gin.SetMode(gin.ReleaseMode)
...
[GIN-debug] GET    /api/nighting-build/release/get_config_content -->
devops_build/api.GetConfigContent (2 handlers)
[GIN-debug] POST   /api/nighting-build/release/set_config -->
devops_build/api.SetConfigContent (2 handlers)
[GIN-debug] POST   /api/nighting-build/release/pub_config -->
```

```
devops_build/api.PubConfig (2 handlers)
    [GIN-debug] GET    /api/nighting-build/list_config_of_pro -->
devops_build/api.ListAllConfigOfProject (2 handlers)
    [GIN-debug] GET    /api/nighting-build/list_config_of_pro_and_nsp -->
devops_build/api.ListConfigOfProjectAndNamespace (2 handlers)
    [GIN-debug] GET    /api/nighting-build/list_config_of_deployment -->
devops_build/api.ListConfigsOfDeployment (2 handlers)
    [GIN-debug] PATCH  /api/nighting-build/config -->
devops_build/api.PatchConfig (2 handlers)
    [GIN-debug] DELETE /api/nighting-build/config -->
devops_build/api.DeleteConfig (2 handlers)
    [GIN-debug] PUT    /api/nighting-build/config -->
devops_build/api.AddConfig (2 handlers)
    [GIN-debug] POST   /api/nighting-build/gitlab_callback -->
devops_build/api.PipeListening (2 handlers)
     NFO[0000] 2023-11-26 22:10:19.009735338 +0800 CST m=+0.037126101 Server
Run http://0.0.0.0:5001/
     NFO[0000] 2023-11-26 22:10:19.009776205 +0800 CST m=+0.037166967 Enter
Control + C Shutdown Server
```

注：这里为了方便调试和更新，直接以二进制和systemd的方式运行；后续的devops-release应用也采用同样的部署方式。

以上输出信息表明，devops-build 服务已经正常启动了，监听本地 5001 端口。

另外，新项目的 CD 发版工作都精简成了两个接口来实现，达到了自动化和简化操作的目的。例如，如果需要发版名为 testci、版本号为 1.0 的应用镜像，则只需先操作 192.10.1.199:5001/api/nighting-build/project/resource/default 接口方法（HTTP 请求方法为 PUT）。这里推荐用 Postman 工具来实现，其 JSON 文件格式如下：

```
{
"project_name":"testci",
"type":"tenant",
"channel":"dev",
"cluster_name":"tenant-devops",
"namespace":"devops",
"language":"go"
}
```

上面的 JSON 文件格式解释如下：

（1）project_name：项目名，如这里是 testci，请注意要保证其唯一性。

（2）type：业务定义的服务类型，tenant 表示租户服务，ops 表示运维服务，这里填 tenant 即可。

（3）channel：产研环境，如 test、dev、stage 等，这里填 dev。

（4）cluster_name：集群名称，如 tenant-devops（注意：这里与数据库中定义的一致即可）。

（5）namespace：集群命名空间，这里是 devops（注意：这里与数据库中定义的一致即可）。

（6）language：选填，因为都是 Go 应用项目，这里填 go 即可。

接口涉及的文件为 /workspace/cloudnative/devops_build/api/projects.go，相关代码如下：

```
    // @Summary 增加项目
    // @Accept application/json
    // @Param project_name body string false "ProjectName"
    // @Param repo_type body string true "仓库类型，应该有个下拉框，目前只有gitlab一种"
    // @Param repo_url body string true "仓库地址，有此参数的话会自动同步其他git相关
    // 的信息（projectname,gitlab_id,topic,descript等）"
    // @Param tags body string true "项目类型 type=ops or tenant or center"
    // @Param enabled body bool true "Enabled"
    // @Param project_token body string true "项目令牌 目前都是 cicd??????"
    // @Param descript body string false "Descript"
    // @Param topic body string  false "Topic"
    // @Param gitlab_id body string false "GitlabId"
    // @Param group body string false "Group"
    // @Success 200 {object} ReturnData
    // @Failure 400 {object} ReturnData
    // @Router /api/nighting-build/project [put]
    func AddProjects(c *gin.Context) {
        //todo
        var project = m.Project{}
        devopsdb := database.GetDevopsDb()
        err := c.ShouldBindJSON(&project)
        fmt.Println(project)
        if err != nil{
                c.JSON(http.StatusBadRequest, &ReturnData{Err_code: 0,
Err_msg: err.Error()})
                return
        }
        if project.RepoType == "gitlab"{
                project.RepoUrl = strings.TrimSpace(project.RepoUrl)
                // 此处代码为硬编码，后续更优化，以变量的方式传参
                repoUrlSplit :=
strings.Split(strings.TrimSuffix(project.RepoUrl, ".git"),
"http://10.1.0.200:30808/")
                if len(repoUrlSplit) != 2{
                        c.JSON(http.StatusBadRequest, &ReturnData{Err_code: 0,
Err_msg: "仓库地址不合法！ "})
```

```
                return
            }
            groupPro := strings.Trim(repoUrlSplit[1], "/")
            groupProSplit := strings.Split(groupPro, "/")
            if len(groupProSplit) < 2{
                c.JSON(http.StatusBadRequest, &ReturnData{Err_code: 0,
Err_msg: "仓库地址不合法！"})
                return
            }

            gitlabPro, err := gitlabapi.GitlabApiObj.GetSingleProject(c,
groupProSplit[0], groupProSplit[1])
            if err != nil{
                logrus.Error(err)
                c.JSON(http.StatusOK, &ReturnData{Err_code: 0,
Err_msg: "从gitlab获取项目信息出错！"})
                return
            }
            project.GitlabId = gitlabPro.ID
            project.Group = groupProSplit[0]
            if len(project.ProjectName) == 0{
                project.ProjectName = gitlabPro.Name
            }
            project.Enabled = true
            topic := ""
            for _, t := range gitlabPro.Topics{
                topic = topic + t +","
            }
            topic = strings.Trim(topic,",")
            project.Topic = topic
            project.Descript = gitlabPro.Description
            project.ProjectToken = "cicd??????"

            fmt.Println(project)
            if project.GitlabId == 0{
                c.JSON(200, gin.H{"err_code": 0, "err_msg": "从gitlab
获取项目失败"})
                return
            }
            exist, err := devopsdb.IfExistProject(c, project)
            if err != nil{
```

```
                        logrus.WithContext(c).Errorf("Error: Judge Project exist
failed,err: %s", err)
                        c.JSON(http.StatusBadRequest, &ReturnData{Err_code:
0, Err_msg: err.Error()})
                        return
              }

              if exist{
                        c.JSON(http.StatusOK, &ReturnData{Err_code: 0,
Err_msg: "记录已经存在，插入失败"})
                        return
              }
         }

         if len(project.Tags) == 0{
               project.Tags = "type=tenant,"
         }
         pro, err := devopsdb.InsertIntoProjects(c, project)
         if err != nil{
               c.JSON(http.StatusBadRequest, &ReturnData{Err_code: 0,
Err_msg: err.Error()})
               logrus.WithContext(c).Errorf("Error: Insert Project failed,
err: %s", err)
              return
         }
         c.JSON(http.StatusOK, &ReturnData{Err_code: 1, Err_msg: "ok", Data:
pro.Id})
    }

    // @Summary 更新项目
    // @Accept application/json
    // @Param Id body int true "Id"
    // @Param ProjectName body string true "ProjectName"
    // @Param RepoType body string true "目前用不到"
    // @Param RepoUrl body string true "目前用不到"
    // @Param Tags body string true "项目类型 type=ops or tenant or center"
    // @Param Enabled body bool true "Enabled"
    // @Param ProjectToken body string true "项目令牌，目前都是 cicd??????"
    // @Param EnabledBranchs body string true "可发版的分支，使用正则表达式"
    // @Param Descript body string true "Descript"
    // @Param Topic body string  true "Topic"
```

```
    // @Param GitlabId body string true "GitlabId"
    // @Param Group body string true "Group"
    // @Success 200 {object} ReturnData
    // @Failure 400 {object} ReturnData
    // @Router /api/nighting-build/project [patch]
    func PatchProjects(c *gin.Context) {
          pro := m.Project{}
          err := c.BindJSON(&pro)
          if err != nil{
                  c.JSON(http.StatusBadRequest, &ReturnData{Err_code: 0,
Err_msg: err.Error()})
                  return
          }
          devopsdb := database.GetDevopsDb()
          result, err := devopsdb.UpdateProject(c, pro)
          if err != nil{
                  c.JSON(http.StatusBadRequest, &ReturnData{Err_code: 0,
Err_msg: err.Error()})
                  logrus.WithContext(c).Errorf("Patch Project failed err:", err)
                  return
          }
          if result{
               c.JSON(http.StatusOK, &ReturnData{Err_code: 1, Err_msg: "ok"})
          }
    }

    func AddNewProResourceDefault(c *gin.Context) {
          newProResourceDefault := project.NewProResourceDefault{}
          err := c.BindJSON(&newProResourceDefault)
          if err != nil{
                  c.JSON(http.StatusBadRequest, &ReturnData{Err_code: 0, Err_msg:
"参数绑定错误！"})
                  return
          }
          res, err := project.AddNewProResourceFromTemp(c,
newProResourceDefault)
          if err != nil{
                  c.JSON(http.StatusOK, &ReturnData{Err_code: 0, Err_msg:
err.Error(), Data: res})
                  return
          }
```

```
        c.JSON(http.StatusOK, &ReturnData{Err_code: 1, Err_msg: "ok", Data:
res})
    }

    func SyncProjectFromGitlab(c *gin.Context) {
        project := &m.Project{}
        err := c.BindJSON(project)
        if err != nil{
                c.JSON(http.StatusBadRequest, gin.H{"err_code": 0, "err_msg":
"参数绑定错误"})
                return
        }
        if project.RepoType == "gitlab"{
                project.RepoUrl = strings.TrimSpace(project.RepoUrl)
                repoUrlSplit :=
strings.Split(strings.TrimSuffix(project.RepoUrl, `.git`), "gitlab.yiban.io/")
            if len(repoUrlSplit) != 2{
                    c.JSON(http.StatusBadRequest, &ReturnData{Err_code:
0, Err_msg: "仓库地址不合法！"})
                    return
            }
            groupPro := strings.Trim(repoUrlSplit[1], "/")
            groupProSplit := strings.Split(groupPro, "/")
            if len(groupProSplit) < 2{
                    c.JSON(http.StatusBadRequest, &ReturnData{Err_code:
0, Err_msg: "仓库地址不合法！"})
                    return
            }

            gitlabPro, err := gitlabapi.GitlabApiObj.GetSingleProject(c,
groupProSplit[0], groupProSplit[1])
            if err != nil{
                    logrus.Error(err)
                    c.JSON(http.StatusOK, &ReturnData{Err_code: 0,
Err_msg: "从gitlab获取项目信息出错！"})
                    return
            }
            project.GitlabId = gitlabPro.ID
            project.Group = groupProSplit[0]
            project.ProjectName = gitlabPro.Name
            project.Enabled = true
```

```
            topic := ""
            for _, t := range gitlabPro.Topics{
                    topic = topic + t +","
            }
            topic = strings.Trim(topic,",")
            project.Topic = topic
            project.Descript = gitlabPro.Description
        }
        devopsdb := database.GetDevopsDb()
        ok, err := devopsdb.UpdateProject(c, *project)
        if err != nil || !ok{
                logrus.Error(err)
                c.JSON(200, gin.H{"err_code": 0, "err_msg": "查询错误"})
                return
        }
        c.JSON(200, gin.H{"err_code": 1, "err_msg": "ok", "data": project})

    }

    // DeleteProject
    // @Summary 删除一个项目
    // @Param proid query string true "项目 ID"
    // @Success 200 {object} ReturnData
    // @Failure 400 {object} ReturnData
    // @Router /api/nighting-build/project [delete]
    func DeleteProject(c *gin.Context) {
        strid := c.Query("proid")
        proid, err := strconv.Atoi(strid)
        if err != nil{
                c.JSON(http.StatusBadRequest, &ReturnData{
                        Err_code: 0,
                        Err_msg:  err.Error(),
            })
            return
        }
        devopsdb := database.GetDevopsDb()
        _, err = devopsdb.DeleteProjectById(c, proid)
        if err != nil {
               c.JSON(http.StatusBadRequest, &ReturnData{Err_code: 0,
Err_msg: err.Error()})
               logrus.WithContext(c).Errorf("Delete Project failed err:", err)
```

```
                return
        }

        c.JSON(http.StatusOK, &ReturnData{
                Err_code: 1,
                Err_msg:  "ok",
        })
}
```

真正起作用的文件是 /workspace/cloudnative/devops_build/internal/project/project.go，其源码如下：

```
package project

import (
        "context"
        "fmt"
        "strings"
        "time"

        "devops_build/database"
        "devops_build/database/model"
        Logr "github.com/sirupsen/logrus"
)

type NewProResourceDefault struct {
        ProjectName   string `json:"project_name"`
        Type          string `json:"type"`
        Channel       string `json:"channel"`
        ClusterName   string `json:"cluster_name"`
        Namespace     string `json:"namespace"`
        Language      string `json:"language"`
}
type CreateProResourceDefaultRes struct {
        ProjectId     int `json:"project_id"`
        DeploymentId  int `json:"deployment_id"`
        ServiceId     int `json:"service_id"`
        RouteId       int `json:"route_id"`
}

func AddNewProResourceFromTemp(ctx context.Context, info
NewProResourceDefault) (*CreateProResourceDefaultRes, error) {
        devopsdb := database.GetDevopsDb()
```

```
        resInfo := CreateProResourceDefaultRes{}
        projectTemp, err := devopsdb.GetProjectByName(ctx, "project_template")
        if err != nil{
                fmt.Println("no projectTemp")
                return nil, err
        }
        deploymentTemp, err := devopsdb.GetDeploymentByName(ctx, "deployment_
template")
        fmt.Println(deploymentTemp[0])
        fmt.Println(deploymentTemp[0].Id,deploymentTemp[0].ChannelName)
        if err != nil || len(deploymentTemp) == 0{
                fmt.Println("no deploymentTemp")
                return nil, err
        }
        serviceTemp, err := devopsdb.GetServicesByDeploymentIdAndChannel(ctx,
deploymentTemp[0].Id, deploymentTemp[0].ChannelName)
        if err != nil{
                fmt.Println("no serviceTemp")
                return nil, err
        }
        routeTemp, err := devopsdb.GetRoutesByProId(ctx, projectTemp.Id)
        if err != nil || len(routeTemp) == 0{
                fmt.Println("no routeTemp")
                return nil, err
        }

        configmapTemp, err := devopsdb.GetRoutesByProId(ctx, projectTemp.Id)
        if err != nil || len(configmapTemp) == 0{
                 fmt.Println("no configmapTemp")
                 return nil, err
        }
        newProject := projectTemp

        newProject.Id = 0
        newProject.ProjectName = info.ProjectName

        proExist, err := devopsdb.IfExistProject(ctx, *newProject)
        var p *model.Project
        if false && (err != nil || proExist){
                 p, err = devopsdb.GetProjectByName(ctx, info.ProjectName)
                 if err != nil{
```

```
                logr.Error(err)
                return nil, err
            }
        } else {
            p, err = devopsdb.InsertIntoProjects(ctx, *newProject)
            if err != nil{
                logr.Error(err)
                return nil, err
            }
        }

        resInfo.ProjectId = p.Id

        newDeployment := deploymentTemp[0]

        newDeployment.Id = 0
        newDeployment.ChannelName = info.Channel
        newDeployment.DeploymentName = info.ProjectName
        newDeployment.ProjectId = p.Id
        newDeployment.Content = strings.ReplaceAll(newDeployment.Content,
"${__project_name}", info.ProjectName)
        fmt.Println(newDeployment.Content)
        deploymentExist, err := devopsdb.IfExistDeployment(ctx,
*newDeployment)
        if err != nil || deploymentExist{
            return &resInfo, fmt.Errorf("资源已经存在! value:%v",
newDeployment)
        }
        d, err := devopsdb.InsertIntoDeployment(ctx, *newDeployment)
        if err != nil{
            logr.Error(err)
            return nil, err
        }
        resInfo.DeploymentId = d.Id

        newService := serviceTemp
        newService.ChannelName = info.Channel
        newService.Id = 0
        newService.DeploymentId = d.Id
        newService.Content = strings.ReplaceAll(newService.Content,
    "${__project_name}", info.ProjectName)
```

```
	serviceExist, err := devopsdb.IfExistService(ctx, *newService)
	if err != nil || serviceExist{
		return &resInfo, fmt.Errorf("资源已经存在！ value:%v", newService)
	}
	s, err := devopsdb.InsertIntoService(ctx, *newService)
	if err != nil{
		logr.Error(err)
		return nil, err
	}
	resInfo.ServiceId = s.Id

	newRoute := routeTemp[0]
	newRoute.Id = 0
	newRoute.Channel = info.Channel

	clusterInfo, err := devopsdb.GetClusterIdByName(ctx, info.ClusterName)
	if err != nil{
		logrus.Error(err)
		return &resInfo, err
	}
	newRoute.ClusterId = clusterInfo.Id
	// 默认允许 main 分支是可以发版，后续再根据实际情况整改
	newRoute.RefRep = "main"
	newRoute.NspName = info.Namespace
	newRoute.ProjectId = p.Id
	newRoute.CreateTime = time.Now()
	newRoute.UpdateTime = time.Now()
	routeExist, err := devopsdb.IfExistRoute(ctx, *newRoute)
	if err != nil || routeExist{
		return &resInfo, fmt.Errorf("资源已经存在！ value:%v", newService)
	}
	r, err := devopsdb.InsertIntoRoutes(ctx, *newRoute)
	if err != nil{
		logr.Error(err)
		return &resInfo, err
	}
	resInfo.RouteId = r.Id
	return &resInfo, nil
}
```

devops-build 会根据预先插入的 project_template 和 deployment_template 等模板文件配置，然

后依次向 postgreSQL 数据库中插入 projects、deployment、service 及 routes 等表的数据。

这里以 deployment_template 文件为例进行说明，其内容如下：

```
apiVersion: apps/v1
kind: Deployment
metadata:
  labels:
    k8s-app: ${__project_name}
  name: ${__project_name}
spec:
  progressDeadlineSeconds: 600
  replicas: 1
  selector:
    matchLabels:
      k8s-app: ${__project_name}
  template:
    metadata:
      labels:
        k8s-app: ${__project_name}
    spec:
      containers:
        - image: 10.1.0.201:8083/${__project_name}:1.0
          imagePullPolicy: Always
          lifecycle:
            preStop:
              exec:
                command:
                  - /bin/bash
                  - -c
                  - sleep 20
          name: ${__project_name}
          env:
            - name: ELASTIC_APM_SERVICE_NAME
              value: ${__project_name}
            - name: ELASTIC_APM_ENVIRONMENT
              value: ${__k8s_namespace}
            - name: ELASTIC_APM_TRANSACTION_IGNORE_URLS
              value: /
          resources:
            limits:
              cpu: 500m
```

```
                memory: 512Mi
              requests:
                cpu: 250m
                memory: 256Mi
            securityContext:
              privileged: false
            volumeMounts:
              - mountPath: /opt/${__project_name}/config
                name: settings
        imagePullSecrets:
          - name: odregistry
        restartPolicy: Always
        terminationGracePeriodSeconds: 30
        volumes:
          - configMap:
              defaultMode: 420
              name: ${__project_name}
            name: settings
```

注意，这里因为有 ConfigMap 资源需要挂载，所以需要提前建立一个名为 testci 的 ConfigMap 资源（内容为空即可，因为最终的 ConfigMap 文件由 Apollo 来生成）。创建命令如下：

```
kubectl create cm testci -n devops
```

这里还是承接上面的内容，192.10.1.199:5001/api/nighting-build/project/resource/default 接口的返回值如下：

```
{
    "err_code": 1,
    "err_msg": "ok",
    "data": {
        "project_id": 13,
        "deployment_id": 7,
        "service_id": 6,
        "route_id": 6
    }
}
```

对于以上这些返回值，尤其是 project_id 和 deployment_id 的值，下面的接口会复用。下面调用另一个接口将 testci 应用部署到指定的 Kubernetes 集群，接口为 10.1.0.199:5001/api/nighting-build/release/callback，HTTP 提交方法为 POST 请求，其 JSON 数据如下：

```
{
     "id":"1",
     "event_type":"build",
```

```
        "tag_id":"1.0",
        "deployments":[6],
        "project_id":12,
        "namespace_id":1
    }
```

事实上，调用上面的接口会在内部调用 devops-release 应用，其相关函数如下：

```
    func (spec *UpdateSpec) applyDeployment(ctx context.Context, deploymentString
string, clientset *kubernetes.Clientset) error {
        if len(deploymentString) != 0{
                deploymentYaml := appsv1.DeploymentApplyConfiguration{}
                d :=
yaml.NewYAMLToJSONDecoder(bytes.NewBufferString(deploymentString))
                e := d.Decode(&deploymentYaml)
                if e != nil{
                        nlog.Error(" 解析 yaml 有问题，请注意 ")
                        nlog.WithContext(ctx).Error(e)
                        return nil
                }
                ctx, ctxCancel := context.WithTimeout(context.Background(),
60*time.Second)
                defer ctxCancel()
                containerIndex := k8sutil.GetContainerIndexByName(*deployment
Yaml.Name, deploymentYaml)
                img := strings.Split(*deploymentYaml.Spec.Template.Spec.Containers
[containerIndex].Image,":")
                //nlog.Infof("img:",img)
                if len(img) < 2{
                     nlog.WithContext(ctx).Infof("skip deployment:%s\n",
spec.Project.ProjectName)

                     return errors.New("Error Deployment:wrong image format")
            }
            // 获取 kubectl apply 操作的镜像名：版本号，这里 Docker 仓库要注意域名或
            //IP 的区别
            //image := img[0] +":" + spec.Version
            image := img[0] +":" + img[1] + ":" + spec.Version
            //nlog.Infof("img:",img)

            if len(spec.Version) != 0{
        deploymentYaml.Spec.Template.Spec.Containers[containerIndex].Image =
```

```
&image
                }
                // 如果 nighting-build 从 job log 中读到了 image 信息，则用此镜像信息
                if strings.Contains(spec.Version,":") {
                        image = spec.Version
        deploymentYaml.Spec.Template.Spec.Containers[containerIndex].Image =
&spec.Version
                }

                nlog.Infof("Applying image:%s to namespace: [%s]",
*deploymentYaml.Spec.Template.Spec.Containers[containerIndex].Image, spec.
K8sNamespace.Name)
                /* _, e = clientset.AppsV1().Deployments(spec.Namespace).Apply
(ctx, &deploymentYaml, v1.ApplyOptions{FieldManager: "nighting-release"})*/

                _, e = clientset.AppsV1().Deployments(spec.K8sNamespace.Name).Apply
(ctx, &deploymentYaml, v1.ApplyOptions{Force: true, FieldManager: "nighting-
release"})

                if e != nil{
                        nlog.WithContext(ctx).Infof("skip deployment:%s\n", spec.
Project.ProjectName)
                        return e
                }
        } else {
                    nlog.WithContext(ctx).Infof("skip deployment:%s\n", spec.
Project.ProjectName)
        }
        return nil
    }
```

由于 Apollo 的相关代码控制在 devops-release 组件中，因此这里也需要提前查看 devops-release 组件的日志：

```
    INFO[1854] releaseInfo:{11 build 1.0 [11] 17 1}
    INFO[1854] 重新部署 devops:testci
    INFO[1854] 依次获取发版资源 :configmap/deployment/service
    INFO[1854] 从 apollo 拉取配置!
    获取项目发版资源 [dev devops testci tenant]
    最终生成的 configMap 为 []:INFO[1854] 配置为空
    INFO[1854] message:
    项目名: testci
```

```
操作类型：更新服务
环境：devops
集群：tenant-devops
命名空间：namespace--devops
重启信息：没有从配置中心读取到配置，将跳过 configmap apply，请检查配置中心是否有相应配置和项目名是否正确!
INFO[1854] Applying image:10.1.0.201:8083/testci:1.0 to namespace: [devops]
[GIN] 2023/12/18 - 14:32:50 | 200 |   45.982951ms |       127.0.0.1 | POST "/api/nighting-release/callback"
```

注意这行日志“最终生成的 configMap 为 []:INFO[1854] 配置为空”，说明从 Apollo 中没有获取到对应名为 testci 的 configMap 资源，然后直接执行了将 10.1.0.201:8083/testci:1.0 部署到 tenant-devops 集群的 devops 命名空间的操作。

4.4.2 devops-release 组件服务

devops-release 组件只与 devops-build 组件交互，然后与 Kubernetes 和 Apollo 等敏感资源打交道，针对 Deployment 资源提供 build/apply、restart 及 patch_image 操作，其工作流如图 4.15 所示。

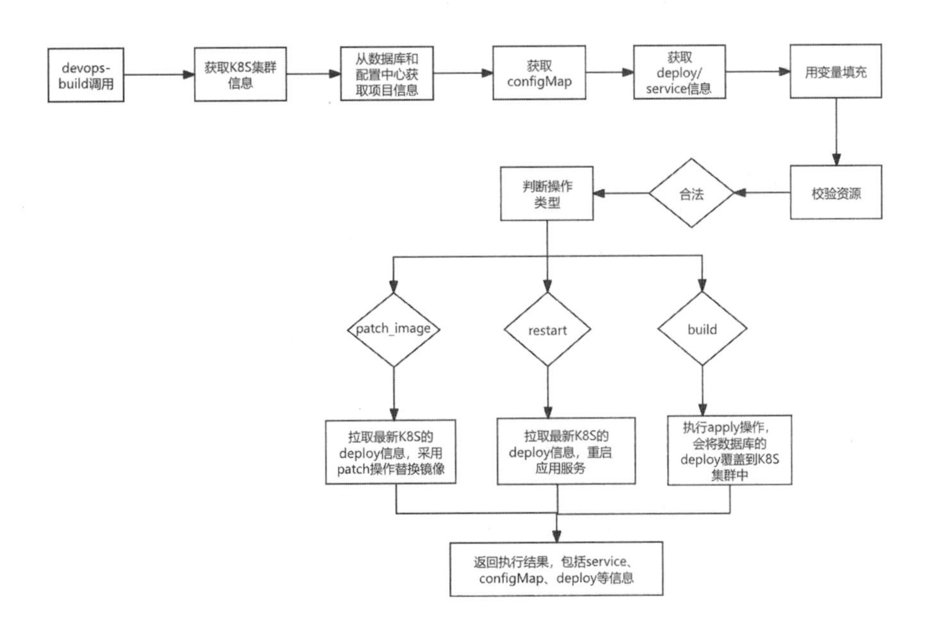

图 4.15 devops-release 组件工作流

devops-release 组件的目录结构可以用下面的命令显示：

```
tree /Users/yuhongchun/repo/workspace/cloudnative/devops_release
```

显示结果如下：

```
├── Dockerfile
├── README.md
├── api
│   └── handle.go
├── cmd
│   ├── cmd.go
│   └── server
│       └── server.go
├── config
│   ├── config.go
│   ├── model.go
│   ├── pri-release-settings.yaml
│   ├── settings.default.yaml
│   └── settings.yaml
├── database
│   ├── etcd.go
│   ├── model
│   │   └── model.go
│   ├── postgres.go
│   └── relational
│       ├── db_test
│       │   └── db_test.go
│       ├── postgres
│       │   ├── postgres.go
│       │   └── utils.go
│       └── relational.go
├── devops.sql
├── devops_release
├── go.mod
├── go.sum
├── internal
│   ├── buildv2
│   │   ├── apply.go
│   │   ├── build.go
│   │   ├── model.go
│   │   ├── plan.go
│   │   └── resourceSet.go
│   ├── middlewares
│   │   └── init.go
```

```
│    └── service
│        ├── add_new_tenant
│        │   └── add_new_tenant.go
│        ├── compare
│        │   └── compare.go
│        ├── convert
│        │   └── convert.go
│        ├── k8s_resource
│        │   ├── k8s_resource.go
│        │   ├── local_k8s.go
│        │   ├── model.go
│        │   ├── remote_k8s.go
│        │   └── sync_resource.go
│        └── k8s_resource_test
│            └── k8s_resource_test.go
├── log
│   └── nlog.log
├── main
├── main.go
├── manage.sh
├── pkg
│   └── env.go
├── rest.http
├── router
│   └── router.go
├── script
│   └── scan_config
│       └── main.go
├── test
│   └── apollo
│       └── main.go
└── util
    ├── aes_util
    │   └── aes_util.go
    ├── apollo
    │   ├── apollo_config.go
    │   ├── apollo_openapi.go
    │   ├── apollo_sdk.go
    │   ├── apollo_test.go
    │   ├── apollo_yaml_parse.go
    │   ├── import_to_apollo.go
```

```
│    ├── log
│    │   └── nlog.log
│    └── yaml_test
│        └── main.go
├── devops_error
│    └── devops_error.go
├── dingtalk_util
│    ├── dingtalk_util.go
│    └── test
│        └── main.go
├── http_util
│    └── http_util.go
├── k8s_util
│    └── k8s_util.go
├── model
│    └── model.go
├── myyaml
│    ├── myyaml.go
│    └── test
│        └── main.go
├── noticer
│    └── noticer.go
└── test
    ├── main.go
    └── settings.yaml

40 directories, 65 files
```

查看配置文件，命令为 cat ./config/settings.yaml。配置文件的内容如下：

```
application:
  host: 0.0.0.0
  ishttps: false
  mode: debug
  name: devops-release
  port: 5000
  readtimeout: 20
  writertimeout: 20
  nighting_build_url: http://192.168.1.199:5001/
  aes_key: 'asptxie12/98ajlrzmknm.ahkjk/jl;k'
postgres:
   dsn: "host=10.1.0.200 user=postgres password=pgsql@123456@
```

```
dbname=devops_build port=32432 sslmode=disable TimeZone=Asia/Shanghai"
    sentry:
      dsn: ""
      source: ""
    apm:
      logLevel: "info"
      filePath: "./log/nlog"
      maxFileSize: 100
      maxBackups: 3
      maxAge: 30
      compress: false
    tecent_cloud:
      secret_id: ""
      secret_key: ""
    apollo:
      channel:
        - key: dev
          address_openapi: http://10.1.0.200:32110
          address_sdk: http://10.1.0.200:32109
          k8s_cluster: tenant-devops
          apps:
            - id: devops-project
              type: tenant
              token:
8fb412e8fc29560144da447190de16b224824610b89b6ff11f8d095d6106a6cf
              secret: 5f0be9e6899046aba31604c35806aa51
```

这里要添加第三方应用 Token 为 Apollo 授权以便查看其应用，界面如图 4.16 所示。

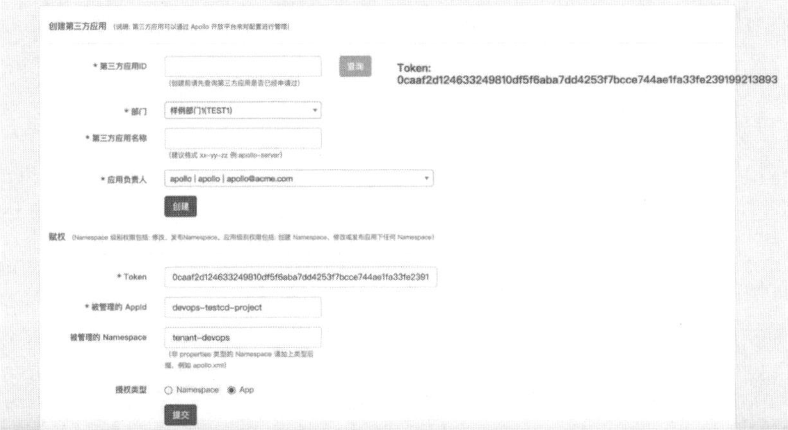

图 4.16　为 Apollo 添加第三方应用授权 Token

下面依次说明图 4.16 中各选项的作用：

（1）第三方应用 ID：设置唯一标识，使用字母和数字即可。

（2）部门：设置应用归属部门，用于筛选。

（3）第三方应用名称：设置应用名称，一个简单的描述。

（4）应用负责人：设置 openapi 操作的用户，该用户必须具有该项目的操作权限，如没有操作权限，则需要在 Apollo 项目中授权，否则调用时会出现“403”错误。

（5）Token：自定义生成，openapi 客户端参数。

（6）被管理的 AppId：已在 Apollo 中存在的需要操作的 Apollo 的应用 id。

（7）被管理的 Namespace：openapi 可操作的 namesapce，授权类型为 namespace 时，最少填一个。

（8）授权类型：选中 Namespace 单选按钮，表示指定 namesapce；选中 App 单选按钮，表示所有 namespace。

然后，可以执行启动 devops-release 服务的命令来启动该应用，命令如下：

```
ELASTIC_APM_ACTIVE=false ./main server -c config/settings.yaml
```

显示结果如下：

```
INFO[0000] starting server
INFO[0000] {20 20 debug devops-release 0.0.0.0 5000 false asptxie12/98ajlrzmknm.ahkjk/jl;k}
INFO[0000] mode:debug
[GIN-debug] [WARNING] Running in "debug" mode. Switch to "release" mode in production.
 - using env:   export GIN_MODE=release
 - using code:  gin.SetMode(gin.ReleaseMode)

[GIN-debug] GET    /ping                     --> devops_release/router.InitRouter.func1 (3 handlers)
[GIN-debug] POST   /api/nighting-release/callback --> devops_release/api.HandleCallback (3 handlers)
[GIN-debug] POST   /api/nighting-release/convert --> devops_release/api.ConvertYamlOrKV (3 handlers)
[GIN-debug] POST   /api/nighting-release/add_new_tenant --> devops_release/api.CreateANewEnv (3 handlers)
[GIN-debug] GET    /api/nighting-release/getstatus --> devops_release/api.GetDeployStatus (3 handlers)
[GIN-debug] GET    /api/nighting-release/image_list --> devops_release/api.GetDockerTagList (3 handlers)
[GIN-debug] POST   /api/nighting-release/compare_config --> devops_release/api.CompareConfigInK8sAndApollo (3 handlers)
[GIN-debug] GET    /api/nighting-release/get_config -->
```

```
devops_release/api.GetApolloConfig (3 handlers)
    [GIN-debug] POST   /api/nighting-release/set_config -->
devops_release/api.SetApolloConfig (3 handlers)
    [GIN-debug] POST   /api/nighting-release/pub_config -->
devops_release/api.PubApolloConfig (3 handlers)
     NFO[0000] 2023-12-09 11:18:12.706156645 +0800 CST m=+0.027505037 Server
Run http://0.0.0.0:5000/
     NFO[0000] 2023-12-09 11:18:12.706180359 +0800 CST m=+0.027528751 Enter
Control + C Shutdown Server
    INFO[0000] 对 dev 通 道 进 行 监 听! address_openapi:http://10.1.0.200:32110
address_sdk:http://10.1.0.200:32109
    INFO[0000] dev 通道内监听 !appid:devops-project
    [{DEV [default devops]}]
    INFO[0000] 对 default 集群进行监听!
    [{devops-project default application default app namespace properties
false []  2023-12-03T21:39:59.000+0800 2023-12-03T21:39:59.000+0800} {devops-
project default testci  properties false [{password 123456 apollo apollo 2023-
12-03T21:44:51.000+0800 2023-12-03T21:44:51.000+0800} {user yuhongchun apollo
apollo 2023-12-03T21:46:25.000+0800 2023-12-03T21:46:25.000+0800}]   2023-12-
03T21:44:03.000+0800 2023-12-03T21:44:03.000+0800}]
    INFO[0000] 对 devops 集群进行监听!
    [{devops-project devops application default app namespace properties
false [] 2023-12-03T22:56:31.000+0800 2023-12-03T22:56:31.000+0800} {devops-
project devops testci  properties false [{username yht apollo apollo 2023-12-
03T22:56:58.000+0800 2023-12-03T22:56:58.000+0800} {password 123456 apollo
apollo 2023-12-09T10:31:00.000+0800 2023-12-09T10:31:00.000+0800}]   2023-12-
03T22:56:31.000+0800 2023-12-03T22:56:31.000+0800}]
    INFO[0002] 监听成功! clustername:default
    INFO[0002] 监听成功! clustername:devops
```

执行到这一步，说明 devops-release 服务是正常启动的。当然，读者肯定会有疑问，Kubernetes 中的关键资源 configMap 究竟是如何生成的，又是如何与前面的这些资源，如 project 和 deployment 关联的呢？

下面介绍 devops-release 应用获取 configMap 的流程，如图 4.17 所示。

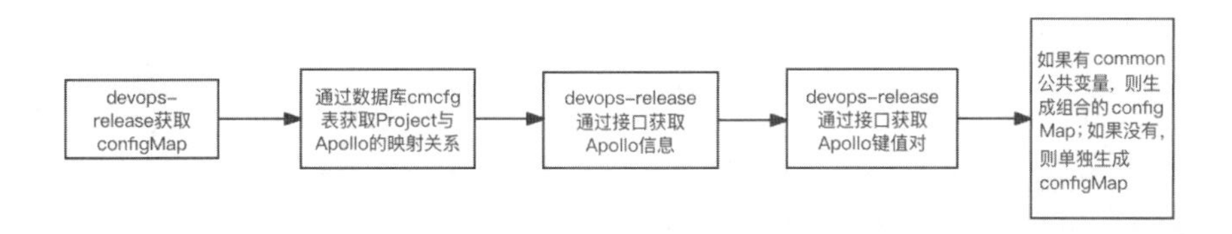

图 4.17　devops-release 应用获取 configMap 的流程

图 4.17 中的 cmcfg 表是很关键的数据，它记录了某个 configMap 变量以及 project_id 与 deployment_id 之间一一对应的关系，也是 devops-release 组件连接 Apollo 的关键，其表结构见表 4.5。

表 4.5　cmcfg 表的结构

名　字	类　型	说　明	是否主键
id	int4	唯一标识	是
project_id	int4	项目 id	否
config_name	text	配置名称（即 Apollo 的配置名）	否
content	text	保留字段（选填内容）	否
deployment_id	int4	绑定的 deployment，决定着配置变更重启的 deployment	否
file_name	text	生成的配置文件名	否
configmpa_name	text	生成的 configMap 名	否
restart_after_pub	bool	配置发布后是否重启	否

不仅这个配置很重要，相应的 API 接口也很多，可以通过 devops-build 的 swagger 接口文档查看：http://10.1.0.199:5001/swagger/index.html。

其中与 config 配置（即 cmcfg 表）相关的 API 接口及接口的代码在 /workspace/cloudnative/devops_build/api/config.go 文件中。其代码内容如下：

```
    // ListConfigOfDeployment
    // @Summary 查询 deployment 下的 config
    // @Produce json
    // @Param deployment_id query int true "deployment Id"
    // @Success 200 {object} ReturnData{data=[]model.Config}
    // @Failure 400 {object} ReturnData{data=[]model.Config}
    // @Router /api/nighting-build/list_config_of_deployment [get]
    func ListConfigsOfDeployment(c *gin.Context) {
         deploymentIdStr := c.Query("deployment_id")
         deploymentId, err := strconv.Atoi(deploymentIdStr)
         if err != nil{
                 c.JSON(http.StatusBadRequest, ReturnData{Err_code: 0,
Err_msg: "参数绑定错误！"})
               return
         }
         devopsdb := database.GetDevopsDb()
         configs, err := devopsdb.GetConfigByDeploymentId(c, deploymentId)
         if err != nil{
                 logrus.Error(err)
```

```
            c.JSON(500, ReturnData{Err_code: 0, Err_msg: "查询错误！"})
            return
        }
        c.JSON(http.StatusOK, ReturnData{Err_code: 1, Err_msg: "ok", Data:
configs})
    }

    // @Summary 更新某个项目发版路径下的config
    // @Accept application/json
    // @Param id body int true "Id"
    // @Param project_id body string true "project_id"
    // @Param config_name body string true "ProjetcId"
    // @Param file_name body string true "file_name"
    // @Param configmap_name body string true "configmap_name"
    // @Param content body string true "content"
    // @Param restart_after_pub body int true "restart_after_pub"
    // @Param namespace_id body int  true "namespace_id"
    // @Success 200 {object} ReturnData
    // @Failure 400 {object} ReturnData
    // @Router /api/nighting-build/config [patch]
    func PatchConfig(c *gin.Context) {
        config := &model.Config{}
        err := c.BindJSON(config)
        if err != nil{
            c.JSON(400, gin.H{"err_code": 0, "err_msg": "参数绑定错误！"})
            return
        }
        devopsdb := database.GetDevopsDb()
        _, err = devopsdb.UpdateConfigById(c, *config)
        if err != nil{
            c.JSON(500, gin.H{"err_code": 0, "err_msg": "修改错误！"})
            return
        }
        c.JSON(http.StatusOK, gin.H{"err_code": 1, "err_msg": "ok"})
    }

    // DeleteConfig
    // @Summary 删除一个配置
    // @Param id query string true "ID"
    // @Success 200 {object} ReturnData
    // @Failure 400 {object} ReturnData
    // @Router /api/nighting-build/config [delete]
```

```
func DeleteConfig(c *gin.Context) {
      idStr := c.Query("id")
      id, err := strconv.Atoi(idStr)
      if err != nil{
              c.JSON(http.StatusBadRequest, ReturnData{Err_code: 0,
Err_msg: "参数错误! "})
            return
      }
      devopsdb := database.GetDevopsDb()
      _, err = devopsdb.DeleteConfigById(c, id)
      if err != nil{
      c.JSON(500, ReturnData{Err_code: 0, Err_msg: err.Error()})
              return
      }
      c.JSON(200, ReturnData{Err_code: 1, Err_msg: "ok"})
}

// @Summary 增加某个项目发版路径下的 config
// @Accept application/json
// @Param project_id body string true "project_id"
// @Param config_name body string true "配置名"
// @Param file_name body string true "文件名"
// @Param configmap_name body string true "configmap_name"
// @Param content body string true "content"
// @Param restart_after_pub body int true "restart_after_pub"
// @Param namespace_id body int  true "namespace_id"
// @Param deployment_id body int  true "deployment_id"
// @Success 200 {object} ReturnData
// @Failure 400 {object} ReturnData
// @Router /api/nighting-build/config [put]
func AddConfig(c *gin.Context) {
      configInfo := &model.Config{}
      err := c.BindJSON(configInfo)
      if err != nil{
              c.JSON(400, ReturnData{Err_code: 0, Err_msg: "参数绑定错误! "})
              return
      }
      devopsdb := database.GetDevopsDb()
      _, err = devopsdb.InsertIntoConfig(c, *configInfo)
      if err != nil{
      c.JSON(500, ReturnData{Err_code: 0, Err_msg: err.Error()})
            return
```

```
    }

    c.JSON(200, ReturnData{Err_code: 1, Err_msg: "ok"})
}
```

通过查看 devops-build 组件的 swagger 文档也能看到相应数据，如图 4.18 所示。

图 4.18　devops-build 组件的 swagger 文档中关于 configMap 的接口文档

除了有新增、修改、查询 configMap 资源的接口外，还有删除 configMap 资源的接口。例如，要新增一个 configMap 资源配置，可以通过 PUT 方法调用以下接口，即 10.1.0.199:5001/api/nighting-build/config，其要传递的 JSON 数据格式如下：

```
{
    "id": 7,
    "project_id": 17,
    "config_name": "testci",
    "file_name": "testci",
    "configmap_name": "testci",
    "content": "apiVersion: v1\nkind: ConfigMap\nmetadata:\n  name: testci\
ndata:\n  setting.yaml: |-\n    user: yhc666",
    "restart_after_pub": true,
    "namespace_id": 1,
    "deployment_id": 11,
    "deployment_name": "",
```

```
        "namespace_name": ""
    }
```

其中的content内容可以自行定义，因为最终的configMap内容生成全部由Apollo来控制；但是需要注意deployment_id和project_id，它们是需要建立一一对应关系的，如果出错，会将名为testci的应用附加到出错的对应deployment上。

执行成功以后，会在PostgreSQL数据库的cmcfg表中生成数据。可以用GET方法查询此数据，如10.1.0.199:5001/api/nighting-build/list_config_of_deployment?deployment_id=11，最后返回结果如下：

```
    {
        "err_code": 1,
        "err_msg": "ok",
        "data": [
            {
                "id": 8,
                "project_id": 17,
                "config_name": "testci",
                "file_name": "testci",
                "configmap_name": "testci",
                "content": "apiVersion: v1\nkind: ConfigMap\nmetadata:\n  name:
testci\ndata:\n  setting.yaml: |-\n    user: yhc666",
                "restart_after_pub": true,
                "namespace_id": 1,
                "deployment_id": 11,
                "deployment_name": "",
                "namespace_name": ""
            }
        ]
    }
```

对于其他接口，如更新及删除cmcfg表的接口就不再一一演示和说明了。

事实上，用devops-release的callback接口部署testci应用后，通常可以通过其日志查看整个configMap的执行及生成过程，其日志如下：

```
INFO[0706] releaseInfo:{11 build 1.0 [11] 17 1}
namespaceId:
1
INFO[0706] 重新部署 devops:testci
INFO[0706] 依次获取发版资源 :configmap/deployment/service
INFO[0706] 从 apollo 拉取配置!
[dev devops testci tenant]
INFO[0706] 通过 Apollo 来重新生成应用配置
```

```
    INFO[0706] 从 apollo 拉取配置 |appid:devops-project cluster:devops name:testci
    http://10.1.0.200:32110/openapi/v1/envs/DEV/apps/devops-project/clusters/
devops/namespaces/testci
    INFO[0706]  从 apollo 拉 取 公 共 配 置 !appid:devops-project cluster:devops
name:testci
    http://10.1.0.200:32110/openapi/v1/envs/DEV/apps/devops-project/clusters/
devops/namespaces/common
    ********************configmap***************************
    apiVersion: v1
    kind: ConfigMap
    metadata:
      name: testci
    data:
      setting.yaml: |-
        mysql_port: 33071
    **********************end******************************
    INFO[0706] config 信息为 :&{8 17 testci testci testci apiVersion: v1
    kind: ConfigMap
    metadata:
      name: testci
    data:
      setting.yaml: |-
        user: yhc666 true 1 11}
    [apiVersion: v1
    kind: ConfigMap
    metadata:
      name: testci
    data:
      setting.yaml: |-
        mysql_port: 33071]INFO[0706] Applying image:10.1.0.201:8083/testci:1.0
to namespace: [devops]
    [GIN] 2023/12/16 - 17:28:54 | 200 |    63.30076ms |       127.0.0.1 | POST
"/api/nighting-release/callback"
```

Apollo 具体生成 ConfigMap 的文件为 /workspace/cloudnative/devops_release/util/apollo/apollo_config.go。

其具体执行函数在 /workspace/cloudnative/devops_release/util/apollo/apollo_openapi.go 文件中。

事实上，还有个场景会经常发生，就是应用的配置被更改，所以，这里需要设置对应的场景。如果监听到应用对应的 Apollo namespace 键值对发生更改，则需要重新生成 ConfigMap 并且重启相应的 Deployment，其流程如图 4.19 所示。

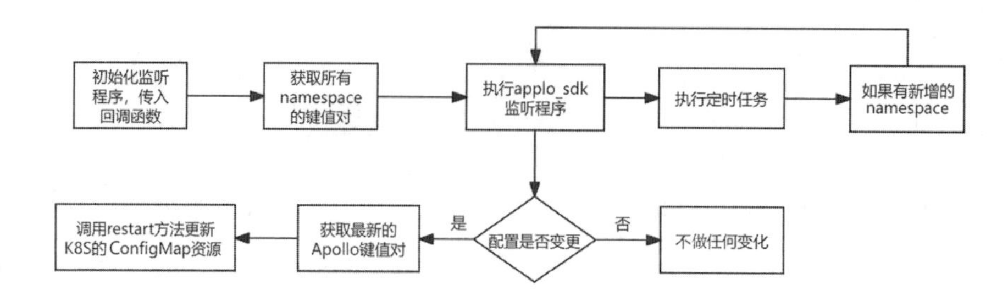

图 4.19　重新生成 ConfigMap 并且重启 Deployment 流程

如果通过 Apollo 界面来修改 testci 的配置并重新发布，则名为 testci 的 Deployment 将重启并重新生成 Pod。devops-release 应用的日志如下：

```
    INFO[0038] 监听到apollo配置中心配置变更! appid:devops-project env:dev cluster:tenant-devops namespace:devops pro:testci
    INFO[0038] 配置变更，重启服务!
    1 devops
    1 testci
    configMap集合为：&{8 17 testci testci testci apiVersion: v1
    kind: ConfigMap
    metadata:
      name: testci
    data:
      setting.yaml: |-
        user: yhc666 true 1 11}
    Deployments: [11]INFO[0038] 依次获取发版资源:configmap/deployment/service
    INFO[0038] 从apollo拉取配置!
    [dev devops testci tenant]
    INFO[0038] 通过apollo来重新生成应用配置
    INFO[0038] 从apollo拉取配置!appid:devops-project cluster:devops name:testci
    http://10.1.0.200:32110/openapi/v1/envs/DEV/apps/devops-project/clusters/devops/namespaces/testci
    INFO[0038] 从apollo拉取公共配置!appid:devops-project cluster:devops name:testci
    http://10.1.0.200:32110/openapi/v1/envs/DEV/apps/devops-project/clusters/devops/namespaces/common
    ********************configmap****************************
    apiVersion: v1
    kind: ConfigMap
    metadata:
      name: testci
    data:
      setting.yaml: |-
```

```
    mysql_port: 3306
**********************end******************************
INFO[0038] config 信息为 :&{8 17 testci testci testci apiVersion: v1
kind: ConfigMap
metadata:
  name: testci
data:
  setting.yaml: |-
    user: yhc666 true 1 11}
INFO[0038] deployMsgId 11
b 的值为 : metadata:
  creationTimestamp: null
  generation: 1
  labels:
    k8s-app: testci
  name: testci
spec:
  progressDeadlineSeconds: 600
  replicas: 1
  revisionHistoryLimit: 10
  selector:
    matchLabels:
      k8s-app: testci
  strategy:
    rollingUpdate:
      maxSurge: 25%
      maxUnavailable: 25%
    type: RollingUpdate
  template:
    metadata:
      creationTimestamp: null
      labels:
        k8s-app: testci
    spec:
      containers:
      - env:
        - name: ELASTIC_APM_SERVICE_NAME
          value: testci
        - name: ELASTIC_APM_ENVIRONMENT
          value: devops
```

```
          - name: ELASTIC_APM_TRANSACTION_IGNORE_URLS
            value: /
          image: 10.1.0.201:8083/testci:1.0
          imagePullPolicy: Always
          lifecycle:
            preStop:
              exec:
                command:
                - /bin/bash
                - -c
                - sleep 20
          name: testci
          resources:
            limits:
              cpu: 500m
              memory: 512Mi
            requests:
              cpu: 250m
              memory: 256Mi
          securityContext:
            privileged: false
          terminationMessagePath: /dev/termination-log
          terminationMessagePolicy: File
          volumeMounts:
          - mountPath: /opt/testci/config
            name: settings
        dnsPolicy: ClusterFirst
        imagePullSecrets:
        - name: odregistry
        restartPolicy: Always
        schedulerName: default-scheduler
        securityContext: {}
        terminationGracePeriodSeconds: 30
        volumes:
        - configMap:
            defaultMode: 420
            name: testci
          name: settings
status: {}
```

```
    INFO[0038] Restart Deployment testci with image:10.1.0.201:8083/testci:1.0
in namespace: [devops], because of the config has changed
    INFO[0038] Restarted deployment testci in cluster[tenant-devops]-
namespace[testci] success
    INFO[0038] 重启服务成功!
    INFO[0038]
    项目名：testci
    操作类型：配置变更
    环境：devops
    集群：tenant-devops
    命名空间：devops
    重启信息：正常
```

重启 Deployments 的逻辑在 /workspace/cloudnative/devops_release/internal/buildv2/build.go 文件中，其相应代码如下：

```
    type UpdateSpec struct {
        K8sClusterName string
        Version        string
        OpsType        string
        ClusterToken   string
        ClusterCa      string
        ClusterUrl     string
        Project        model.Project
        Deployments    []int
        ConfigMap      string
        K8sNamespace   model.K8sNamespace
    }

    // restartProject: restart deployment if config file has changed.
    // stage1: get config.
    // stage2: get apply configmap
    // stage3: restart deployment
    // stage4: update svc
    func (s *UpdateSpec) restart(ctx context.Context) error {
        fmt.Printf("重启的 Deployments 名字为：%v",s.Deployments)
        opsMsg := dingtalkutil.GetOpsInfoWithContext(ctx)
        var err error
        if s == nil{
                return errors.New("restartProject faild: nil restartProject
pointer!")
        }
```

```
        devopsdb := database.GetDevopsDb()
        // connect k8s
        clientset, err := s.getK8sClient()
        if err != nil{
                return err
        }
        resource, err := s.GetProjectResource(ctx, devopsdb)
        if err != nil{
                return err
        }
         if len(resource.Configmaps) == 0 && len(resource.Services) == 0 &&
len(resource.Deployments) == 0{
                return errors.New("not a valid k8s resource")
        }

        // // replace template
        // resource.Configmap, err = s.getRealConfig(resource.Configmap, client)
        // if err != nil{
        //      return err
        //}

        // apply configmap
        if len(resource.Configmaps) == 0{
                 logr.Info("配置为空")
                 message := fmt.Sprintf(noticer.ConfigEmptyMsg,
s.Project.ProjectName, "重启服务", s.K8sNamespace.Name, s.K8sClusterName,
s.K8sNamespace.Description)
                 logr.Info("message:",message)
                 //sendToWatchers(ctx, message, s.Project.Id,
s.K8sNamespace.Name)
        }

        // 找出 configMap 关联的 Deployment 并重启
        for i, cfg := range resource.Configs{
             if !cfg.RestartAfterPub{
                    continue
             }
             configMsg := dingtalkutil.ConfigInfo{
                    Id:   cfg.Id,
                    Name: cfg.ConfigName,
```

```
            }


            err = s.applyConfigmap(ctx, resource.Configmaps[i], clientset)
            if err != nil{
                    opsMsg.Status = "Warning"
                    configMsg.Err = err.Error()
                    opsMsg.Configs = append(opsMsg.Configs, configMsg)
                    continue
            }

            opsMsg.Configs = append(opsMsg.Configs, configMsg)
            deployMsg := dingtalkutil.DeploymentInfo{
                    Id: cfg.DeploymentId,
            }
            // 通过前面定义的 PostgreSQL 的 cmcfg 表记录查找 configMap 对应的
            // deployment_id 记录，从而得知要重启的 Deployment
            deployment, err := devopsdb.GetDeploymentById(ctx,
cfg.DeploymentId)
            if err != nil{

                    deployMsg.Err = "未获取到 deployment"
                    logr.Error("未获取到 deployment")
                    opsMsg.Deployments = append(opsMsg.Deployments,
deployMsg)
                continue
            }

            deployMsg.Id = deployment.Id
            deployMsg.Name = deployment.DeploymentName
            logr.Infof("deployMsgId %v",deployMsg.Id)
            //同步并重启 deployment，防止本地覆盖线上已经更改的资源
            d, err :=
clientset.AppsV1().Deployments(s.K8sNamespace.Name).Get(ctx,
deployment.DeploymentName, metav1.GetOptions{})
            if err == nil && d != nil && len(d.Name) != 0{
                    err = k8sresource.SyncDeploymentToLocal(ctx,
deployment.Id, *d)
                    if err != nil{
                         logr.Errorf("同步 deployment 失败！")
                    }
```

```
            } else {
                    logr.Infof("同步 deployment 失败! K8S 端 deployment 未部署")
                    deployMsg.Err = "同步 deployment 失败! K8S 端 deployment 未部署"
                    opsMsg.Status = "Warnging"
                    opsMsg.Deployments = append(opsMsg.Deployments, deployMsg)
                    continue
            }

            //变量替换
            deployment.Content = s.getRealDeployment(deployment.Content)
            err = s.restartDeployment(ctx, deployment.Content, clientset)
            if err != nil{
                     opsMsg.Status = "Warning"
                     deployMsg.Err = err.Error()
                     logr.WithContext(ctx).Errorf("restartdeployment
pro:%s deployment:%s  err:%v", s.Project.ProjectName,
deployment.DeploymentName, err)
                     opsMsg.Deployments = append(opsMsg.Deployments, deployMsg)
                     continue
            }

        }
        // apply service
        for _, service := range resource.Services{
            serviceMsg:= dingtalkutil.ServiceInfo{
                    Id: service.Id,
            }
            err = s.applyService(ctx, service.Content, clientset)
            if err != nil{
                     serviceMsg.Err = err.Error()
                     opsMsg.Status = "Warning"
                     opsMsg.ServiceInfo = append(opsMsg.ServiceInfo, serviceMsg)
                     logr.WithContext(ctx).Errorf("apply service pro:%s
    service:%s  err:%v", s.Project.ProjectName, service.Name, err)
            }
            opsMsg.ServiceInfo = append(opsMsg.ServiceInfo, serviceMsg)
        }
        return nil
    }
```

接下来，回顾前面的接口 10.1.0.199:5001/api/nighting-build/release/callback，针对 Kubernetes

资源，它能够执行 build、patch_image 及 restart 应用等操作。这些功能的具体实现源码主要包含在 buildv2 包中，其文件目录明细如下：

```
tree /workspace/cloudnative/devops_release/internal/buildv2
```

显示结果如下：

```
├── apply.go
├── build.go
├── model.go
├── plan.go
└── resourceSet.go
0 directories, 5 files
```

devops-release 针对 Kubernetes 资源的绝大部分操作都是通过 buildv2 包来实现的，所以这里完整展示相应文件内容。

model.go 文件定义了与发版相关的结构体，代码如下：

```
package buildv2

type K8s_cluster_info struct {
    Name      string `yaml:"cluster_name"`
    Url       string
    Token     string
    Ca        string
    TLK_ID    string
    Namespace []string `yaml:"namespaces"`
}

type OpsInfo struct {
    ProjectName    string           `json:"project_name"`
    Type           string           `json:"type"`
    OpsProResults  []OpsProResult   `json:"ops_pro_results"`
    Channel        string           `json:"channel"`
    Status         string           `json:"status"`
}
type OpsProResult struct {
    Namespace        string    `json:"namespace"`
    ClusterName      string    `json:"cluster_name"`
    Status           string    `json:"status"`
    Message          string    `json:"message"`
    DeploymentErrs   []string  `json:"deployment_errs"`
    ServiceErrs      []string  `json:"service_errs"`
    ConfigMapErrs    []string  `json:"configmap_errs"`
```

```
}
```

plan.go 文件会检查项目在 router 表中是否指定了发版路径。如果已指定，则通过；如果未指定，则拒绝。其中的源码内容如下：

```
package buildv2

import (
    "context"
    "devops_release/database"
    "devops_release/database/model"
    logr "github.com/sirupsen/logrus"
)

type ClusterInfo struct {
    ClusterName string   `yaml:"cluster_name"`
    Namespaces  []string `yaml:"namespaces"`
}
type ChannelMapping struct {
    Channel string              `yaml:"channel_name"`
    Cluster []K8s_cluster_info `yaml:"clusters"`
}
type ChannelMappingYaml struct {
    ChannelMapping []*ChannelMapping `yaml:"channel_mapping"`
}

// 在router表中确定project_id和deployment_id唯一的资源有没有发版路由
func getReleaseRoutes(ctx context.Context, projectid int, namespaceId int)
[]*model.Route {
    devopsdb := database.GetDevopsDb()
    routes := []*model.Route{}
    //fmt.Println("namespaceId:")
    //fmt.Println(namespaceId)

    routes, err := devopsdb.GetRoutesByProIdAndNspId(ctx, projectid,
namespaceId)
    if err != nil || routes == nil || len(routes) == 0{
        logr.Errorf("未在系统中找到%d namespaceId:%s 的发版路径！请添加！
err: ", projectid, namespaceId, err)
    }
    return routes
}
```

resourceSet.go 文件是发版流程中首个执行环节，它会依次获取 configmap、deployment 和 service，最终生成 ProjectResource 结构体。其源码内容如下：

```
package buildv2

import (
      "context"
      "fmt"

      "devops_release/database/model"
      "devops_release/database/relational"
      "devops_release/util/apollo"
      logr "github.com/sirupsen/logrus"
)

type ProjectResource struct {
      Configmaps    []string
      Deployments []*model.Deployment
      Services      []*model.Service
      Configs       []*model.Config
}

// 获取项目发版资源 (deployment,configmap,service)
func (s *ProjectResource) setConfig(ctx context.Context, devopsdb
relational.DevopsDb, clusterName string, project model.Project, k8sNamespace
model.K8sNamespace, class string) error {
      // env cluster namespace projectname，从 apollo 获取配置
      logr.Info("依次获取发版资源 :configmap/deployment/service")
      logr.Infof("从 apollo 拉取配置！")
      ch := apollo.GetApChannelFromConfigByClusterName(clusterName)
      if ch == nil{
            logr.Errorf("%s 不在 apollo 配置中！", clusterName)
            return fmt.Errorf("%s 不在 apollo 配置中！", clusterName)
      }
      fmt.Println("获取项目发版资源",[]string{ch.Key, k8sNamespace.Name,
project.ProjectName, class})

      // 获取 deployment 与配置的映射关系
      for _, deployment := range s.Deployments{
            // 通过数据库查询 deployment 相应的 config 信息
            configs, err := devopsdb.GetConfigByDeploymentId(ctx,
```

```
deployment.Id)
                if err != nil{
                        logr.Errorf("未获取到%s的配置，忽略。",
deployment.DeploymentName)
                        continue
                }
                s.Configs = append(s.Configs, configs...)
        }

        //opsMsg := dingtalkutil.GetOpsInfoWithContext(ctx)
        for _, config := range s.Configs{
                logr.Infof("通过Apollo来重新生成应用配置")
                //configmap, err := apollo.GetConfigFromApolloV2(ch.Key,
k8sNamespace.Name, *config, class)
                configmap, err := apollo.GetConfigFromApolloV2(ch.Key,
k8sNamespace.Name, project.ProjectName, class)
                logr.Info("config信息为:",config)
                if err != nil{
                        //opsMsg.Status = "Warning"
                        //opsMsg.ErrMsg = "配置" + config.ConfigName + "获取失败！" +
err.Error()
                        logr.Errorf("获取apollo配置失败！ err: %v", err)
                }
                if len(configmap) == 0{
                        //TODO:
                        logr.Info("configmap is empty,skip apply
info!deploymentId:", project.ProjectName)
                }
                s.Configmaps = append(s.Configmaps, configmap)

        }

        // set configmap

        // set deployment
        if len(s.Deployments) == 0{
                fmt.Println("projectId:", project.Id)
                logr.WithContext(ctx).Errorf("系统中未找到%s-%s的deployments，
或查询出错！", project.ProjectName, k8sNamespace.Name)
                return fmt.Errorf("系统中未找到%s-%s的deployments，或查询出错！",
project.ProjectName, k8sNamespace.Name)
```

```
        }
        // set service
        services := []*model.Service{}
        for _, deployment := range s.Deployments{
                servicesT, err := devopsdb.GetServicesByDeploymentId(ctx,
deployment.Id)
                if err != nil || len(servicesT) == 0{
                        logr.Infof("未找到%s-%s的service，跳过... err:%v",
deployment.DeploymentName, k8sNamespace.Name, err)
                        continue
                }
                services = append(services, servicesT...)
        }
        s.Services = services
        return nil
    }
```

build.go 文件主要用于判断操作类型。如果是 eventType=build，则会执行 build 操作；如果是 eventType=restart，则会执行 restart 操作；如果是 eventType=patch_image，则会执行 patch_image 操作。

apply.go 文件对应的就是相关的 apply 及 restart deployment 操作，其源码如下：

```
    package buildv2

    import (
        "bytes"
        "context"
        "encoding/json"
        "fmt"
        "strings"
        "time"

        "devops_release/database/model"
        k8sresource "devops_release/internal/service/k8s_resource"
        k8sutil "devops_release/util/k8s_util"
        "github.com/pkg/errors"
        logr "github.com/sirupsen/logrus"
        apiappsv1 "k8s.io/api/apps/v1"
        v1 "k8s.io/apimachinery/pkg/apis/meta/v1"
        "k8s.io/apimachinery/pkg/types"
        "k8s.io/apimachinery/pkg/util/yaml"
        appsv1 "k8s.io/client-go/applyconfigurations/apps/v1"
```

```
        corev1 "k8s.io/client-go/applyconfigurations/core/v1"
        "k8s.io/client-go/kubernetes"
        "k8s.io/client-go/util/retry"
    )

    func (spec *UpdateSpec) applyConfigmap(ctx context.Context, configmapString
string, clientset *kubernetes.Clientset) error {
        if len(configmapString) != 0{
                ctx, ctxCancel := context.WithTimeout(context.Background(),
60*time.Second)
                defer ctxCancel()
                configmap := corev1.ConfigMapApplyConfiguration{}
                d :=
yaml.NewYAMLToJSONDecoder(bytes.NewBufferString(configmapString))

                err := d.Decode(&configmap)
                if err != nil{
                        logr.WithContext(ctx).Error(err)
                        return err
                }
            _, err =
clientset.CoreV1().ConfigMaps(spec.K8sNamespace.Name).Apply(ctx, &configmap,
v1.ApplyOptions{
                        Force:        true,
                        FieldManager: "nighting-release"})
                if err != nil{
                        logr.WithContext(ctx).Error(err)
                        return err
                }
        } else {
                logr.WithContext(ctx).Infof("skip configmap:%s\n",
spec.Project.ProjectName)
        }
        return nil
    }
    func (spec *UpdateSpec) applyDeployment(ctx context.Context,
deploymentString string, clientset *kubernetes.Clientset) error {
        if len(deploymentString) != 0{
              deploymentYaml := appsv1.DeploymentApplyConfiguration{}
              d :=
yaml.NewYAMLToJSONDecoder(bytes.NewBufferString(deploymentString))
```

```
            e := d.Decode(&deploymentYaml)
            if e != nil{
                    logr.Error(" 解析 yaml 有问题，请注意 ")
                    logr.WithContext(ctx).Error(e)
                    return nil
            }
            ctx, ctxCancel := context.WithTimeout(context.Background(),
60*time.Second)
            defer ctxCancel()
            containerIndex :=
k8sutil.GetContainerIndexByName(*deploymentYaml.Name, deploymentYaml)
            img :=
strings.Split(*deploymentYaml.Spec.Template.Spec.Containers[containerIndex].
Image,":")
            if len(img) < 2{
                    logr.WithContext(ctx).Infof("skip deployment:%s\n",
spec.Project.ProjectName)
                    return errors.New("Error Deployment:wrong image format")
            }
            // 获取 kubectl apply 操作的镜像名：版本号，这里 Docker 仓库要注意域名或
            //IP 的区别
            //image := img[0] +":" + spec.Version
            image := img[0] +":" + img[1] + ":" + spec.Version

            if len(spec.Version) != 0{
        deploymentYaml.Spec.Template.Spec.Containers[containerIndex].Image = &image
             }
             // 如果 nighting-build 从 job log 中读到了 image 信息，则用此镜像信息
             if strings.Contains(spec.Version,":") {
                     image = spec.Version
        deploymentYaml.Spec.Template.Spec.Containers[containerIndex].Image =
&spec.Version }
            logr.Infof("Applying image:%s to namespace: [%s]",
*deploymentYaml.Spec.Template.Spec.Containers[containerIndex].Image,
spec.K8sNamespace.Name)
            _, e =
clientset.AppsV1().Deployments(spec.K8sNamespace.Name).Apply(ctx,
&deploymentYaml, v1.ApplyOptions{Force: true, FieldManager:
"nighting-release"})

            if e != nil{
```

```
                    logr.WithContext(ctx).Infof("skip deployment:%s\n",
spec.Project.ProjectName)
                    return e
            }
        } else {
                logr.WithContext(ctx).Infof("skip deployment:%s\n", spec.
Project.ProjectName)
        }
        return nil
    }
    func (spec *UpdateSpec) patchDeployment(ctx context.Context, deployment
*apiappsv1.Deployment, localDeployment model.Deployment, clientset
*kubernetes.Clientset) error {
        k8sContaniers := deployment.Spec.Template.Spec.Containers
        if len(k8sContaniers) == 0{
                return fmt.Errorf("deployment 中没有镜像信息！ ")
        }
        imageUrl := k8sContaniers[0].Image
        imageUrlSplit := strings.Split(imageUrl,":")
        if len(imageUrlSplit) < 2{
                logr.WithContext(ctx).Infof("skip deployment:%s\n", spec.
Project.ProjectName)
                fmt.Println(imageUrl)
                return errors.New("Error Deployment:wrong image format")
        }
        // 这里 Docker 仓库要注意域名或 IP 的区别
        imageUrl = imageUrlSplit[0] +":" + imageUrlSplit[1] +":" + spec.Version
        // 如果是 devops-build 应用从 job log 中读到了 image 信息，则用此镜像信息
        if strings.Contains(spec.Version,":") {
                imageUrl = spec.Version
        }
        containerIndex := k8sutil.GetContainerIndexByName(deployment.Name,
*deployment)
        deployment.Spec.Template.Spec.Containers[containerIndex].Image =
imageUrl
        if len(k8sContaniers) == 0{
                return errors.New("k8s containers is empty!")
        }
        //初始化 imagePatch，将其镜像设置为更新后的值
        imagePatch := []k8sresource.ImageJsonPatch{
            {
```

```
                    Op:    "replace",
                    Patch:
fmt.Sprintf("/spec/template/spec/containers/%d/image", containerIndex),
                    Value: imageUrl,
            },
        }
        imagePatchBytes, err := json.Marshal(imagePatch)
        if err != nil{
                logr.Error("patch marshal json failed!err:", err)
                return fmt.Errorf("patch marshal json failed!err:%v", err)
        }
       logr.Infof("Patching image:%s to namespace: [%s]", imageUrl,
spec.K8sNamespace.Name)
        _, err =
clientset.AppsV1().Deployments(spec.K8sNamespace.Name).Patch(ctx,
deployment.Name, types.JSONPatchType, imagePatchBytes,
v1.PatchOptions{FieldManager: "nighting-release"})
        if err != nil{
                logr.Error("patch image failed!err:", err)
                return fmt.Errorf("patch image failed!err:%v", err)
        }
        k8sresource.SyncDeploymentToLocal(ctx, localDeployment.Id,
*deployment)
        return nil
    }

    func (spec *UpdateSpec) applyService(ctx context.Context, serviceString
string, clientset *kubernetes.Clientset) error {
        if len(serviceString) != 0{
                serviceYaml := corev1.ServiceApplyConfiguration{}
                d :=
yaml.NewYAMLToJSONDecoder(bytes.NewBufferString(serviceString))
                e := d.Decode(&serviceYaml)
                if e != nil{
                        logr.WithContext(ctx).Infof("skip service:%s\n",
spec.Project.ProjectName)

                        return e
                }
                ctx, ctxCancel := context.WithTimeout(context.Background(),
60*time.Second)
```

```
                defer ctxCancel()
                _, e =
clientset.CoreV1().Services(spec.K8sNamespace.Name).Apply(ctx, &serviceYaml,
v1.ApplyOptions{FieldManager: "nighting-release"})
                if e != nil{
                        logr.WithContext(ctx).Infof("skip service:%s\n",
spec.Project.ProjectName)
                        return e
                }
        } else {
                logr.WithContext(ctx).Infof("skip service:%s\n",
spec.Project.ProjectName)
        }
        return nil
    }
    func (s *UpdateSpec) restartDeployment(ctx context.Context,
deploymentString string, clientset *kubernetes.Clientset) error {
        if len(deploymentString) != 0{
                deploymentYaml := appsv1.DeploymentApplyConfiguration{}
                d :=
yaml.NewYAMLToJSONDecoder(bytes.NewBufferString(deploymentString))
                e := d.Decode(&deploymentYaml)
                if e != nil{
                        logr.WithContext(ctx).Error(e)

                        return nil
                }
                ctx, ctxCancel :=context.WithTimeout(context.Background(),
60*time.Second)
                defer ctxCancel()

                // Start to Restart!
                retryErr := retry.RetryOnConflict(retry.DefaultRetry, func()
error {
                        if deploymentYaml.Name == nil{
                                return errors.Errorf("键值对不合法！")
                        }
                        result, getErr :=
clientset.AppsV1().Deployments(s.K8sNamespace.Name).Get(context.TODO(),
*deploymentYaml.Name, v1.GetOptions{})
                        if getErr != nil{
```

```
                    logr.WithContext(ctx).Errorf("Failed to get latest
version of Deployment: %v", getErr)
                    return getErr
                }
                // 设置重启 Deployment 输出日志注解
                t := time.Now().Format("2006-01-02 15:04:05")
                m := make(map[string]string)
                m["devops.com/restartedAt"] = t
                result.Spec.Template.Annotations = m
                containerIndex :=
k8sutil.GetContainerIndexByName(result.Name, *result)
                image :=
result.Spec.Template.Spec.Containers[containerIndex].Image
                logr.WithContext(ctx).Infof(
                    "Restart Deployment %s with image:%s in namespace:
[%s], because of the config has changed",
                    *deploymentYaml.Name,
                    image,
                    s.K8sNamespace.Name)
                _, updateErr :=
clientset.AppsV1().Deployments(s.K8sNamespace.Name).Update(context.TODO(),
result, v1.UpdateOptions{FieldManager: "nighting-release"})
                return updateErr
            })
            if retryErr != nil{
                logr.WithContext(ctx).Errorf("Update failed: %v", retryErr)
                return retryErr
            }
            logr.WithContext(ctx).Infof("Restarted deployment %s in
cluster[%s]-namespace[%s] success", s.Project.ProjectName, s.K8sClusterName,
s.Project.ProjectName)
        } else {
            logr.WithContext(ctx).Infof("skipdeployment:%s\n", s.Project.
ProjectName)
        }
        return nil
    }
```

此外，devops-release 应用还支持重启 Kubernetes 的 deploy 应用，其接口还是 10.1.0.199:5001/api/nighting-build/release/callback，HTTP 方法还是 POST，主要是 JSON 数据格式发生了更新，其内容如下：

```
{
    "id":"11",
    "event_type":"restart",
    "tag_id":"1.0",
    "deployments":[11],
    "project_id":17,
    "namespace_id":1
}
```

另外，Kubernetes 的版本更新的操作与上面也一样，只是 JSON 格式不一样，其 JSON 格式如下：

```
{
    "id":"11",
    "event_type":"patch_image",
    "tag_id":"1.1",
    "deployments":[11],
    "project_id":17,
    "namespace_id":1
}
```

其实就是名为 testci 的 Deployment 文件发生了变更，这个通过 devops-release 应用日志可以看到：

```
INFO[0536] releaseInfo:{11 patch_image 1.1 [11] 17 1}
INFO[0536] 重新部署 devops:testci
INFO[0536] 依次获取发版资源 :configmap/deployment/service
INFO[0536] 从 apollo 拉取配置!
获取项目发版资源 [dev devops testci tenant]
INFO[0536] 通过 apollo 来重新生成应用配置
INFO[0536] 从 apollo 拉取配置 |appid:devops-project cluster:devops name:testci
http://10.1.0.200:32110/openapi/v1/envs/DEV/apps/devops-project/clusters/devops/namespaces/testci
INFO[0536] 从 apollo 拉取公共配置 !appid:devops-project cluster:devops name:testci
http://10.1.0.200:32110/openapi/v1/envs/DEV/apps/devops-project/clusters/devops/namespaces/common
********************configMap****************************
apiVersion: v1
kind: ConfigMap
metadata:
  name: testci
data:
```

```
  setting.yaml: |-
    user: yhc
**********************end*******************************
INFO[0536] config 信息为 :&{9 17 testci testci testci apiVersion: v1
kind: ConfigMap
metadata:
  name: testci
data:
  setting.yaml: |-
    user: yhc666 true 1 11}
最终生成的 configMap 为 [apiVersion: v1
kind: ConfigMap
metadata:
  name: testci
data:
  setting.yaml: |-
  user:  yhc]:INFO[0536]  Patching  image:10.1.0.201:8083/testci:1.1  to
  namespace: [devops]
b 的值为 : metadata:
  creationTimestamp: null
  generation: 9
  labels:
    k8s-app: testci
  name: testci
spec:
  progressDeadlineSeconds: 600
  replicas: 1
  revisionHistoryLimit: 10
  selector:
    matchLabels:
      k8s-app: testci
  strategy:
    rollingUpdate:
      maxSurge: 25%
      maxUnavailable: 25%
    type: RollingUpdate
  template:
    metadata:
      creationTimestamp: null
      labels:
        k8s-app: testci
```

```
    spec:
      containers:
      - env:
        - name: ELASTIC_APM_SERVICE_NAME
          value: testci
        - name: ELASTIC_APM_ENVIRONMENT
          value: devops
        - name: ELASTIC_APM_TRANSACTION_IGNORE_URLS
          value: /
        image: 10.1.0.201:8083/testci:1.1
        imagePullPolicy: Always
        lifecycle:
          preStop:
            exec:
              command:
              - /bin/bash
              - -c
              - sleep 20
        name: testci
        resources:
          limits:
            cpu: 500m
            memory: 512Mi
          requests:
            cpu: 250m
            memory: 256Mi
        securityContext:
          privileged: false
        terminationMessagePath: /dev/termination-log
        terminationMessagePolicy: File
        volumeMounts:
        - mountPath: /opt/testci/config
          name: settings
      dnsPolicy: ClusterFirst
      imagePullSecrets:
      - name: odregistry
      restartPolicy: Always
      schedulerName: default-scheduler
      securityContext: {}
      terminationGracePeriodSeconds: 30
      volumes:
```

```
      - configMap:
          defaultMode: 420
          name: testci
        name: settings
status: {}

[GIN] 2023/12/23 - 17:41:19 | 200 |    86.79864ms |       127.0.0.1 | POST
"/api/nighting-release/callback"
```

整个 SasS 项目主要是在腾讯云上运行，其中测试环境、开发环境、预发布环境、各项目环境都依赖 Apollo 切换配置来区分。在试运行阶段，系统较稳定，出现的 bug 也较少。后续线上环境（也是利用 Apollo 来切换环境）还加入了工单审核环境及测试介入，在一定程度上减轻了研发及运维的重复发版工作。

4.5 小　　结

本章主要介绍了使用 Go 语言开发 CD 自动化发版工具，发版工具 devops-build 和 devops-release 应用的流程及重要核心源码，以及其中用到的重要开源组件，如 gitlab CI、gitlab-runner 及 Apollo，读者可以结合自己团队或业务的实际情况考虑是否投入使用。

第 5 章　云原生 MySQL 架构选型

关系型数据库支持事务机制，可以提供对结构化数据的增删改查等功能，是一种不可或缺的关键数据服务。MySQL 是目前最流行的开源关系型数据库，目前业界主流的部署方式均为一主一从 + 多只读的模式。主从之间半同步复制，主与只读之间进行异步复制。

数据库高可用的基本要求：数据不丢失、服务持续可用和自动的主备切换。

传统的 MySQL 集群很多技术上的选择，如 MHA、DRBD+Heartbeat 等，主要是依赖 HaVip。

> **注**：高可用虚拟IP（High-Availability Virtual IP Address，HaVip）是一种可以独立创建和释放的私网IP资源。HaVip可以与高可用软件（Keepalived）配合使用，搭建高可用主备服务，提高业务的可用性。

5.1　云原生高可用数据库选型难点

项目背景：某大型汽车集团项目，云原生集群选用的是 Mesos 分布式集群，核心业务已经在该集群上运行了较长时间；希望 MySQL 数据库也能以云原生容器化的方式，在 Mesos 分布式集群上大规模地以自动化部署和自动化运维的方式进行交付。

难点：云原生数据库的高可用性和取代烦琐的数据库运维。

优势：①一键安装数据库服务，实现自动化运维；②采用主流云原生技术，降低数据库成本；③提升资源利用率，降低 40% 计算资源和 80% 运维人员成本。

MySQL 高可用方案在云原生应用中有如下难点：

（1）传统的 MySQL 高可用方案都要摈弃，因为有 HaVip 的硬性要求。

（2）MySQL NDB 架构不适合实际业务（需要将所有的业务数据库引擎由 InnoDB 引擎转成 NDB 引擎，理论上不太现实）。

（3）敏捷弹性伸缩：基于副本的弹性横向扩展，扩缩容过程不中断业务访问。

（4）自行开发的 Failover Monitor 中间件程序也存在数据不一致的风险：Failover Monitor 其实并不具备实时动态 MySQL 路由的能力也无法保证 MySQL 从库（Slave）的只读性；LB 与 Failover Monitor 之间是异步刷新，理论上存在数据写到 MySQL 从库的风险，这样会导致数据分叉。其工作流程如图 5.1 所示。

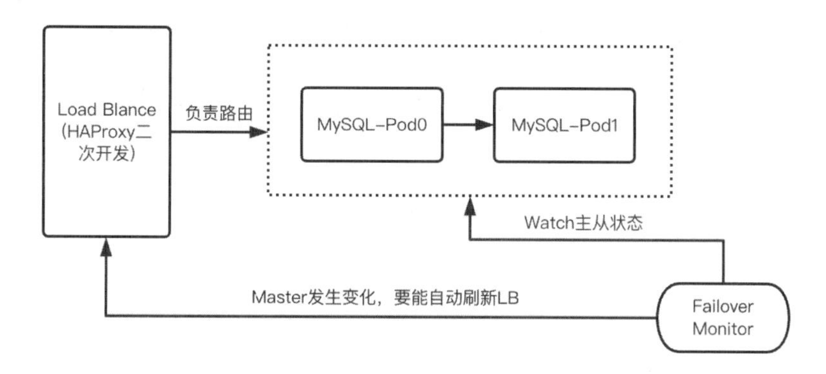

图 5.1 Failover Monitor 工作流程

针对云原生 MySQL 高可用方案的需求如下：

（1）数据要实现强一致性。

（2）减少 HaVip 的依赖性，表示要放弃传统物理环境下的 DRBD+Heartbeat、MHA 等 MySQL 高可用方案。

（3）减少商用共享存储依赖，数据存储采取本地磁盘的方式，然后采取多副本的方式。

（4）尽量实现自动化运维和简化运维动作。

（5）通过混乱测试，因为项目中的很多环境都基于虚拟机 / 云主机，运维重启机器也是考虑因素之一。

（6）故障转移的敏捷性，主库发生宕机时，集群能够自动选主并快速转移故障，并且确保转移前后的数据一致。

（7）敏捷弹性伸缩，即基于副本的弹性横向扩展，扩缩容过程不能中断业务访问。

最终决定，采用 MySQL 的 MGR 作为此云原生 MySQL 高可用交付方案。

5.2 GTID 的工作方式

MGR 集群技术是以 MySQL GTID 为基础的，所以在介绍 MGR 之前得先了解和熟悉 GTID。

5.2.1 MySQL 官方引入 GTID 的目的

MySQL 5.6 版本新增了一种主从复制方式 GTID（Global Transaction Identifier，全局事务标识）。GTID 由 server_uuid 和事务 ID 组成，格式为 GTID=server_uuid:transaction_id。其中，server_uuid 是数据库启动时自动生成的，存放在数据库目录下的 auto.cnf 文件中；transaction_id 是事务提交时由系统顺序分配的序列号。一个 GTID 在一个服务器上只执行一次，从而避免重复执行导致数据混乱或主从不一致。

GTID 的功能主要有以下两点：

（1）根据 GTID 可以知道事务最初是在哪个实例上提交的。

（2）GTID 方便了 Replication 的 Failover（主从故障切换，下略）。

这里详细解释下第（2）点。在 MySQL 5.6 的 GTID 出现之前，Replication Failover 的操作过程如图 5.2 所示。

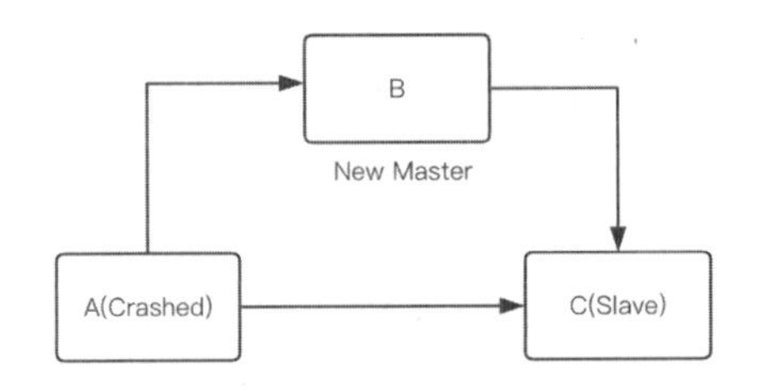

图 5.2　Replication Failover 的操作过程

此时，如果 Server A 的服务器宕机，需要将业务切换到 Server B 上。同时，又需要将 Server C 的复制源改成 Server B。修改复制源的命令语法如下：

```
CHANGE MASTER TO MASTER_HOST='xxx', MASTER_LOG_FILE='xxx', MASTER_LOG_POS = NNNN
```

而难点在于，由于同一个事务在每台机器上的 binlog 文件名和位置都不同，因此找到 Server C 的当前同步停止点，对应 Server B 的 master_log_file 和 master_log_pos 的时间就成了难题。这也是 M-S 复制集群需要使用 MMM 或 MHA 这样的额外管理工具的一个重要原因。这个问题在 MySQL 5.6 引入 GTID 之后，就变得非常简单了。由于同一个事务的 GTID 在所有节点上的值一致，因此根据 Server C 当前停止点的 GTID，就能唯一定位到 Server B 上的 GTID。甚至由于 MASTER_AUTO_POSITION 功能的出现，都不需要知道 GTID 的具体值，直接使用以下命令：

```
CHANGE MASTER TO MASTER_HOST='xxx', MASTER_AUTO_POSITION
```

就可以完成 Failover 的工作。

与传统复制相比，GTID 用于主从复制的优势如下：

- 更简单地实现 Failover，不用像以前那样需要查找 log_file 和 log_Pos。
- 更简单地搭建主从复制。
- 比传统复制更加安全。
- GTID 是连续没有空洞的，因此主从库出现数据冲突时，可以用添加空事物的方式跳过。

从架构设计的角度看，GTID 是一种很好的分布式 ID 实践方式，通常来说，分布式 ID 有两个基本要求：全局唯一性和趋势递增。

因为这个 ID 是全局唯一，所以在分布式环境中很容易识别，又因为趋势递增，所以 ID 具有相应的趋势规律，在必要时方便进行顺序提取。行业内使用较多的是基于 Twitter 的 ID 生成算法 snowflake，所以换一个角度来理解 GTID，就是一种优雅的分布式设计。

MySQL 官方是从 5.6.5 版本开始支持 GTID 的，在 5.6.10 版本之后开始完善；如果要将其用于生产环境，可以考虑采用 MySQL 5.7 或以上版本。

5.2.2　GTID 与 binlog 日志的关系

MySQL 的二进制日志 binlog 是事务安全型的，这个二进制日志可以说是 MySQL 最重要的日志。它记录了所有的 DDL 和 DML（除了数据查询）语句，以事件形式记录，还包含语句执行所消耗的时间。开启 GTID 之后，MySQL 的 binlog 日志结构如图 5.3 所示。

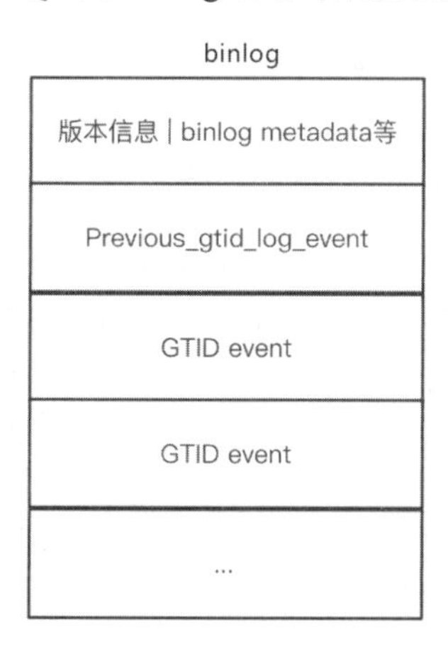

图 5.3　MySQL 的 binlog 日志结构（开启 GTID）

每个 binlog 文件头部都有 Previous_gtid_log_event。在每次轮询 binlog 时，这个信息会存储在 binlog 头部。需要注意的是，Previous-GTIDS 只会记录这台机器上执行过的所有 binlog 文件，而不包括手动设置的 gtid_purged 值。换句话说，如果手动执行了 SET GLOBAL gtid_purged=xxx，则 xxx 不会出现在 Previous_gtid_log_event 中。

那么，GTID 和 binlog 之间的关系是怎么对应的呢？如何才能找到 GTID=? 对应的 binlog 文件呢?

假设有 4 个 binlog：bin.001、bin.002、bin.003 和 bin.004。

bin.001：Previous-GTIDs=empty；binlog_event 有 1-40。

bin.002：Previous-GTIDs=1-40；binlog_event 有 41-80。

bin.003：Previous-GTIDs=1-80；binlog_event 有 81-120。

bin.004：Previous-GTIDs=1-120；binlog_event 有 121-160。

假设现在要找到 GTID=$A，那么 MySQL 的扫描顺序如下：

（1）从最后一个 binlog 开始扫描（即 bin.004）。

（2）bin.004 的 Previous-GTIDs=1-120，如果 $A=140 > Previous-GTIDs，那么结果肯定在 bin.004 中。

（3）bin.004 的 Previous-GTIDs=1-120，如果 $A=88 包含在 Previous-GTIDs 中，则继续对比上一个 binlog 文件 bin.003，然后循环前面两个步骤，直到找到为止。

5.2.3　GTID 的重要参数说明

GTID 的重要参数说明见表 5.1。

表 5.1 GTID 的重要参数说明

GTID 的重要参数	参数说明
gtid_executed	执行过的所用 GTID
gtid_purged	丢弃过的 GTID
gtid_mod	GTID 模式
gtid_next	Session 级别的变量，下一个 GTID
gtid_owned	正在运行的 GTID
enforce_gtid_consistency	保证 GTID 安全的参数

下面重点介绍 MySQL 主从复制中涉及的几个重要的 GTID 参数。

（1）gtid_executed。在当前实例上执行过的 GTID 集合。gtid_executed 是全局参数，GTID 集合包含所有在该服务器上执行过的事务编号和使用 set gtid_purged 语句设置过的事务编号。使用 SHOW MASTER STATUS 和 SHOW SLAVE STATUS 命令得到的 Executed_Gtid_Set 列值就取自全局参数 gtid_executed。

实际上，gtid_executed 包含所有记录到 binlog 中的事务。设置 set sql_log_bin=0 后执行的事务不会生成 binlog 事件，也不会被记录到 gtid_executed 中。执行 RESET MASTER 可以将该变量置空。可以通过以下命令来查看 mysql-bin.000002 中的相关信息。

```
show binlog events in 'mysql-bin.000002';
```

（2）gtid_purged。gtid_purged 是全局参数，GTID 集合包含从 binlog 中清除掉的事务 ID，该集合是全局参数 gtid_executed 的子集。binlog 不可能永远驻留在服务上，需要定期进行清理（通过 expire_logs_days 可以控制定期清理），否则它迟早会把磁盘用尽。gtid_purged 用于记录本机上执行过，但是已经被清除的 binlog 事务集合。它是 gtid_executed 的子集，只有 gtid_executed 为空时才能手动设置 gtid_purged 变量，此时会同时更新 gtid_executed 和 gtid_purged 相同的数值。

gtid_executed 为空表示要么之前没有启动过基于 GTID 的复制，要么执行过 RESET MASTER。执行 RESET MASTER 时同样也会把 gtid_purged 置空，即始终保持 gtid_purged 是 gtid_executed 的子集。

（3）gtid_next。会话级变量，指示如何产生下一个 GTID。

gtid_next 可能的取值如下。

- AUTOMATIC：自动生成下一个 GTID，实际上是分配一个当前实例上尚未执行过的序号最小的 GTID。
- ANONYMOUS：设置后执行事务不会产生 GTID。
- 显式指定的 GTID：可以指定任意形式合法的 GTID 值，但不能是当前 gtid_executed 中已经包含的 GTID，否则下次执行事务时会报错。

GTID 的生成受 gtid_next 控制。在主服务器上，gtid_next 是默认的 AUTOMATIC，即在每次事务提交时自动生成新的 GTID。它从当前已执行的 GTID 集合（gtid_executed）中，寻找一个大于 0 的、未使用的最小值作为下一个事务的 GTID。同时在 binlog 的实际更新事务事件前面插入一个 set gtid_next 事件。

5.2.4 MySQL 容器 GTID 主从复制

GTID 主从复制的主要流程如下。

（1）当一个事务在主机器执行并提交时，会生成一个 GTID，并被一同记录到 binlog 日志中。

（2）binlog 传输并存储到从机器的 relaylog 后，读取 GTID 的值设置 gtid_next 变量，即告诉从机器，下一个要执行的 GTID 值。

（3）SQL 线程从 relaylog 中获取 GTID，然后对比从机器端的 binlog 是否有该 GTID。

（4）如果有记录，说明该 GTID 的事务已经执行，从机器会忽略。

（5）如果没有记录，从机器就会执行该 GTID 事务，并将该 GTID 记录到自身的 binlog，在读取执行事务前会先检查其他 session 是否持有该 GTID，以确保不被重复执行。

（6）在解析过程中会判断是否有主键，如果没有，就用二级索引，如果有就用全部扫描。

作者的线上环境用得最多的是 MySQL 5.7，所以这里就用 MySQL 5.7 进行演示，在 my.cnf 配置文件中需要添加以下配置：

```
gtid_mode= ON
enforce-gtid-consistency= ON
```

下面是 MySQL 主从机器的 Docker 启动脚本及对应的 my.cnf 配置文件。主从机器的 Docker 启动脚本如下：

```
#!/bin/bash
CNAME=mysql-docker-0
CIMAGE=mysql:5.7
CPORT=3306
MYSQL_SECRET=123456

docker rm -f ${CNAME} >/dev/null 2>&1 || true
docker run \
-d --restart=always \
--name ${CNAME} \
--memory=${CMEM} \
-p ${CPORT}:${CPORT} \
-v /data/mysql/docker0/my.cnf:/etc/my.cnf:ro \
-v /data/mysql/docker0/mysqldb:/var/lib/mysql \
-e MYSQL_ROOT_PASSWORD=${MYSQL_SECRET} \
${CIMAGE}
```

主从机器对应的 my.cnf 配置文件内容如下：

```
# For advice on how to change settings please see
# http://dev.mysql.com/doc/refman/5.7/en/server-configuration-defaults.html
```

```
[mysqld]
#
# Remove leading # and set to the amount of RAM for the most important data
# cache in MySQL. Start at 70% of total RAM for dedicated server, else 10%.
# innodb_buffer_pool_size = 128M
#
# Remove leading # to turn on a very important data integrity option: logging
# changes to the binary log between backups.
# log_bin
#
# Remove leading # to set options mainly useful for reporting servers.
# The server defaults are faster for transactions and fast SELECTs.
# Adjust sizes as needed, experiment to find the optimal values.
# join_buffer_size = 128M
# sort_buffer_size = 2M
# read_rnd_buffer_size = 2M
skip-host-cache
skip-name-resolve
datadir=/var/lib/mysql
socket=/var/lib/mysql/mysql.sock
secure-file-priv=/var/lib/mysql-files
user=mysql

# 这里注意 MySQL 主从机器的 id 号，不要重复了
server-id = 1
binlog_format = row
expire_logs_days = 7
max_binlog_size  = 100M
gtid_mode = ON
enforce_gtid_consistency = ON
binlog-checksum = CRC32
master-verify-checksum = 1
log-bin = /var/lib/mysql/mysql-bin
log_bin_index = /var/lib/mysql/mysql-bin.index
#read_only=on
#super_read_only=on
```

```
    # Disabling symbolic-links is recommended to prevent assorted security
risks
    symbolic-links=0

    log-error=/var/log/mysqld.log
    pid-file=/var/run/mysqld/mysqld.pid
```

在 MySQL 主机上执行以下操作，为用户分配主从复制权限。

```
    grant replication slave on *.* to 'repl'@'%' identified by '123456';
    flush privileges;
```

在 MySQL 从机上执行以下操作。

```
    mysql> reset master;
    mysql> reset slave;
    mysql> stop slave;
    mysql> CHANGE MASTER TO  MASTER_HOST='10.1.0.201', MASTER_USER='repl',
MASTER_PASSWORD='123456', MASTER_PORT=3306, MASTER_AUTO_POSITION = 1;
    mysql> start slave;
```

最后用 show slave status \G; 命令来验证，显示结果如下：

```
    *************************** 1. row ***************************
                   Slave_IO_State: Waiting for master to send event
                      Master_Host: 10.1.0.201
                      Master_User: repl
                      Master_Port: 3306
                    Connect_Retry: 60
                  Master_Log_File: mysql-bin.000003
              Read_Master_Log_Pos: 629
                   Relay_Log_File: 4b4f033d3922-relay-bin.000003
                    Relay_Log_Pos: 842
            Relay_Master_Log_File: mysql-bin.000003
                 Slave_IO_Running: Yes
                Slave_SQL_Running: Yes
                  Replicate_Do_DB:
              Replicate_Ignore_DB:
               Replicate_Do_Table:
           Replicate_Ignore_Table:
          Replicate_Wild_Do_Table:
      Replicate_Wild_Ignore_Table:
                       Last_Errno: 0
                       Last_Error:
                     Skip_Counter: 0
```

```
             Exec_Master_Log_Pos: 629
                 Relay_Log_Space: 2948886
                 Until_Condition: None
                  Until_Log_File:
                   Until_Log_Pos: 0
              Master_SSL_Allowed: No
              Master_SSL_CA_File:
              Master_SSL_CA_Path:
                 Master_SSL_Cert:
               Master_SSL_Cipher:
                  Master_SSL_Key:
           Seconds_Behind_Master: 0
   Master_SSL_Verify_Server_Cert: No
                   Last_IO_Errno: 0
                   Last_IO_Error:
                  Last_SQL_Errno: 0
                  Last_SQL_Error:
     Replicate_Ignore_Server_Ids:
                Master_Server_Id: 1
                     Master_UUID: 1666e7d8-341d-11ee-943e-0242ac110003
                Master_Info_File: /var/lib/mysql/master.info
                       SQL_Delay: 0
             SQL_Remaining_Delay: NULL
         Slave_SQL_Running_State: Slave has read all relay log; waiting for
more updates
              Master_Retry_Count: 86400
                     Master_Bind:
         Last_IO_Error_Timestamp:
        Last_SQL_Error_Timestamp:
                  Master_SSL_Crl:
              Master_SSL_Crlpath:
              Retrieved_Gtid_Set: 1666e7d8-341d-11ee-943e-0242ac110003:1-7
               Executed_Gtid_Set: 1666e7d8-341d-11ee-943e-0242ac110003:1-7
                   Auto_Position: 1
            Replicate_Rewrite_DB:
                    Channel_Name:
              Master_TLS_Version:
   1 row in set (0.00 sec)
```

MySQL 主从复制的 IO_Running 和 SQL_Running 线程目前均显示为 Yes 状态，证明 GTID 主从复制是成功的。

5.3　MySQL 组复制

MySQL Group Replication（MGR）是 MySQL 5.7 版本出现的新特性，提供高可用、高扩展、高可靠（强一致性）的 MySQL 集群服务。MGR 技术在保证数据一致性的基础上，可自动进行故障检测、自动切换，具备防脑裂机制，兼具多节点写入等优点，是一个很好的技术发展方向。

5.3.1　MGR 配置

MGR 由多个实例节点共同组成一个数据库集群，系统提交事务必须经过半数以上节点同意，集群中的每个节点都维护一个数据库状态机，以保证节点间事务的一致性。

MGR 可配置为单主模型和多主模型两种工作模式。以下是两种模式的简介：

（1）单主模型：从复制组中的多个 MySQL 节点中自动选举一个 master（主）节点，只有 master 节点可以写，其他节点自动设置为 ReadOnly 状态。当 master 节点故障时，会自动选举一个新的 master 节点，选举成功后设置为可写，其他 slave（从）节点将指向这个新的 master。

（2）多主模型：复制组中的任何一个节点都可以写，因此没有 master 和 slave 的概念，只要突然故障的节点数量不多，这个多主模型就能继续可用。

MGR 使用 Paxos 分布式算法提供节点间的分布式协调。正因如此，它要求组中大多数节点在线才能达到法定票数，从而对一个决策做出一致的决定。MySQL 官方文档中规定 MGR 目前最多支持 9 个节点，其中，容忍的故障节点数 F 和总集群节点数 N 的计算公式为

$N=2\times F+1$

如果 $F=1$，那么 $N=3$；如果 $F=2$，那么 $N=5$。

MGR 配置主要还是用 my.cnf 文件实现，内容如下：

```
datadir=/data
socket=/data/mysql.sock

server-id=1                              # 注意每个节点的 server-id 都要不同
gtid_mode=on                             # 开启 MGR 必需项
enforce_gtid_consistency=on              # 开启 MGR 必需项
log-bin=/data/master-bin                 # 必需项
binlog_format=row                        # 开启 MGR 必需项
binlog_checksum=none                     # 必需项
master_info_repository=TABLE             # 必需项
relay_log_info_repository=TABLE          # 必需项
relay_log=/data/relay-log
log_slave_updates=ON
sync-binlog=1
log-error=/data/error.log
pid-file=/data/mysqld.pid
```

```
    transaction_write_set_extraction=XXHASH64         # 必需项
    loose-group_replication_group_name="aaaaaaaa-aaaa-aaaa-aaaa-aaaaaaaaaaaa"
                                                      # 必需项
    loose-group_replication_start_on_boot=off         # 建议设置为 off
    loose-group_replication_local_address = "192.110.103.41:31061"
                                                      # 必需项，下一行也是必需项
    loose-group_replication_group_seeds =
"192.110.103.41:31061,192.110.103.42:31061,192.110.103.43:31061"
    loose-group_replication_bootstrap_group = OFF
    loose-group_replication_single_primary_mode = FALSE # = multi-primary
    loose-group_replication_enforce_update_everywhere_checks=ON # =
multi-primary
```

下面分析上面的配置选项：

（1）因为组复制基于 GTID，所以必须开启 gtid_mode 和 enforce_gtid_consistency。

（2）组复制必须设置 log-bin 和 binlog_format。这样才能从日志记录中收集信息且保证数据一致性。

（3）由于 MySQL 对复制事件校验的设计缺陷，组复制不能对它们进行校验，因此设置 binlog_checksum=none。

（4）组复制要将 master 和 relay log 的元数据写入 mysql.slave_master_info 和 mysql.slave_relay_log_info 中。

（5）组中的每个节点都保留了完整的数据副本，它是无共享的模式。因此所有节点上都必须开启 log_slave_updates，这样新节点随便选哪个作为提供者都可以进行异步复制。

（6）sync-binlog=1 是为了保证提交每次事务都立刻将 binlog 刷盘，即使 MySQL 发生故障也不丢失日志。

（7）最后的 6 行是组复制插件的配置。以 loose- 开头表示即使启动组复制插件，MySQL 也可以继续正常运行。这个前缀是可选的。

（8）transaction_write_set_extraction 项表示写集合以 XXHASH64 的算法进行哈希。所谓写集合，是指对事务中所修改的行进行唯一标识，在后续检测并发事务是否修改特定的数据行冲突时使用。它基于主键生成，所以使用组复制，表中必须有主键。

（9）loose_group_replication_group_name 项表示 MGR 复制组的名称。

5.3.2　MGR 工作流简介

MGR 是一种可用于实现容错系统的技术。复制组是一个通过消息传递交互的 Server 集群，由多个 Server 成员组成，所有成员独立完成各自的事务。

当客户端发起一个更新事务时，该事务首先在本地执行，执行完成后会尝试提交。在真正提交之前，需要将产生的复制写集广播给其他成员。如果冲突探测成功，组内会决定该事务可以提交，其他成员可以应用该事务；如果冲突探测失败，则该事务会被回滚。最终，所有组内成员以相同的顺序接收同一组事务。因此，组内成员以相同的顺序应用相同的修改，保证组内数据的强一致性。

MGR 是 MySQL 官方开发的一个开源插件，与其他的异步复制和半同步复制不同，它是利用 MySQL 的组复制技术来实现高可用的一种解决方案，保证了数据的强一致性。

这里可以用 performance_schema.replication_group_member_stats 表来监视组复制成员，其代码如下：

```
select * from performance_schema.replication_group_member_stats\G ;
```

显示结果如下：

```
*************************** 1. row ***************************
                      CHANNEL_NAME: group_replication_applier
                           VIEW_ID: 15718289704352993:39
                         MEMBER_ID: 509810ee-f3d7-11e9-a7d5-a0369fac2de4
       COUNT_TRANSACTIONS_IN_QUEUE: 0
        COUNT_TRANSACTIONS_CHECKED: 10
          COUNT_CONFLICTS_DETECTED: 0
COUNT_TRANSACTIONS_ROWS_VALIDATING: 0
TRANSACTIONS_COMMITTED_ALL_MEMBERS: aaaaaaaa-aaaa-aaaa-aaaa-aaaaaaaaaa
aa:1-60:1000003-1000006:2000003-2000006
LAST_CONFLICT_FREE_TRANSACTION: aaaaaaaa-aaaa-aaaa-aaaa-aaaaaaaaaaaa:58
```

Group 视图 ID 由前缀时间戳 + 序号部分组成。

（1）前缀时间戳：在 Group 初始化时产生，为当时的时间戳，Group 存活期间该值不变。

（2）序号部分：Group 初始化时，第一个视图序号为 1，以后任何成员加入或退出，序号都将增一（上面的例子表示经过了 39 次视图切换）。

MGR 以组视图（Group View，简称视图）为基础进行成员管理。视图是指 Group 在一段时间内的成员状态，如果在这段时间内没有成员变化，也就是没有成员加入或退出，则这段连续的时间为一个视图；如果发生了成员加入或退出变化，则视图也会发生变化。MGR 使用视图 ID（View ID）来跟踪视图的变化并区分视图的先后顺序，如图 5.4 所示。

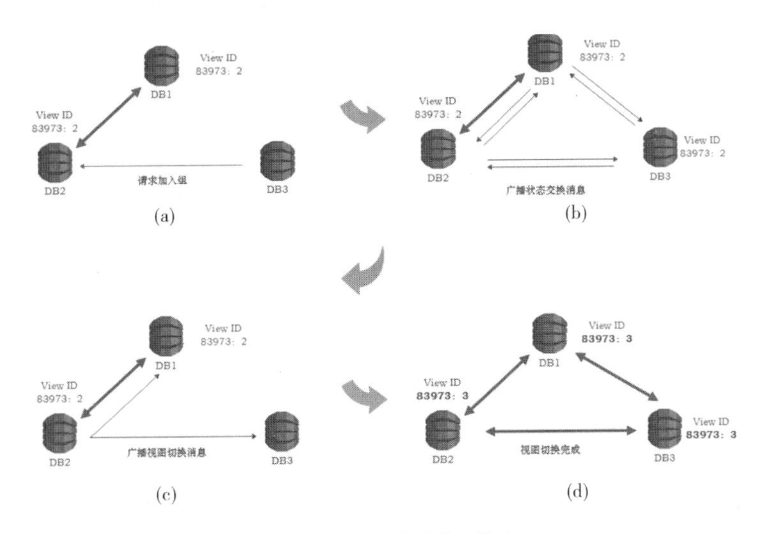

图 5.4　MGR 视图工作流

当一个节点请求加入 Group 时，会首先根据配置的 group_replication_group_seeds 参数与 Group 的种子成员建立 TCP 连接（Gcs_xcom_control::do_join()）。该种子成员会根据自己的 group_replication_ip_whitelist（IP 白名单）检查是否允许新节点加入，MGR 默认不限制新节点的 IP。连接建立后，新节点发送请求申请加入组。收到请求后，种子成员广播视图变化的消息给 Group 中的所有节点，包括申请加入的节点，如图 5.4（b）所示；各节点收到消息后开始做视图切换。每个节点都会广播一个状态交换消息，每个状态交换消息包含节点的当前状态和信息，如图 5.4（c）所示。发送交换消息后，各个节点开始接收其他节点广播的消息，将其中的节点信息更新到本节点所维护的成员列表中。

完成视图切换只是成员加入 Group 要做的第一步，只是说明该成员可以接收到 Group 中通过 Paxos 协议达成共识的消息，并不意味着可以将成员设置为 ONLINE（上线）对外提供服务。原因是新成员还需要进行数据同步，建立起正确的数据版本（recovery_module --> start_recovery），然后才能执行 Paxos 协议消息，进而上线提供正常的用户访问服务。

实质上，MGR 是使用 Paxos 分布式协议构建了一个分布式的状态机复制机制。这是实现多主复制的核心技术，它的存在使 MGR 不会出现脑裂现象，而且只要同时当家的成员不超过半数，数据便不会丢失，也保证了只要 binlog event 没有被传输到半数以上的成员，本地成员就不会将事务的 binlog event 写入 binlog。这样就避免了在故障机器重启时，该故障机器上有其他组内成员不存在的数据。

5.3.3 MGR 技术

1. MGR 技术的优缺点

（1）MGR 技术的优点。

1）高一致性保障。该技术融合了原生复制与 Paxos 协议，以插件形式提供，确保了数据的一致性安全。通过分布式协议和分布式恢复机制，该技术能够保障组内数据的最终一致性。组复制功能强大，内置了组成员管理、数据一致性校验、冲突检测与处理、节点故障监测以及数据库故障转移等机制，从而实现了无须人工干预或额外工具的本地高可用性。此外，组复制还支持自动选主的单主模式和允许任意更新的多主模式。借助一个先进的组通信系统，多主组复制（MGR）内部实现了 Paxos 算法，以自动协调数据复制、一致性和成员管理。

2）卓越的容错性能。组复制采用流行的 Paxos 分布式算法，实现服务器间的分布式协调。为确保小组持续运作，要求大多数成员保持在线，并就每个变更达成共识。这样，MySQL 数据库在发生故障时仍能安全运行，无须人工干预，且不存在数据丢失或损坏的风险。只要大多数节点未发生故障，系统就能继续工作。同时，系统具备自动检测机制，当节点间出现资源冲突时，会按照先到先服务的原则进行处理，并内置了自动化脑裂防护机制，确保系统稳定运行。

3）插件化设计。支持插件检测，如果新增节点小于集群当前节点主版本号，则拒绝该节点加入集群；大于则加入，但无法作为主节点。

4）没有第三方组件依赖。

5）支持全链路 SSL 通信。

6）支持 IP 白名单。

7）不依赖网络多播。

8）强大的自愈能力，当服务器重新加入组时，会自动重新与组同步。

9）有单主和多主模式，从维护的角度推荐单主模式。

10）支持多节点写入，具备冲突检测机制，可以适应多种应用场景需求。

（2）MGR 技术的缺点。

1）仅支持 innodb 引擎表，并且每张表必须要有主键。

2）必须开启 GTID 和 binlog，并且要设置 binlog 为 row 格式。

3）DDL 语句不支持原子性，不能检测冲突，执行后需自行校验是否一致；不支持对表进行锁操作。

4）多主模式不支持外键。

5）一个 MGR 集群最多支持 9 个节点。

2. MGR 适用的业务场景

MGR 适用的业务场景有以下几种：

（1）对主从延迟比较敏感的场景。

（2）希望对写服务提供高可用，并且不想用第三方软件的场景。

（3）数据保持一致性的场景。

下面介绍在业务场景中如何解决读写负载大的问题。

（1）写负载大的问题：解决方案为分库分表和优化复杂的 SQL 语句。

（2）读负载大的问题：解决方案为增加 Slave 节点数量和读写分离（配合 MySQL Router 一起使用）。

MySQL Router 是轻量级的中间件，它不检查过滤数据包，只将数据请求重定向到正确的后端节点。众所周知，中间件的引入会产生一定的开销（官方公布的性能消耗约为 1%），MySQL Router 在后面的章节再重点介绍。

目前，银行应用 MySQL 的比例较高，并且也已开始推广上线 MGR 架构；G 行数据库规划秉持传统数据库和开源数据库并行使用模式，MySQL 线上应用也有上百套，其中的 A 类系统中的分布式企业总线开始应用实践 MGR 技术。

5.3.4　使用 Docker 搭建 MGR 集群

这里用 3 台 CentOS 7.9 x86_64 物理机进行演示，Docker 网络采取 host 主机模式，对外提供的服务端口为 3306。MGR 的 Docker 物理环境见表 5.2。

表 5.2　MGR 的 Docker 物理环境

容器名	主机 IP	服务端口
mysql-mgr-server1	10.1.0.201	3306
mysql-mgr-server2	10.1.0.202	3306
mysql-mgr-server3	10.1.0.203	3306

容器的启动脚本如下：

```
docker run -d --hostname=mysql-mgr-server1 --network=host  -p 3306:3306 -p
33061:33061  --volume=/data/mysql/docker0/my.cnf:/etc/my.cnf
--volume=/data/mysql/docker0/mysqldb:/var/lib/mysql  --name=mysql-mgr-server1
-e MYSQL_ROOT_PASSWORD=123456  mysql:5.7
```

容器对应的my.cnf配置文件如下：

```
[mysqld]
datadir=/var/lib/mysql
pid-file=/var/run/mysqld/mysqld.pid
port=3306
socket=/var/run/mysqld/mysqld.sock
### 注意每个节点此处的 server_id 不能相同
server_id=2
gtid_mode=ON
enforce_gtid_consistency=ON
master_info_repository=TABLE
relay_log_info_repository=TABLE
binlog-format=ROW
binlog_checksum=NONE
log-slave-updates=1
log_bin=binlog
relay-log=bogon-relay-bin
### 本地的对外地址
report_host = 10.1.0.201

# Group Replication
transaction_write_set_extraction = XXHASH64
loose-group_replication_group_name="aaaaaaaa-aaaa-aaaa-aaaa-aaaaaaaaaaaa"
loose-group_replication_start_on_boot = off
loose-group_replication_local_address = '10.1.0.201:33061'
loose-group_replication_group_seeds ='10.1.0.201:33061,10.1.0.202:33061,
10.1.0.203:33061'

loose-group_replication_bootstrap_group=off
loose-group_replication_single_primary_mode = on
loose-group_replication_enforce_update_everywhere_checks = off
```

以上只是一个MySQL实例的配置文件和启动脚本，其他MySQL实例的配置文件和启动脚本需要改动一下，注意，其中的server_id不能重复。

下面启动第一个 MySQL 实例，具体执行命令如下：

```
mysql> SET SQL_LOG_BIN=0;
Query OK, 0 rows affected (0.00 sec)

mysql> CREATE USER rpl_user@'%';
Query OK, 0 rows affected (0.00 sec)

mysql> GRANT REPLICATION SLAVE ON *.* TO rpl_user@'%' IDENTIFIED BY 'rpl_pass';
Query OK, 0 rows affected, 1 warning (0.00 sec)

mysql> FLUSH PRIVILEGES;
Query OK, 0 rows affected (0.00 sec)

mysql> SET SQL_LOG_BIN=1;
Query OK, 0 rows affected (0.00 sec)

mysql> CHANGE MASTER TO MASTER_USER='rpl_user', MASTER_PASSWORD='rpl_pass'
FOR CHANNEL 'group_replication_recovery';
Query OK, 0 rows affected, 2 warnings (0.03 sec)

mysql> INSTALL PLUGIN group_replication SONAME 'group_replication.so';
Query OK, 0 rows affected (0.01 sec)

mysql> SET GLOBAL group_replication_bootstrap_group=ON;
Query OK, 0 rows affected (0.01 sec)

mysql> START GROUP_REPLICATION;
Query OK, 0 rows affected (0.01 sec)

mysql> SET GLOBAL group_replication_bootstrap_group=OFF;
Query OK, 0 rows affected (0.01 sec)
```

第二个和第三个 MySQL 实例的执行命令相同，分别如下：

```
mysql> SET SQL_LOG_BIN=0;
Query OK, 0 rows affected (0.00 sec)

mysql> CREATE USER rpl_user@'%';
Query OK, 0 rows affected (0.00 sec)

mysql> GRANT REPLICATION SLAVE ON *.* TO rpl_user@'%' IDENTIFIED BY 'rpl_pass';
Query OK, 0 rows affected, 1 warning (0.00 sec)

mysql> FLUSH PRIVILEGES;
```

```
    Query OK, 0 rows affected (0.00 sec)

    mysql> SET SQL_LOG_BIN=1;
    Query OK, 0 rows affected (0.00 sec)

    mysql> CHANGE MASTER TO MASTER_USER='rpl_user', MASTER_PASSWORD='rpl_pass'
FOR CHANNEL 'group_replication_recovery';
    Query OK, 0 rows affected, 2 warnings (0.03 sec)

    mysql> INSTALL PLUGIN group_replication SONAME 'group_replication.so';
    Query OK, 0 rows affected (0.01 sec)

    mysql> START GROUP_REPLICATION;
    Query OK, 0 rows affected (0.01 sec)
```

若在执行 START GROUP_REPLICATION 命令时报错，查看 Docker 容器日志发现如下内容：

```
    2023-08-14T02:54:09.350559Z 0 [ERROR] Plugin group_replication reported:
'This member has more executed transactions than those present in the
group. Local transactions: 926beb75-3a4d-11ee-a34b-005056991f7d:1-5 > Group
transactions: aaaaaaaa-aaaa-aaaa-aaaa-aaaaaaaaaaaa:1-4,
    fa1615ca-3a47-11ee-b2c3-00505699077c:1-5'
    2023-08-14T02:54:09.350590Z 0 [ERROR] Plugin group_replication reported:
'The member contains transactions not present in the group. The member will
now exit the group.'
    2023-08-14T02:54:09.350594Z 0 [Note] Plugin group_replication reported:
'To force this member into the group you can use the group_replication_allow_
local_disjoint_gtids_join option'
    2023-08-14T02:54:09.350610Z 0 [Note] Plugin group_replication reported:
'Group membership changed to 10.1.0.203:3306, 10.1.0.201:3306 on view
16919793187165110:8.'
    2023-08-14T02:54:09.350637Z 2 [Note] Plugin group_replication reported:
'Going to wait for view modification'
    2023-08-14T02:54:12.904489Z 0 [Note] Plugin group_replication reported:
'Group membership changed: This member has left the group.'
    2023-08-14T02:54:17.904740Z 2 [Note] Plugin group_replication reported:
'auto_increment_increment is reset to 1'
    2023-08-14T02:54:17.904768Z 2 [Note] Plugin group_replication reported:
'auto_increment_offset is reset to 1'
    2023-08-14T02:54:17.905138Z 7 [Note] Error reading relay log event for
channel 'group_replication_applier': slave SQL thread was killed
    2023-08-14T02:54:17.905149Z 7 [Note] Slave SQL thread for channel 'group_
replication_applier' exiting, replication stopped in log 'FIRST' at position 0
    2023-08-14T02:54:17.907026Z 4 [Note] Plugin group_replication reported:
```

```
'The group replication applier thread was killed'
```

处理办法如下：

```
mysql> show variables like 'group_replication_allow_local_disjoint_gtids_join';
+---------------------------------------------------+-------+
| Variable_name                                     | Value |
+---------------------------------------------------+-------+
| group_replication_allow_local_disjoint_gtids_join | OFF   |
+---------------------------------------------------+-------+
1 row in set (0.00 sec)

mysql> set global group_replication_allow_local_disjoint_gtids_join=1;
Query OK, 0 rows affected (0.00 sec)

mysql> START GROUP_REPLICATION;
Query OK, 0 rows affected (2.12 sec)
```

上面的步骤完全执行成功以后，可以在任意 MGR 实例上执行以下命令。例如，通过以下命令查看 MGR 集群及实例情况：

```
SELECT * FROM performance_schema.replication_group_members;
```

命令显示结果如图 5.5 所示。

CHANNEL_NAME	MEMBER_ID	MEMBER_HOST	MEMBER_PORT	MEMBER_STATE
group_replication_applier	926beb75-3a4d-11ee-a34b-005056991f7d	10.1.0.203	3306	ONLINE
group_replication_applier	fa1615ca-3a47-11ee-b2c3-00505699077c	10.1.0.201	3306	ONLINE
group_replication_applier	facc8726-3a4f-11ee-b88e-005056993b87	10.1.0.202	3306	ONLINE

图 5.5　MGR 集群实例分布情况

下面查看 MGR 集群的主节点，命令如下：

```
SELECT ta.* ,tb.MEMBER_HOST,tb.MEMBER_PORT,tb.MEMBER_STATE FROM performance_schema.global_status ta,performance_schema.replication_group_members tb WHERE ta.VARIABLE_NAME='group_replication_primary_member' and ta.VARIABLE_VALUE=tb.MEMBER_ID;
```

命令显示结果如图 5.6 所示。

VARIABLE_NAME	VARIABLE_VALUE	MEMBER_HOST	MEMBER_PORT	MEMBER_STATE
group_replication_primary_member	fa1615ca-3a47-11ee-b2c3-00505699077c	10.1.0.201	3306	ONLINE

图 5.6　MGR 主节点结果显示

从上面的命令显示结果可以看出，MGR 已经搭建成功了，不过也能发现一个很明显的问题：因为容器机器或者网络的问题，如果 MySQL 实例重启，则 MGR 的主节点一定会发生变化，这样会导致客户端的连接程序无法随着主节点的改变更新配置文件，因此，下面引入前面提到的中间件—— MySQL Router。

基于MGR的机制，当主节点宕机离开集群时，剩余的其他节点会基于Paxos协议选举一个新的主节点。这里有一个问题，如果应用程序端连接到了主节点，这时主节点宕机离开集群，可用的数据库IP地址也会发生变化，但应用程序端还是会向失败的主节点尝试连接，虽然可以修改客户端应用程序的连接配置，但是这个操作基本是不现实的。

MySQL Router是MySQL官方提供的一个轻量级中间件，是MySQL InnoDB Cluster的一部分，可以在应用程序端和后端MySQL服务器之间提供透明路由，主要用于解决MySQL主从库集群的高可用、负载均衡、易扩展等问题。MySQL Router作为一个流量转发层，位于应用程序端与MySQL服务器之间，其功能类似于LVS（即起负载均衡的作用）。MySQL InnoDB Cluster的物理架构如图5.7所示。

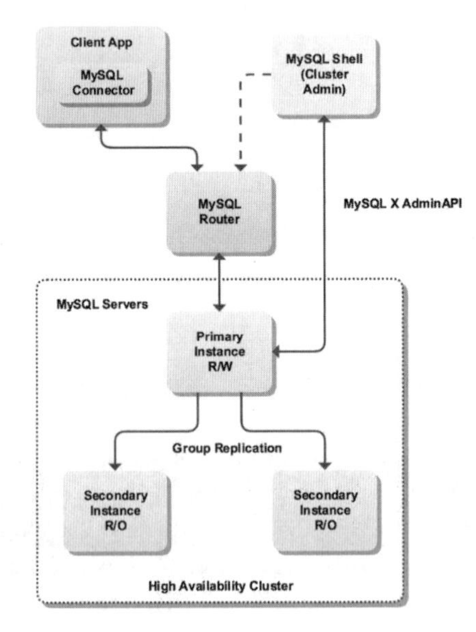

图5.7　MySQL InnoDB Cluster的物理架构

MySQL InnoDB Cluster通过MySQL Shell可以配置一个高可用自动进行故障转移的Cluster。Cluster是这个高可用方案中的一个虚拟节点，它会在MGR的所有成员上创建一个名为MySQL_innodb_cluster_metadata的数据库，存储集群的元数据信息，包括集群信息、集群成员、组复制信息、连接的MySQL Router等信息，以提供MySQL Router查询。它相当于对组复制上的成员做了一层逻辑上的封装，以一个集群的模式展现出来，各节点的状态与对应实例在组复制中成员的状态实时同步，但是集群的节点与组复制的成员只在创建集群时同步，后期组复制的成员变更并不会自动同步到集群中，可以在集群中做手动的节点增减，这样使得面向应用程序端的具体实例实现了更可控、更灵活的高可用。

5.3.5　MySQL MGR生产环境下的监控

数据库管理的关键是要能提前发现问题，若要想提前发现问题，必须要有监控系统，这里推荐Prometheus云原生监控方案。如果MGR要应用于生产环境，强大详细的监控指标是必不可少的。下面详细介绍部分监控指标。

（1）MySQL 的常用监控指标。

1）非功能指标。① QPS：数据库每秒处理的请求数量，包括 DML 和 DDL，这样才能体现数据库的性能。② TPS：数据库每秒处理的事务数量。③并发数：数据库当前并行处理的会话数量。④连接数：连接到数据库的会话数量。⑤缓存命中率：InnoDB 的缓存命中率。

这些都是常见的 MySQL 监控事件，像 Prometheus 这样成熟的开源监控，也提供了 exporter 监控组件，这里采用开源的 mysql_exporter 方案即可。

2）功能指标。①可用性：数据库是否正常对外提供服务，周期性地连接服务器执行 select @@version。②阻塞：当前是否有阻塞的会话，锁住了别人需要的资源。如果 MySQL 版本大于 5.7 版本，查询 sys.innodb_lock_waits 表，wait_started 大于多少秒。③死锁：当前事务是否产生了死锁，相互锁住了对方的资源，可以使用下面的语句查询：

```
set global innodb_print_all_deadlocks=on
```

（2）打印的日志比较多，需要自行分析。

1）慢查询。实时慢查询监控，可以查 information_schema.processlist 表或者用下面的语句查询：

```
show full processlist;
```

2）主从延迟。用于异步复制架构中，重要业务场景需要监控主从延迟的时间。

```
show slave status;
```

重要的指标，如 MGR 服务的可用性、主从复制延迟时间等；监控和告警是业务稳定的基础，尤其是数据库这种较敏感的使用场景。这里需要用 Go 语言编写类似的监控组件，然后通过 AlertManager + Webhook 中间件向企业微信、飞书、钉钉业务群推送相应的告警通知。

5.4　小　　结

云原生数据库就是指在云原生架构上的数据库，与传统的数据库系统相比，都需要存储与计算，当数据量过大时，数据库就需要进行扩容，传统的扩容过程比较慢，而业务高峰过后缩容也很困难，通常会造成极大的资源浪费，也很难应对业务层需要的快速变化，这是传统架构非常大的弊端之一。而在云原生的计算存储分离架构下，业务节点可以根据需要自由地对计算、存储进行快速的扩缩容等操作。云原生数据库具备高性能、高可扩展、一致性、标准、容错、易于管理和多云支持等特性。

本章介绍了基于云原生高可用数据库 MGR 集群的原理和详细的搭建流程。在做好备份和监控的前提下，可以考虑在某些业务场景适当使用 MGR 集群方案，以体验云原生数据库带来的便利和产品交付能力。

第 6 章　用 DC/OS SDK 开发 Framework

6.1　DC/OS 系统简介

在解释 DC/OS SDK 之前，先了解一下 DC/OS 系统。

D2IQ（原名 Mesosphere）公司提出了基于 Mesos 的 DC/OS（数据中心操作系统）的概念，希望企业能像使用一台计算机一样使用整个数据中心。

DC/OS 是一种新型的横跨数据中心或者云机器的操作系统，能够将全部机器资源池化，使它们像一台大计算机。Apache 的开源项目 Mesos 是这个操作系统的内核。这与 Linux 世界的工作模式类似。例如，Linux 内核本身并没有多大用处，也可能以产品的形式对外提供服务，像 Ubuntu 那样的发行版，其中已经围绕内核增加了所有的系统服务和工具，使之成为一个完整的产品。目前已经为数据中心做了同样的事情，使用一系列组件包装了 Mesos 内核，如初始化系统（Marathon）、文件系统（HDFS）、应用打包和部署系统、图形化界面和命令行。所有这些组件组成了 DC/OS 系统。

Mesos 项目主要由 C++ 语言编写，代码仍在快速演化中，稳定版本目前为 1.11.0。

Mesos 拥有许多引人注目的特性，例如：

（1）支持数万个节点的大规模场景（Apple、Twitter、eBay 等公司实践），它的诞生甚至比 YARN 还早几年，并于 2010 年很快被应用到 Twitter，成为 Twitter 自定义 PaaS 的实现基础，管理着 Twitter 超过 300000 台服务器上的应用部署。

（2）支持多种应用框架，包括 Marathon、Singularity 和 Aurora 等。

（3）支持 HA（基于 ZooKeeper 实现）。

（4）支持 Docker、UCR 等容器机制进行任务隔离。

（5）提供了多个流行语言的 API，包括 Python、Java、C++ 等。

（6）自带简洁易用的 WebUI，方便用户直接进行操作。

Mesos 系统架构如图 6.1 所示。

IaaS 层抽象的是机器，PaaS 则更多考虑部署、管理应用和服务。在交互方面，PaaS 可能是和开发者直接交互，而 Mesos 则是以 API 的形式和软件程序交互，它在 PaaS 层与 IasS 层之间工作，更多的是作为 PaaS 实现，如图 6.2 所示。

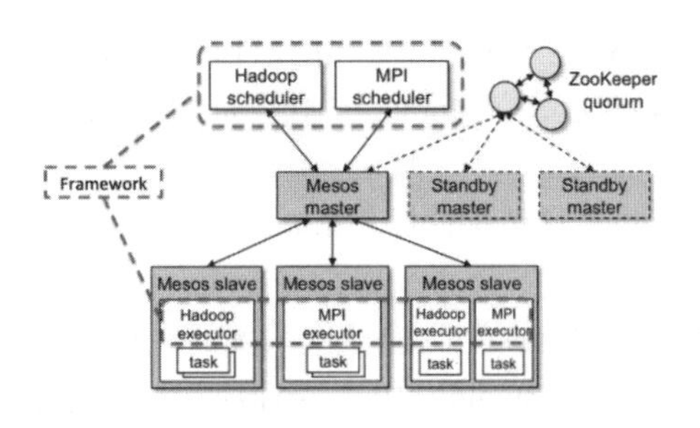

图 6.1　Mesos 系统架构

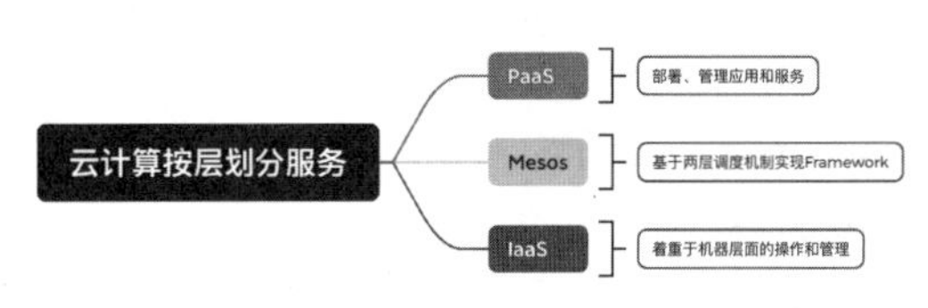

图 6.2　Mesos 可以实现 PaaS 服务

Mesos 的亮点一：两级调度机制。Mesos 以 Framework 的形式提供了两级调度机制，将任务的调度和执行分离。面对各种类型的任务，在调度阶段，由 Framework 的 scheduler（调度器）以资源邀约的形式向 Mesos master 申请资源；在执行阶段，由 Framework 的 executor（执行器）执行任务。

Mesos 的亮点二：围绕 Framework 建设的生态系统。在 Apache Mesos 的术语中，使用 Mesos API 在集群中调度任务的 Mesos 应用程序称为 Framework（框架）。一旦将任务调度委托给 Framework 应用程序，以及采用插件架构，就能直接打造 Mesos 问鼎数据中心资源管理的生态系统。因为每接入一种新的 Framework，master 无须为此编码，Agent 模块可以复用，使得在 Mesos 所支持的宽泛领域中业务迅速增长。而开发者就可以专注于他们的应用和 Framework 的选择。

Mesos 典型的应用场景有以下几种：

（1）Spark on Mesos。

（2）Jenkins on Mesos。

（3）基于 Marathon Docker 的编排服务等。

站在调度的角度，Mesos 与 Kubernetes 集群的区别如下：

（1）开源的 Kubernetes 系统是单体式调度：单体式调度器使用复杂的调度算法结合集群的全局信息，计算出高质量的放置点，不过延迟较高。

（2）Apache Mesos 是两级调度：两级调度器通过将资源调度和作业调度分离，解决单体式调度器的局限性。两级调度器允许根据特定的应用做不同的作业调度逻辑，同时保持了不同作业之间共享集群资源的特性，但是无法实现高优先级应用的抢占。

6.2　DC/OS SDK 简介

DC/OS 除了内核 Mesos，还有两个关键组件 Marathon 和 Chronos。其中，Marathon（又称分布式 init）是一个用于启动长时间运行应用程序和服务的框架，Chronos（又称分布式 cron）是一个在 Mesos 上运行和管理计划任务的框架。

Marathon 按照官方的定义，是一个基于 Mesos 的私有 PaaS，它实现了 Mesos 的 Framework 功能。Marathon 提供了服务发现和负载平衡、部署所需的 REST API 服务、授权机制和 SSL 支持、配置约束等功能。

Marathon 支持通过 Shell 命令和 Docker 部署应用，提供 Web 界面，并允许设置 CPU/ 内存、实例数等参数，支持单应用的部署但不支持复杂的集群定义。Marathon 本身是用 Scala 语言实现的。

Marathon 支持运行长服务，如 Web 应用等，并且能够原样运行任何 Linux 二进制发布版本，如 Tomcat 等。

Mesos 和 Marathon 之间的关系可以理解为，如果将 Mesos 类比为操作系统的内核，负责资源调度，则 Marathon 可以类比为服务管理系统，如 init、systemd，用来管理应用的状态信息。Marathon 将应用程序部署为长时间运行的 Mesos 任务。

DC/OS 系统架构如图 6.3 所示。

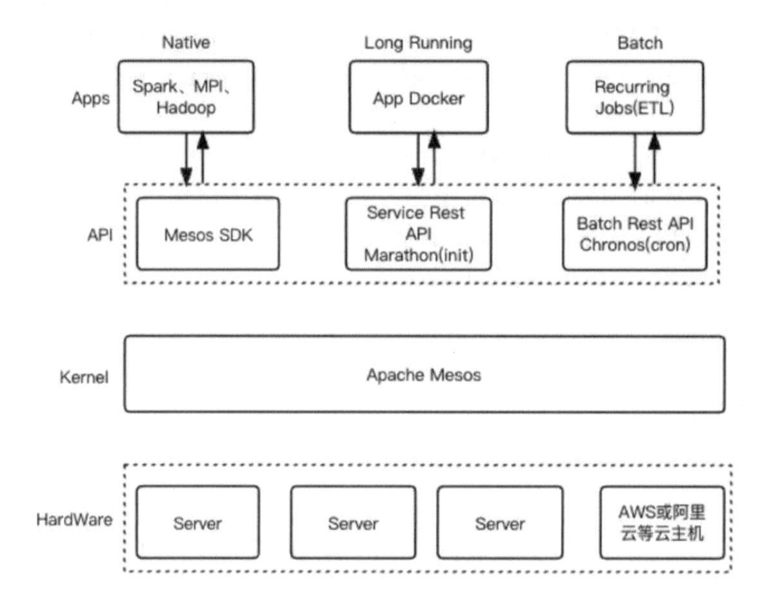

图 6.3　DC/OS 系统架构

D2IQ DC/OS 可以在任意的现代 Linux 环境、公有或私有云、虚拟机甚至裸机环境中运行，当前支持的平台有亚马逊 AWS、谷歌 GCE、微软 Azure 和 OpenStack 等。据 D2IQ 官网显示，DC/OS 在其公有仓库上提供了 40 多种服务（Framework）组件，如 Hadoop、Spark、Cassandra、Jenkins、Kafka 和 MemSQL 等。

DC/OS 的优势如下。

（1）生产验证：基于 Mesos 和 Marathon，DC/OS 是业界成熟、企业级的容器编排平台。

（2）两级调度：Mesos 具有两级调度设计，使平台服务能够智能地通过自动分发任务和容器来调度工作负载，从而提高平台利用率。

（3）快速部署有状态服务：复杂的分布式系统可以在几分钟内完成部署。

（4）自动故障恢复：针对所有类型的应用程序，服务和工作负载，DC/OS 内置了高可用性和容错能力。

（5）资源效率：容器实现了性能隔离，消除了静态分区环境并提高了服务器利用率。在 DC/

OS 上运行 Spark、Kafka 和 Cassandra，可以动态地扩缩容各种计算资源。

（6）简化操作：通过基于 GUI/CLI 的监控和管理来控制整个数据中心资源。DC/OS 提供了一个统一的 GUI 界面和互操作接入，用于管理应用程序和有状态服务的整个生命周期。

（7）跨平台一致性：DC/OS 提供了一个抽象层，无论部署在裸机、虚拟机、私有云或公共云上，都能提供标准的用户体验。

D2IQ 的 DC/OS SDK（软件开发套件）是为 DC/OS 开发者提供的，使他们能够更轻松地使用不同的语言（Java、Go 和 Python）创建新的数据中心服务，打包部署这些应用，并且使用命令行工具中的一条命令即可完成安装。DC/OS SDK 开放了 Mesos API，同时还提供了 DC/OS 的扩展包，构建服务（调度器的开发、执行器等）库，Alpha、Beta 和 Production 阶段的认证清单和开发者手册。此外，D2IQ 还有一个非常强大的社区成长计划，这是确保应用程序开发成功的重要一环。

对于构建和运行分布式应用程序来说，按机器级别的抽象是错误的，因为会迫使用户去推断机器的具体细节信息，如 IP 地址和本地存储。通过将所有的机器，包括虚拟的或物理的，组成一个资源池，DC/OS 为任务、服务发现和传递任务协调消息的自动配置，提供了内置的原语。因为开发人员无须亲自写（测试和调试）这样的代码，所以效率得到很大的提升，可以更快速地提供更好的用户体验。通过构建数据中心操作系统的抽象，可以将新的可伸缩数据中心的产品，快速推向市场。这就是操作系统的催化剂作用。

但是，之所以使用数据中心操作系统的抽象也有经济因素。操作系统提供了可以在其上分发跨基础设施运行的应用程序的平台。这使开发者得以进入一个宽广、无摩擦的市场，在这里开发人员可以分发这些应用程序。例如，在 DC/OS 中构建一个非常复杂的打包系统，可以打包一个复杂的分布式系统（如 Cassandra、Kafka 和 Spark 等），运维人员或者开发人员使用一个命令，即可在几分钟内完成安装、配置和运行这个系统。这是前所未有的，而这些复杂的分布式系统的定义框架（Framework），均可以利用 DC/OS SDK 来完成。

在对接客户的云原生需求时，发现很多客户，如某省联通、某大型汽车供应商、某省卫健核酸项目等，都在生产环境下广泛使用 DC/OS 系统大规模部署集群，并且 DC/OS 在高并发流量下有不俗的表现。客户都有类似的云原生业务开发需求，所以笔者后续还是系统地介绍了 DC/OS SDK，并且基于此开发了一些相应的基于云原生的框架产品，如 Hadoop Framework 和 MySQL InnoDB Cluster Framework 等。

> 注：框架支持UI界面安装和命令行安装。在开发阶段，UI界面安装方便，在生产环境上线以后，命令行安装的方式更为简便。例如，dcos package install hdfs --cli --yes。

6.3 DC/OS SDK 开发工作涉及的网络和存储

在很多微服务架构中，有南北流量（NORTH-SOUTH traffic）和东西流量（EAST-WEST traffic）两个术语；同样，DC/OS 也需要对应的系统组件来处理南北流量和东西流量。

南北流量和东西流量是数据中心环境中的网络流量模式。下面通过一个例子来理解这两个术语。

1. 南北流量

客户端和服务器之间的流量称为南北流量。简而言之，南北流量是 server-client 流量，也可以简单理解为集群内访问。

2. 东西流量

不同服务器之间的流量与数据中心或不同数据中心之间的网络流量称为东西流量。简而言之，东西流量是 server-server 流量，也可以简单理解为集群外访问。

目前，在大数据生态系统中，如 Hadoop 生态系统（大量 server 驻留在数据中心，用 MapReduce 处理），东西流量远大于南北流量。

参考应用部署在 Kubernetes DC/OS 中的难点，要关心的其实也是网络、存储部分。

（1）推荐用 Mesos 网络的 Overlay 模式（Calico CNI），其性能方面比较优异。

（2）DC/OS 集群内 DNS 解析由 DC/OS DNS 提供，包括内网地址解析和自动发现功能。

（3）集群内（东西流量）访问可以用 DC/OS DNS 域名，集群外（南北流量）访问则可以用 Marathon-LB。

（4）由于分布式文件系统 GlusterFS 在高负载下不是太稳定，因此这块倾向于采取本地 Local PV 的方式，其作用在于 Pod 重启以后不更换主机，从而达到持久化的目的。

DC/OS 如何提供东西流量和南北流量呢？参考 D2IQ 官方文档，其主要实现如下：

DC/OS DNS 是在每个 DC/OS 代理节点和管理节点上运行的分布式 DNS 服务器，作为被称为 dcos-net 的 Erlang 虚拟机的一部分。在领导管理节点上运行的实例定期轮询领导管理节点的状态，并为由 DC/OS 启动的每个应用程序生成 FQDN，然后，它将此信息发送给 DC/OS 中的同类服务器。所有这些 FQDN 的后缀都为 directory。

DC/OS 会拦截发源于代理节点的所有 DNS 查询。如果查询以 directory 结尾，则 DC/OS DNS 会在本地解析；如果以 mesos 结尾，则 DC/OS DNS 会把查询转发给在管理节点上运行的 Mesos DNS 实例；否则，它会根据 TLD（顶级域名）将查询转发给已配置的上游 DNS 服务器。

DC/OS DNS 是 DC/OS 集群的主 DNS 服务器，适合利用名为 dcos-l4lb（内部 4 层）的 DC/OS 内部负载均衡器进行负载均衡的任何服务。通过 dcos-l4lb 进行负载均衡的任何服务均可获得 virtual-ip-address（即 VIP 地址，集群内服务一般由 IPVS 转发实现，VIP 地址可以对应多个实例）以及 "*.l4lb.thisdcos.directory" 域中的 FQDN。

> 注：dcos-l4lb是默认安装的分布式第4层东西向负载均衡器。它具有高度可扩展性和高可用性，提供零跃负载均衡，没有单个阻塞点，并容忍主机故障。dcos-l4lb作为Erlang虚拟机中的应用程序运行dcos-net，在DC/OS集群中的所有代理节点和管理节点上运行。

DC/OS 中有两个软件包，即 Edge-LB 和 Marathon-LB，为 DC/OS 服务提供第 7 层负载均衡。这两个软件包均使用 HAProxy 作为其数据平面，对进入群集的南北向流量（即集群外访问）进行负载均衡。虽然这些软件包主要用于提供第 7 层负载均衡（支持 HTTP 和 HTTPS），但也可为

TCP 和 SSL 流量提供第 4 层负载均衡。Edge-LB 仅适用于 DC/OS 企业版本。

Marathon-LB 简单得多，只管理一个 HAProxy 实例。它只能对 Marathon 框架启动的应用执行负载均衡。Marathon-LB 适用于 DC/OS 的开源版和企业版。在实际工作中，由于 DC/OS 开源版的使用较为广泛，因此 Marathon-LB 也相应地得到了较多的应用。

DC/OS SDK 整体是用 Java 开发的，成熟度较高；在 DC/OS SDK 下开发分布式系统应用，其实只需熟悉 YAML 语法和 Shell 脚本即可，但重点还是要关注分布式系统应用本身。例如，如果要在 DC/OS SDK 下开发支持 Kerberos 认证的高可用 HDFS，则要熟悉高可用 HDFS 的部署和运维，了解如何让 HDFS 支持 Kerberos 认证等，这些才是应该关注的重难点，也是设计 DC/OS SDK 的初衷。

6.4 DC/OS SDK 的基础核心概念

在最高抽象级别，DC/OS 服务分为要启动的任务以及如何启动它们。ServiceSpec 定义了什么是服务，Plan 定义了如何在部署、更新和故障场景中控制服务。将 ServiceSpec 和 Plan 打包后，就可以将服务部署在 Universe 的 DC/OS 集群上。如果熟悉 ServiceSpec、Plan 及包定义用法，就可以通过 DC/OS SDK 写出业务框架（服务）了。下面分别介绍 ServiceSpec、Plan 及包定义。

6.4.1 DC/OS 服务规范定义

有两种方法可以生成有效的 ServiceSpec（服务规范），即创建 YAML 文件或编写 Java 代码。两者都可以生成 Java 接口的有效实现 ServiceSpec。ServiceSpec 可用于在 DC/OS 集群内启动同一服务的一个或多个实例，这里主要讨论创建 YAML 文件的方式。

例如，编写一个 ServiceSpec 描述部署 Kafka 集群的 DC/OS 服务，然后可以在 DC/OS 集群中安装一个或多个 Kafka 集群实例。从这个意义上说，ServiceSpec 类似于类定义，可用于创建作为类实例的许多对象。

这里举个简单的 DC/OS 服务 YAML 定义的例子来说明，它的作用是，每 1000 秒将“hello world”打印到容器沙箱中的标准输出，其代码如下：

```
name: "hello-world"
scheduler:
  principal: "hello-world-principal"
  user: {{SERVICE_USER}}
pods:
  hello-world-pod:
    count: 1
    tasks:
      hello-world-task:
        goal: RUNNING
        cmd: "echo hello world && sleep 1000"
```

```
            cpus: 0.1
            memory: 512
```

其中的关键组件解释如下。

（1）name：DC/OS 服务实例的名称。在同一集群中，任何两个服务的实例都不能具有相同的名称。

（2）scheduler：调度程序管理服务并保持其运行。本小节包含适用于调度程序的设置。可以省略 scheduler 部分以对所有这些设置使用合理的默认值。

1）principal：这是注册框架时使用的 DC/OS 服务账户。在安全的企业版集群中，此账户必须具有执行调度程序操作所需的权限。如果省略此设置，则默认为 <svcname>-principal。

2）user：在主机上运行进程时使用的账户，推荐的默认值为 nobody。

（3）pods：约定写法，用于定义以下各种类型的 pod。

1）hello-world-pod：pod 类型的名称。可以为 pod 类型选择任何名称。在本示例中，定义了一个 pod，其名称为 hello-world-pod。

2）count：pod 的实例数。

3）tasks：pod 中的任务列表。

4）hello-world-task：在本示例中，单个 pod 定义包含一个任务。该任务的名称是 hello-world-task。

5）goal：每个任务都必须有一个目标状态。存在三种可能的目标状态：RUNNING、FINISH 和 ONCE。其中，RUNNING 表示任务应该始终运行，因此如果它退出，则应该重新启动；FINISH 表示如果任务成功完成，则不需要重新启动，除非更新其配置；ONCE 表示如果任务成功完成，则在 Pod 的生命周期内不需要重新启动。

6）cmd：运行启动任务的命令。在这里，任务将向标准输出打印“hello world”并休眠 1000 秒。因为它的目标状态是 RUNNING，所以退出时会再次启动。

7）cpus：定义将分配给任务容器的 CPU 数量。

8）memory：定义将分配给任务容器的内存量。

6.4.2 DC/OS 计划中的相关概念

在上面的简单示例中，如何部署示例服务是显而易见的。它由启动单个任务组成。然而，对于具有多个 Pod 的更复杂服务，DC/OS SDK 允许定义计划（Plan）来编排任务的部署。

下面的示例展示了一个定义了两种类型 Pod 的服务，且每种 Pod 都部署了两个实例。其代码如下：

```
name: "hello-world"
pods:
  hello-pod:
    count: 2
    tasks:
      hello-task:
```

```
        goal: RUNNING
        cmd: "echo hello && sleep 1000"
        cpus: 0.1
        memory: 512
  world-pod:
    count: 2
    tasks:
      world-task:
        goal: RUNNING
        cmd: "echo world && sleep 1000"
        cpus: 0.1
        memory: 512
```

有多种可能的部署策略：并行（parallel）或串行（serial），以及一种 Pod 类型在部署之前等待另一种 Pod 类型成功部署。

默认情况下，DC/OS SDK 将串行部署所有 Pod 实例。在上面的示例中，默认部署顺序如下：

```
hello-pod-0-hello-task
hello-pod-1-hello-task
world-pod-0-world-task
world-pod-1-world-task
```

运行 Pod 的任务能正常启动的前提是，上一个任务必须正常运行完成。这是默认部署策略最简单、最安全的方法。但是，这种默认部署策略无法提供编写丰富服务所需的灵活性。因此，DC/OS SDK 还允许定义编排任务部署的计划。

在本小节中，重点关注如何使用计划来定义服务的初始部署。此外，计划还可以用于协调配置更新、软件升级以及从复杂的故障场景中恢复。

1. 复杂的计划示例一

下面考虑更为复杂的场景：希望并行部署 hello-pod 任务，等待它们达到 RUNNING 状态，然后串行部署 world-pod 任务。可以将 YAML 文件修改如下：

```
name: "hello-world"
pods:
  hello-pod:
    count: 2
    tasks:
      hello-task:
        goal: RUNNING
        cmd: "echo hello && sleep 1000"
        cpus: 0.1
        memory: 512
  world-pod:
    count: 2
```

```
      tasks:
        hello-task:
          goal: RUNNING
          cmd: "echo world && sleep 1000"
          cpus: 0.1
          memory: 512
plans:
  deploy:
    strategy: serial
    phases:
      hello-phase:
        strategy: parallel
        pod: hello-pod
      world-phase:
        strategy: serial
        pod: world-pod
```

计划（Plan）是一个简单的三层层次结构。计划（Plan）由阶段（Phase）组成，阶段又由步骤（Step）组成。每层都可以定义部署其组成元素的策略（strategy）。最高层的策略定义了如何部署阶段。每个阶段的策略定义了如何部署步骤。如果未指定，则默认策略是串行（serial），Plan 最终会生成对应的服务框架，如图 6.4 所示。

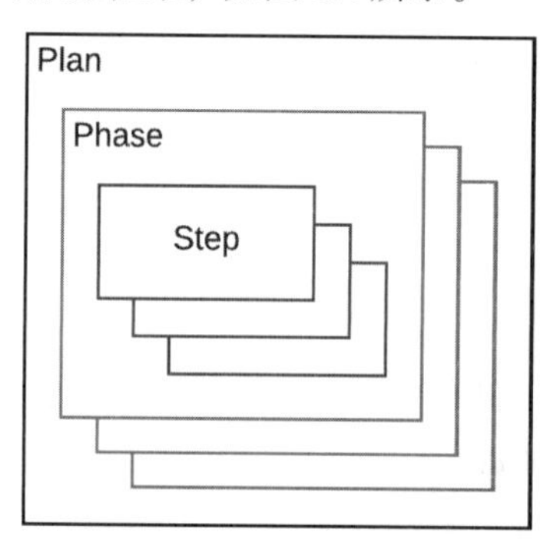

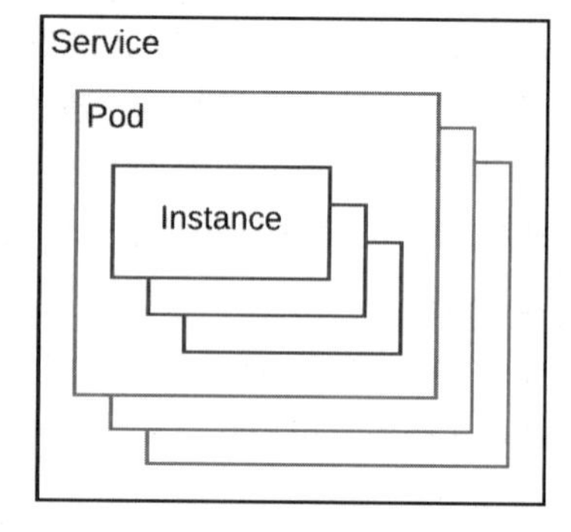

图 6.4　DC/OS SDK 中的 Plan 生成对象模型

阶段封装了 Pod 类型，步骤封装了 Pod 实例。因此在这种情况下有两个阶段：hello-phase 和 world-phase。它们显然与 ServiceSpec 中的特定 Pod 定义相关联。在上面的示例中，不需要具体定义完成部署策略目标的步骤，因此省略了。

该示例的 hello-phase 有两个元素：strategy 和 pod。

```
plans:
  deploy:
    strategy: serial
    phases:
```

```
      hello-phase:
        strategy: parallel
        pod: hello-pod
      world-phase:
        strategy: serial
        pod: world-pod
```

pod 参数引用了 ServiceSpec，即 YAML 中 Plan 字段上面的内容；该策略声明了如何部署 pod 的实例。这里表示即将进行并行部署，但要注意的是，整个部署计划相关的策略是串行的，因此，应一次部署一个阶段。此依赖关系说明了部署步骤，如图 6.5 所示。

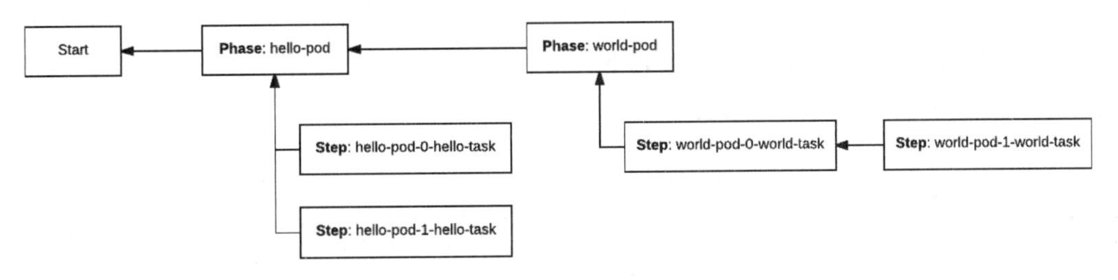

图 6.5　名为 hello 的计划生成实例的阶段演示

world-pod 阶段的部署可以分成两个阶段执行，如下：

串行依赖性：world-pod 必须等待 hello-pod 完成之后才开始执行，这意味着它们是串行化执行的。

并行依赖性：两个 hello 步骤虽然都依赖 hello-pod，但它们彼此之间没有依赖，因此可以并行执行。

2. 复杂的计划示例二

Plans 是 DCOS SDK 的保留关键字，利用它可以编写更强大的自定义计划，例如，Pod 需要在运行 Pod 的主要任务之前运行初始化步骤，可以定义一个这样的任务计划。其 YAML 文件内容如下：

```
name: "hello-world"
pods:
  hello:
    [...]
    tasks:
      init:
        [...]
      main:
        [...]
plans:
  deploy:
    strategy: serial
    phases:
      hello-phase:
```

```
          strategy: serial
          pod: hello
          steps:
            - default: [[init], [main]]
```

该计划表明，默认情况下，hello pod 的每个实例都应生成两个步骤：一个代表 init 任务，另一个代表 main 任务。ServiceSpec 指示 hello 应启动两个 pod，以便按顺序启动以下任务：

```
hello-0-init
hello-0-main
hello-1-init
hello-1-main
```

考虑这样的场景：第一个 pod 的 init 任务只应发生一次，并且所有后续 pod 只应启动其 main 任务。这样的计划可以写成如下形式：

```
pods:
  hello:
    [...]
    tasks:
      init:
        [...]
      main:
        [...]
plans:
  deploy:
    strategy: serial
    phases:
      hello-phase:
        strategy: serial
        pod: hello
        steps:
          - 0: [[init], [main]]
          - default: [[main]]
```

该计划将导致生成以下任务实例：

```
hello-0-init
hello-0-main
hello-1-main
```

6.4.3 DC/OS 包定义

DC/OS 服务必须提供包定义（主要是通过 Universe 包文件）才能安装在 DC/OS 集群上。包定义至少由 4 个文件组成：marathon.json.mustache、config.json、resource.json 和 package.json。

下面介绍每个文件的格式和用途。

（1）marathon.json.mustache 提供 Marathon 应用程序定义的 Mustache 模板文件。它的 mustache 元素由 config.json 和 resource.json 文件中存在的值呈现。

（2）config.json 文件可以简单理解为 marathon.json.mustache 的字典 K/V 值的具体呈现。

（3）resource.json 是所有下载元素的 URI 列表。服务所需的任何文件都必须在此处列出，否则服务将无法安装到 DC/OS 集群。当 DC/OS 集群安装软件包时，只有此处列出的文件可用。此列表包含服务运行所需的一些项目文件，如 bootstrap.zip。

下载 URI 可以结合实际情况来定，如果 DC/OS 集群部署在阿里云 / 华为云集群上，则下载地址可以是 OSS 公共下载地址；如果 DC/OS 集群部署在私有云环境中，则需要私有云内部提供 HTTP 下载或自行搭建 HTTP 下载地址，可以根据实际情况来部署。

（4）package.json 文件包含 Universe 的版本元数据，其中也包括可以部署服务的 DC/OS 的版本号。

建议参考 DC/OS 服务的 helloworld 服务（框架）示例，其文件内容如下：

```
name: {{FRAMEWORK_NAME}}
scheduler:
  principal: {{FRAMEWORK_PRINCIPAL}}
  user: {{FRAMEWORK_USER}}
pods:
  hello:
    count: {{HELLO_COUNT}}
    placement: '{{{HELLO_PLACEMENT}}}'
    rlimits:
      RLIMIT_NOFILE:
        soft: {{HELLO_RLIMIT_NOFILE_SOFT}}
        hard: {{HELLO_RLIMIT_NOFILE_HARD}}
    tasks:
      server:
        goal: RUNNING
        cmd: env && echo hello >> hello-container-path/output && sleep
$SLEEP_DURATION
        cpus: {{HELLO_CPUS}}
        memory: {{HELLO_MEM}}
        volume:
          path: hello-container-path
          type: ROOT
          size: {{HELLO_DISK}}
        env:
          SLEEP_DURATION: {{SLEEP_DURATION}}
```

```
        health-check:
          cmd: stat hello-container-path/output
          interval: 5
          grace-period: 30
          delay: 0
          timeout: 10
          max-consecutive-failures: 3
        labels: {{HELLO_LABELS}}
  world:
    count: {{WORLD_COUNT}}
    allow-decommission: true
    placement: '{{{WORLD_PLACEMENT}}}'
    rlimits:
      RLIMIT_NOFILE:
        soft: {{WORLD_RLIMIT_NOFILE_SOFT}}
        hard: {{WORLD_RLIMIT_NOFILE_HARD}}
    tasks:
      server:
        goal: RUNNING
        cmd: |
               # for graceful shutdown
               # 用于优雅关闭陷阱 SIGTERM 并模拟清理时间范围
               terminated () {
                 echo "$(date) received SIGTERM, zzz for 3 ..."
                 sleep 3
                 echo "$(date) ... all clean, peace out"
                 exit 0
               }
               trap terminated SIGTERM
               echo "$(date) trapping SIGTERM, watch here for the signal..."

               echo 'world1' >>world-container-path1/output &&
               echo 'world2' >>world-container-path2/output &&
               # 执行无限循环的语句以便于下面的测试顺利进行
               while true; do
                 sleep 0.1
               done
        cpus: {{WORLD_CPUS}}
        memory: {{WORLD_MEM}}
```

```
        volumes:
          vol1:
            path: world-container-path1
            type: ROOT
            size: {{WORLD_DISK}}
          vol2:
            path: world-container-path2
            type: ROOT
            size: {{WORLD_DISK}}
        env:
          SLEEP_DURATION: {{SLEEP_DURATION}}
        readiness-check:
          # 执行 readiness-check 健康检测脚本
          # so send the error to /dev/null, BUT also zero-left-pad the  variable
BYTES to ensure that it is zero
          # on empty for comparison sake.
          cmd: BYTES="$(wc -c world-container-path2/output 2>/dev/null| awk
'{print $1;}')" && [ 0$BYTES -gt 0 ]
          interval: {{WORLD_READINESS_CHECK_INTERVAL}}
          delay: {{WORLD_READINESS_CHECK_DELAY}}
          timeout: {{WORLD_READINESS_CHECK_TIMEOUT}}
        kill-grace-period: {{WORLD_KILL_GRACE_PERIOD}}
```

在 helloworld 服务示例的总控制文件 svc.yml 中，所有的变量均来自 Universe 包中的相关文件，最终经过 Mustache 模板会渲染成所需要的值。

> 注：Mustache是一个轻逻辑模板解析引擎，它的优势在于可以应用在JavaScript、PHP、Python、Perl等多种编程语言中。如果{{#keyName}} {{/keyName}}中的keyName 值为null、undefined和false，则表示不渲染输出任何内容。

6.5　使用 DC/OS SDK 开发 MIC 框架

前面章节提到了基于 MGR 的云原生设计方案，本节介绍如何使用 DC/OS SDK 来开发一个基于云原生的 MySQL InnoDB Cluster（以下简称 MIC）框架服务。

这里为了方便查阅相关代码，可以远程克隆 DC/OS Git 仓库到本地工作目录下，其命令为 git clone https://github.com/mesosphere/dcos-commons.git。

然后，在 frameworks 目录下面新建 mysql-innodb-cluster 目录，目录树明细如图 6.6 所示。

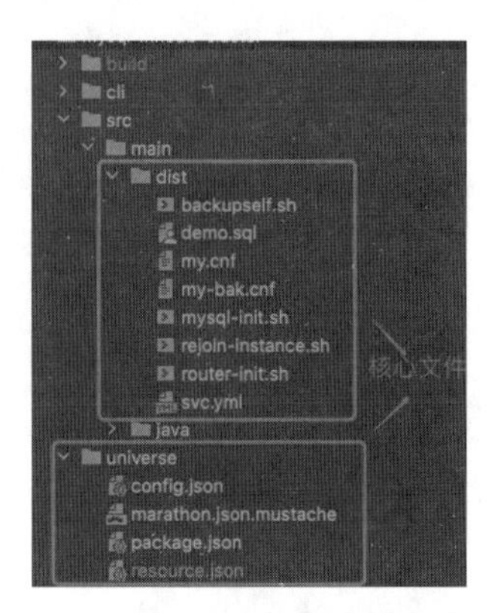

图 6.6 mysql-innodb-cluster 目录树明细

mysql-innodb-cluster 目录树中的部分文件的作用如下。

（1）backupself.sh：利用 xbackup 备份 MySQL 容器脚本。

（2）demo.sql：初始化 MGR 集群时用于插入 MySQL 数据，其中有些参数做了最基础的优化动作。

（3）my.cnf：创建 MGR 集群的 MySQL 文件。

（4）mysql-init.sh：初始化 MGR 集群的脚本文件。

（5）rejoin-instance.sh：MGR 集群中有 MySQL 实例发生机器重启时，重新加入 MGR 集群的自动化脚本。

（6）router-init.sh：加入 MySQL-router 的脚本文件。

（7）svc.yml: 框架的主核心配置文件。

（8）universe：主要是与 MIC 相关的配置和模板渲染文件。

6.5.1 MIC 框架的主核心配置文件

下面分析 MIC 框架的主核心配置文件。其内容如下：

```
name: {{FRAMEWORK_NAME}}
scheduler:
  principal: {{FRAMEWORK_PRINCIPAL}}
  user: {{FRAMEWORK_USER}}
pods:
  mysql:
    count: {{NODE_COUNT}}
    uris:
      - {{BOOTSTRAP_URI}}
      - {{MYSQL_SHELL_URI}}
      - {{DEMO_SQL_URI}}
      - {{MYSQL_XTRABACKUP_URI}}
    rlimits:
      RLIMIT_NOFILE:
```

```
          soft: 65535
          hard: 65535
      image: {{MYSQL_IMAGE}}
      placement: '{{{NODE_PLACEMENT}}}'
      {{#ENABLE_VIRTUAL_NETWORK}}
      networks:
        {{VIRTUAL_NETWORK_NAME}}:
          labels: {{VIRTUAL_NETWORK_PLUGIN_LABELS}}
      {{/ENABLE_VIRTUAL_NETWORK}}
      volumes:
        mysql-data:
          path: "MYSQL_DATA"
          type: {{NODE_DISK_TYPE}}
          size: {{NODE_DISK}}
        mysqlbp-data:
          path: "BKP_DATA"
          type: {{BACKUP_DISK_TYPE}}
          size: {{BACKUP_NODE_DISK}}
      resource-sets:
        server:
          cpus: {{NODE_CPUS}}
          memory: {{NODE_MEM}}
      tasks:
        node:
          goal: RUNNING
          cmd: |
            #export
PATH=${MESOS_SANDBOX}/percona-xtrabackup-2.4.21-Linux-x86_64.glibc2.12/
bin:$PATH
            chmod +x ./config-templates/mysql-init.sh
            ./config-templates/mysql-init.sh
          readiness-check:
            cmd: |
                v_check=$(mysql  -B   -uroot  -p"${MYSQL_ROOT_PASSWORD}"  -h
127.0.0.1 --disable-column-names -e "select 1")
              if [ $v_check = 1 ]; then
                echo "mysql is ready"
                exit 0
              else
                echo "mysql is not yet ready"
                exit 1
```

```
            fi
            interval: 35
            delay: 100
            timeout: 120
          resource-set: server
          env:
            MYSQL_ROOT_PASSWORD: {{MYSQL_ROOT_PASSWORD}}
            NODE_COUNT: {{NODE_COUNT}}
            NODE_MEM: {{NODE_MEM}}
            MYSQL_CHAR_SET_SERVER: {{MYSQL_CHAR_SET_SERVER}}
            MYSQL_INNODB_BUFFER_POOL_SIZE_RATIO:
{{MYSQL_INNODB_BUFFER_POOL_SIZE_RATIO}}
            MYSQL_MAX_CONNECTIONS: {{MYSQL_MAX_CONNECTIONS}}
            MYSQL_MAX_USER_CONNECTIONS: {{MYSQL_MAX_USER_CONNECTIONS}}
            MYSQL_TIME_HOUR: {{MYSQL_TIME_HOUR}}
            MYSQL_EXPIRE_LOGS_DAYS: {{MYSQL_EXPIRE_LOGS_DAYS}}
            MYSQL_TIMEOUT: {{MYSQL_TIMEOUT}}
            MYSQL_CONNECT_ERRORS: {{MYSQL_CONNECT_ERRORS}}
          configs:
            my.cnf:
              template: my.cnf
              dest: my.cnf
            mysql-init.sh:
              template: mysql-init.sh
              dest: mysql-init.sh
            rejoin-instance.sh:
              template: rejoin-instance.sh
              dest: rejoin-instance.sh
            backup-mysqldump.sh:
              template: backup.sh
              dest: backup.sh
            backup-xtrabackup.sh:
              template: backupself.sh
              dest: backupself.sh


    backup:
      count: 1
      {{#ENABLE_VIRTUAL_NETWORK}}
      networks:
        {{VIRTUAL_NETWORK_NAME}}:
```

```
        labels: {{VIRTUAL_NETWORK_PLUGIN_LABELS}}
    {{/ENABLE_VIRTUAL_NETWORK}}
    volumes:
      mysqlbp-data:
        path: "BKP_DATA"
        type: {{NODE_DISK_TYPE}}
        size: {{NODE_DISK}}
    image: {{MYSQL_IMAGE}}
    resource-sets:
      side-cars:
        cpus: 0.1
        memory: 128
    tasks:
      localbackup:
        goal: ONCE
        cmd: |
          chmod +x ./config-templates/backup-xtrabackup.sh
          ./config-templates/backup-xtrabackup.sh
          #tail -f /dev/null
        resource-set: side-cars
        env:
          MYSQL_ROOT_PASSWORD: {{MYSQL_ROOT_PASSWORD}}
          BACKUP_DATABASENAME: {{BACKUP_DATABASENAME}}
        configs:
          backup-xtrabackup.sh:
            template: backupself.sh
            dest: backupself.sh

{{#EXPORTER_ENABLED}}
  exporter:
    count: 3
    uris:
      - {{BOOTSTRAP_URI}}
    image: {{EXPORTER_IMAGE}}
    placement: '{{{EXPORTER_PLACEMENT}}}'
    {{#ENABLE_VIRTUAL_NETWORK}}
    networks:
      {{VIRTUAL_NETWORK_NAME}}:
        labels: {{VIRTUAL_NETWORK_PLUGIN_LABELS}}
    {{/ENABLE_VIRTUAL_NETWORK}}
    resource-sets:
```

```
        exporter:
          cpus: {{EXPORTER_CPUS}}
          memory: {{EXPORTER_MEM}}
          ports:
            exporter:
              port: 9104
              advertise: true
              vip:
                prefix: exporter
                port: 9104
      tasks:
        daemon:
          goal: RUNNING
          cmd: |
            ./bootstrap
            ln -sf /usr/share/zoneinfo/${TZ} /etc/localtime
            export
DATA_SOURCE_NAME="root:${MYSQL_ROOT_PASSWORD}@(mysql-${POD_INSTANCE_INDEX}-
node.${FRAMEWORK_NAME}.autoip.dcos.thisdcos.directory:3306)/"
            /bin/mysqld_exporter
          resource-set: exporter
          env:
            MYSQL_ROOT_PASSWORD: {{MYSQL_ROOT_PASSWORD}}
  {{/EXPORTER_ENABLED}}


    router:
      count: 2
      uris:
        - {{BOOTSTRAP_URI}}
        - {{MYSQL_ROUTER_URI}}
      rlimits:
        RLIMIT_NOFILE:
          soft: 12800
          hard: 12800
      image: {{MYSQL_IMAGE}}
      placement: '{{{ROUTER_PLACEMENT}}}'
      {{#ENABLE_VIRTUAL_NETWORK}}
      networks:
        {{VIRTUAL_NETWORK_NAME}}:
          labels: {{VIRTUAL_NETWORK_PLUGIN_LABELS}}
```

```
{{/ENABLE_VIRTUAL_NETWORK}}
resource-sets:
  mysql-router:
    cpus: {{ROUTER_CPUS}}
    memory: {{ROUTER_MEM}}
    ports:
      master:
        port: 6446
        advertise: true
        vip:
          prefix: master
          port: 6446
      slave:
        port: 6447
        advertise: true
        vip:
          prefix: slave
          port: 6447
tasks:
  node:
    goal: RUNNING
    cmd: |
      chmod +x ./config-templates/router-init.sh
      ./config-templates/router-init.sh
    readiness-check:
      cmd: |
        mysqladmin -uroot -p123456  -h127.0.0.1 -P6446 ping
        if [ $? = 0 ]; then
          echo "mysql-router is ready"
          exit 0
        else
          echo "mysql-router is not yet ready"
          exit 1
        fi
      interval: 45
      delay: 10
      timeout: 60
    resource-set: mysql-router
    env:
      MYSQL_ROOT_PASSWORD: {{MYSQL_ROOT_PASSWORD}}
      CLUSTERNAME: {{ROUTER_CLUSTERNAME}}
```

```
        configs:
          router-init.sh:
            template: router-init.sh
            dest: router-init.sh

plans:
  deploy:
    strategy: serial
    phases:
      mysql-deploy:
        strategy: serial
        pod: mysql
        steps:
          - default: [[node]]
{{#EXPORTER_ENABLED}}
      exporter-deploy:
        strategy: parallel
        pod: exporter
        steps:
          - default: [[daemon]]
{{/EXPORTER_ENABLED}}
      router-deploy:
        strategy: serial
        pod: router
        steps:
          - default: [[node]]
  localbackup:
    strategy: serial
    phases:
      node:
        strategy: serial
        pod: backup
        steps:
          - default: [[localbackup]]
```

这里解释配置文件中的几个关键部分：

（1）configs 部分是利用 DC/OS SDK 的模板渲染功能将文件中的变量用框架系统变量代替。

（2）{{#ENABLE_VIRTUAL_NETWORK}} 部分是由用户选择是否启用 DC/OS 虚拟 Calico 网络。如果不选择，则启动容器的默认 host 模式（即会占用主机端口）。

（3）{{#EXPORTER_ENABLED}} 部分是由用户选择是否启用 mysql-exporter 监控组件。如果选择，则并发安装 mysql-exporter 组件；如果不选择，则忽略 mysql-exporter 组件。

（4）plans部分配置了两个动作：一个是deploy动作，另一个是localbackup动作。其中，deploy动作在安装框架执行安装命令或者利用DC/OS图形化安装时触发；localbackup在执行相应的命令时才会触发。两者是并列关系，相互之间没有完全的依赖关系。例如，可以用以下命令来启动localbackup的Plan：

```
dcos innodb-cluster plan start localbackup
```

mysql-init.sh的脚本内容如下：

```
#!/usr/bin/env bash
set -m

declare nodeSelf="localhost"

function mysqlExec () {
    local mysqlHost=$1
    local options=$2
    local execSql=${@: 3}

    mysql --connect-timeout 3 -h${mysqlHost} -uroot -p${MYSQL_ROOT_PASSWORD}
${options} -e "$execSql" 2>/dev/null
}


./bootstrap
ln -sf /usr/share/zoneinfo/${TZ} /etc/localtime
serverID=$(( ${POD_INSTANCE_INDEX} + 1 ))
echo -e "MySQL Server ID: $serverID"
#hostIP=${MESOS_CONTAINER_IP}


# Calculate the mysql pool buffer size
#MYSQL_INNODB_BUFFER_POOL_SIZE=$(echo "${NODE_MEM} *
#${MYSQL_INNODB_BUFFER_POOL_SIZE_RATIO}"|bc)
MYSQL_INNODB_BUFFER_POOL_SIZE=`echo | awk "{print $NODE_MEM * $MYSQL_
INNODB_BUFFER_POOL_SIZE_RATIO}"`

# my.cnf for related parameter tuning
cp -v ./config-templates/my.cnf  ./my.cnf

sed -i "s@serverid@$serverID@g" ./my.cnf
sed -i "s@INDEX@${POD_INSTANCE_INDEX}@g" ./my.cnf
sed -i "s@FRAMEWORK_NAME@${FRAMEWORK_NAME}@g" ./my.cnf
```

```
    sed -i "s@maxconn@${MYSQL_MAX_CONNECTIONS}@g" ./my.cnf
    sed -i "s@maxuserconn@${MYSQL_MAX_USER_CONNECTIONS}@g" ./my.cnf
    sed -i "s@buffersize@${MYSQL_INNODB_BUFFER_POOL_SIZE}@g" ./my.cnf
    sed -i "s@timehour@${MYSQL_TIME_HOUR}@g" ./my.cnf
    sed -i "s@expiredays@${MYSQL_EXPIRE_LOGS_DAYS}@g" ./my.cnf
    sed -i "s@timeoutconn@${MYSQL_TIMEOUT}@g" ./my.cnf
    sed -i "s@maxerr@${MYSQL_CONNECT_ERRORS}@g" ./my.cnf

    echo
"report_host=mysql-${POD_INSTANCE_INDEX}-node.${FRAMEWORK_NAME}.autoip.dcos.
thisdcos.directory" >> ./my.cnf
    rm -rf /etc/my.cnf
    cp ./my.cnf /etc/my.cnf
    chown -R mysql:mysql /mnt && chmod -R 777 /tmp && chmod -R 777 /mnt

    # start mysqld as background job 1
    pStatus=$?
    /entrypoint.sh mysqld --log-error &

    if [[ ${pStatus} -ne 0 ]]; then
        echo -e "Failed to start mysqld, process exit code: $pStatus"
        exit ${pStatus}
    else
        until mysqlExec ${nodeSelf} -sN "SELECT 1"; do sleep 3; done
    fi

    #sleep 30
    echo -e "Post-flight status: START"
    #rtnCode=0


    # update the password of user root
    # mysql -uroot -p -e "update user set password=password('123456');"


    # install group_replication plugins

    mysqlExec ${nodeSelf} -sN "\
    SET SQL_LOG_BIN=0;\
    GRANT ALL PRIVILEGES ON *.* TO 'root'@'%' IDENTIFIED BY
'${MYSQL_ROOT_PASSWORD}' WITH GRANT OPTION;\
```

```
    GRANT ALL PRIVILEGES ON mysql_innodb_cluster_metadata.* TO root@'%' WITH
GRANT OPTION;\
    GRANT RELOAD, SHUTDOWN, PROCESS, FILE, SUPER, REPLICATION SLAVE,
REPLICATION CLIENT, CREATE USER ON *.* TO root@'%' WITH GRANT OPTION;\
    FLUSH PRIVILEGES;\
    SET SQL_LOG_BIN=1;"

    # Install MySQL Group Replication MGR
    echo -e "install MySQL Group Replication MGR"

    if [[ ${POD_INSTANCE_INDEX} -eq 0 ]]; then
        echo "POD":${POD_INSTANCE_INDEX}
        mysqlExec ${nodeSelf} -sN "\
        SET SQL_LOG_BIN=0;\
        CREATE USER rpl_user@'%';\
        GRANT REPLICATION SLAVE ON *.* TO rpl_user@'%' IDENTIFIED BY 'rpl_pass';\
        SET SQL_LOG_BIN=1;\
        CHANGE MASTER TO MASTER_USER='rpl_user', MASTER_PASSWORD='rpl_pass' FOR
CHANNEL 'group_replication_recovery';\
        INSTALL PLUGIN group_replication SONAME 'group_replication.so';\
        set global group_replication_bootstrap_group = ON;\
        START GROUP_REPLICATION;\
        set global group_replication_bootstrap_group = OFF;"
    else
        echo "POD":${POD_INSTANCE_INDEX}
        mysqlExec ${nodeSelf} -sN "\
        SET SQL_LOG_BIN=0;\
        CREATE USER rpl_user@'%';\
        GRANT REPLICATION SLAVE ON *.* TO rpl_user@'%' IDENTIFIED BY 'rpl_pass';\
        SET SQL_LOG_BIN=1;\
        CHANGE MASTER TO MASTER_USER='rpl_user', MASTER_PASSWORD='rpl_pass' FOR
CHANNEL 'group_replication_recovery';\
        INSTALL PLUGIN group_replication SONAME 'group_replication.so';\
        set global group_replication_allow_local_disjoint_gtids_join=ON;\
        START GROUP_REPLICATION;"
    fi


    #Only print failure status without exiting, to prevent scenarios where Pod
    #restart cannot start
    if [[ $? -ne 0 ]]; then
```

```
        echo -e "install Group Replication MGR status: FAIL"
        source ./config-templates/rejoin-instance.sh
        #exit $rtnCode
        #fi
    else
        echo -e "install Group Replication MRG status: sucesses"

    fi



    #bring background job 1 to foreground
    echo -e "Bring mysqld process to foreground"
    fg 1
```

mysql-init.sh 文件的主要作用是将创建 MGR 的动作以代码的方式自动化，除此之外，它还有以下几个功能：

（1）set-m 主要的作用是启用作业监视，让容器中的 mysqld 以独立后台进程组的方式运行。

（2）自动判断当前节点是否是 MGR 主节点，并根据 MGR 角色重写 my.cnf 文件。

（3）my.cnf 文件的简单优化。

（4）根据当前容器的运行状态，如容器是否是第一次运行或者是否是重启，来判断是否调用 rejoin-instance.sh 脚本（自动重新加入 MGR 节点）。

rejoin-instance.sh 脚本的功能就是在有网络抖动或 DC/OS 节点重启的情况下，使 MGR 实例自动重新加入 MGR 集群。其脚本内容如下：

```
    #!/usr/bin/env bash

    function mysqlExec () {
        local mysqlHost=$1
        local options=$2
        local execSql=${@: 3}

        mysql  --connect-timeout  3  -h${mysqlHost}  -uroot  -p${MYSQL_ROOT_
PASSWORD} ${options} -e "$execSql" 2>/dev/null
    }



    for n in {0..2};do
      nodeSelf=mysql-$n-node.${FRAMEWORK_NAME}.autoip.dcos.thisdcos.directory
      PRIMARY=`mysqlExec ${nodeSelf} -sN "SHOW STATUS LIKE 'group_replication_
primary_member';" | awk '{print $2}'`
      UUID=`mysqlExec  ${nodeSelf}  -sN  "SHOW  GLOBAL  VARIABLES  LIKE  'server_
```

```
uuid';" | awk '{print $2}'`
      echo -e "PRIMARY:"$PRIMARY
      echo -e "UUID":$UUID
      if [[ $PRIMARY = $UUID ]];then
          MASTERNODE=mysql-$n-node.${FRAMEWORK_NAME}.autoip.dcos.thisdcos.
directory
        echo "MASTERNODE:"$MASTERNODE
        break
      fi
    sleep 1
    done


    cat << EOF > rejoin_cluster.js
    shell.connect('$MASTERNODE', '${MYSQL_ROOT_PASSWORD}')
    var cluster=dba.getCluster()
    cluster.removeInstance('root@mysql-${POD_INSTANCE_INDEX}-node.${FRAMEWORK_
NAME}.autoip.dcos.thisdcos.directory:3306',{'force':'true'})
    cluster.addInstance('root@mysql-${POD_INSTANCE_INDEX}-node.${FRAMEWORK_
NAME}.autoip.dcos.thisdcos.directory:3306', {'password': '123456'})
    EOF

    mysqlsh --no-password --js --file=rejoin_cluster.js
```

router-init.sh 脚本的作用是加入 mysql-router 组件，以代码的方式自动化封装。其脚本内容如下：

```
    #!/usr/bin/env bash

    ./bootstrap
    ln -sf /usr/share/zoneinfo/${TZ} /etc/localtime

    mv mysql-router-8.0.21-el7-x86_64 mysql-router

    # Create MySQL InnoDB Cluster
    echo -e "Create MySQL InnoDB Cluster"


    # Select cluster_name from mysql_innodb_cluster_metadata.cluster
    mysql  -hmysql-0-node.${FRAMEWORK_NAME}.autoip.dcos.thisdcos.directory
-uroot -p${MYSQL_ROOT_PASSWORD} -e "select cluster_name  from mysql_innodb_
cluster_metadata.clusters;" | grep ${CLUSTERNAME}
```

```
    if [[ $? -ne 0 ]] && [[ ${POD_INSTANCE_INDEX} = 0 ]];then
    cat << EOF > init_cluster.js
    shell.connect('root@mysql-0-node.${FRAMEWORK_NAME}.autoip.dcos.thisdcos.
directory:3306', '${MYSQL_ROOT_PASSWORD}')
    var cluster = dba.createCluster('${CLUSTERNAME}', {adoptFromGR: true});
    EOF
    mysqlsh --no-password --js --file=init_cluster.js
    else
        echo -e "mysql cluster has been established or Not router-0-node"
    fi

    echo -e "Generate mysql-router configuration file"
    echo -e "POD_INSTANCE_INDEX:"${POD_INSTANCE_INDEX}

    # If the specified second node (read-only) cannot be operated, it will
    #automatically try to connect to the primary node, Add HealthCheck First

    for n in {0..2};do
        nodeSelf=`mysql
-hmysql-$n-node.${FRAMEWORK_NAME}.autoip.dcos.thisdcos.directory -uroot
-p123456 -e "SELECT ta.* ,tb.MEMBER_HOST,tb.MEMBER_PORT,tb.MEMBER_STATE FROM
performance_schema.global_status
ta,performance_schema.replication_group_members tb  WHERE
ta.VARIABLE_NAME='group_replication_primary_member' and
ta.VARIABLE_VALUE=tb.MEMBER_ID;" | grep "autoip.dcos.thisdcos" | awk '{print
$3}'`
        echo $nodeSelf
        if [ ! -n $nodeSelf ]; then
            echo "MIC Master Node:"$nodeSelf
        fi
    done
    echo -e "run mysql-router"
    ./mysql-router/bin/mysqlrouter --bootstrap root:${MYSQL_ROOT_
PASSWORD}@$nodeSelf:3306 --user=root
    ./mysql-router/bin/mysqlrouter -c /mnt/mesos/sandbox/mysql-router/
mysqlrouter.conf
```

my.cnf 文件中的很多配置都与 MGR 相关，变量会从 Universe 的 config.json 文件获取。其内容如下：

```
    [mysqld]
```

```
skip-host-cache
skip-name-resolve
socket=/var/lib/mysql/mysql.sock
secure-file-priv=/var/lib/mysql-files
user=mysql

# Disabling symbolic-links is recommended to prevent assorted security risks
symbolic-links=0
max_connections = maxconn
max_user_connections = maxuserconn
max_connect_errors = maxerr
innodb_buffer_pool_size = buffersizem
wait_timeout=timeoutconn
interactive_timeout=timeoutconn

datadir=/mnt/mesos/sandbox/MYSQL_DATA


server_id=serverid
gtid_mode=ON
enforce_gtid_consistency=ON
binlog_checksum=NONE

log_bin=binlog
log_slave_updates=ON
binlog_format=ROW
expire_logs_days = expiredays
default-time_zone = '+timehour:00'
master_info_repository=TABLE
relay_log_info_repository=TABLE
slave-skip-errors = 1032,1062

transaction_write_set_extraction=XXHASH64

loose-group_replication_group_name="5c7975ec-0000-11e9-a8c9-0800273906ff"

loose-group_replication_start_on_boot=off

loose-group_replication_local_address="mysql-INDEX-node.FRAMEWORK_NAME.
autoip.dcos.thisdcos.directory:24901"
```

```
    loose-group_replication_group_seeds="mysql-0-node.FRAMEWORK_NAME.autoip.
dcos.thisdcos.directory:24901,mysql-1-node.FRAMEWORK_NAME.autoip.dcos.
thisdcos.directory:24901,mysql-2-node.FRAMEWORK_NAME.autoip.dcos.thisdcos.
directory:24901"
    loose-group_replication_bootstrap_group=off
    loose-group_replication_ip_whitelist='mysql-0-node.FRAMEWORK_NAME.autoip.
dcos.thisdcos.directory,mysql-1-node.FRAMEWORK_NAME.autoip.dcos.thisdcos.
directory,mysql-2-node.FRAMEWORK_NAME.autoip.dcos.thisdcos.directory'

    loose-group_replication_single_primary_mode = on
    disabled_storage_engines = MyISAM,BLACKHOLE,FEDERATED,CSV,ARCHIVE
    report_port=3306
```

demo.sql 文件的内容比较简单，这里省略。

6.5.2 MIC 框架的 Universe 包文件

Universe 包文件主要包括 config.json、marathon.json.mustache、package.json 和 resource.json 文件。

config.json 文件的内容如下：

```
{
  "type": "object",
  "properties": {
    "service": {"type":"object"...},
    "mysql": {"type":"object"...},
    "exporter": {"type":"object"...},
    "route": {"type":"object"...},
    "backup": {"type":"object"...}
  }
}
```

marathon.json.mustache 文件的内容如下：

```
{
  "id": "{{service.name}}",
  "cpus": 1.0,
  "mem": 1024,
  "instances": 1,
  "user": "{{service.user}}",
  "cmd": "export
LD_LIBRARY_PATH=$MESOS_SANDBOX/libmesos-bundle/lib:$LD_LIBRARY_PATH; export
MESOS_NATIVE_JAVA_LIBRARY=$(ls
$MESOS_SANDBOX/libmesos-bundle/lib/libmesos-*.so); export JAVA_HOME=$(ls -d
```

```
$MESOS_SANDBOX/jdk*); export JAVA_HOME=${JAVA_HOME%%/}; export PATH=$(ls -d
$JAVA_HOME/bin):$PATH && export JAVA_OPTS=\"-Xms256M -Xmx512M
-XX:-HeapDumpOnOutOfMemoryError\" && ./bootstrap -resolve=false -template=true
&& ./operator-scheduler/bin/mysql ./operator-scheduler/svc.yml",
      "labels": {
        "DCOS_COMMONS_API_VERSION": "v1",
        "DCOS_COMMONS_UNINSTALL": "true",
        "DCOS_PACKAGE_FRAMEWORK_NAME": "{{service.name}}",
        "MARATHON_SINGLE_INSTANCE_APP": "true",
        "DCOS_SERVICE_NAME": "{{service.name}}",
        "DCOS_SERVICE_PORT_INDEX": "0",
        "DCOS_SERVICE_SCHEME": "http"
      },
      {{#service.service_account_secret}}
      "secrets": {
        "serviceCredential": {
          "source": "{{service.service_account_secret}}"
        }
      },
      {{/service.service_account_secret}}
      "env": {
        "PACKAGE_NAME": "{{package-name}}",
        "PACKAGE_VERSION": "5.0.5-2",
        "PACKAGE_BUILD_TIME_EPOCH_MS": "1552562293172",
        "PACKAGE_BUILD_TIME_STR": "Thu Mar 14 2019 11:18:13 +0000",
        "FRAMEWORK_NAME": "{{service.name}}",
        "SLEEP_DURATION": "{{service.sleep}}",
        "FRAMEWORK_USER": "{{service.user}}",
        "FRAMEWORK_PRINCIPAL": "{{service.service_account}}",
        "FRAMEWORK_LOG_LEVEL": "{{service.log_level}}",
        "MESOS_API_VERSION": "V1",

        {{#service.virtual_network_enabled}}
        "ENABLE_VIRTUAL_NETWORK": "yes",
        "VIRTUAL_NETWORK_NAME": "{{service.virtual_network_name}}",
        "VIRTUAL_NETWORK_PLUGIN_LABELS":
"{{service.virtual_network_plugin_labels}}",
        {{/service.virtual_network_enabled}}

        {{#service.service_account_secret}}
        "DCOS_SERVICE_ACCOUNT_CREDENTIAL": { "secret": "serviceCredential" },
```

```
        "MESOS_MODULES":
"{\"libraries\":[{\"file\":\"libmesos-bundle\/lib\/mesos\/libdcos_security.
so\",\"modules\":[{\"name\":
\"com_mesosphere_dcos_ClassicRPCAuthenticatee\"},{\"name\":\"com_mesosphere_
dcos_http_Authenticatee\",\"parameters\":[{\"key\":\"jwt_exp_timeout\",
\"value\":\"5mins\"},{\"key\":\"preemptive_refresh_duration\",\"value\":
\"30mins\"}]}]}]}",
        "MESOS_AUTHENTICATEE": "com_mesosphere_dcos_ClassicRPCAuthenticatee",
        "MESOS_HTTP_AUTHENTICATEE": "com_mesosphere_dcos_http_Authenticatee",
        {{/service.service_account_secret}}

        "BOOTSTRAP_URI": "{{resource.assets.uris.bootstrap-zip}}",
        "EXECUTOR_URI": "{{resource.assets.uris.executor-zip}}",
        "JAVA_URI": "{{resource.assets.uris.jre-tar-gz}}",
        "LIBMESOS_URI": "{{resource.assets.uris.libmesos-bundle-tar-gz}}",
        "MYSQL_SHELL_URI": "{{resource.assets.uris.mysql-shell-uri}}",
        "MYSQL_INIT_URI": "{{resource.assets.uris.mysql-init-uri}}",
        "ROUTER_INIT_URI": "{{resource.assets.uris.router-init-uri}}",
        "DEMO_SQL_URI": "{{resource.assets.uris.demo-sql-uri}}",
        "MYSQL_ROUTER_URI": "{{resource.assets.uris.mysql-router-uri}}",
        "MYSQL_CNF_URI":"{{resource.assets.uris.mysql-cnf-uri}}",
        "MYSQL_XTRABACKUP_URI":
"{{resource.assets.uris.mysql-xtrabackup-tar-gz}}",
        "MYSQL_XTRABACKUP_SHELL_URI":
"{{resource.assets.uris.mysql-xtrabackup-shell-uri}}",
        "MYSQL_MYSQLDUMP_SHELL_URI":
"{{resource.assets.uris.mysql-mysqldump-shell-uri}}",
        "REJOIN_SHELL_URI": "{{resource.assets.uris.rejoin-shell-uri}}",
        "CRONYUM_URI": "{{resource.assets.uris.cronyum-tar-gz}}",
        "MYSQL_IMAGE": "{{resource.assets.container.docker.mysql_cluster}}",
        "EXPORTER_IMAGE":
"{{resource.assets.container.docker.mysql_exporter}}",

        "NODE_COUNT": "{{mysql.common.count}}",
        "NODE_CPUS": "{{mysql.common.cpus}}",
        "NODE_MEM": "{{mysql.common.memory}}",
        "NODE_DISK": "{{mysql.common.disk}}",
        "NODE_DISK_TYPE": "{{mysql.common.disk_type}}",
        "NODE_PLACEMENT": "{{mysql.common.placement}}",

        "ROUTER_CPUS": "{{router.cpus}}",
```

```
        "ROUTER_MEM": "{{router.memory}}",
        "ROUTER_CLUSTERNAME": "{{router.clustername}}",
        "ROUTER_PLACEMENT": "{{router.placement}}",

        "BACKUP_CPUS": "{{backup.cpus}}",
        "BACKUP_MEM": "{{backup.memory}}",
        "BACKUP_DATABASENAME": "{{backup.databasename}}",

        "BACKUP_RESTORE_CPUS": "{{mysql.backup_restore.cpus}}",
        "BACKUP_RESTORE_MEM": "{{mysql.backup_restore.memory}}",
        "STHREE_COMPATIBLE_URI": "{{mysql.backup_restore.endpoint}}",
        "BACKUP_STORE_SERVICENAME": "{{mysql.backup_restore.servicename}}",
        "ACCESS_KEY_ID": "{{mysql.backup_restore.s3_username}}",
        "SECRET_ACCESS_KEY": "{{mysql.backup_restore.s3_password}}",
        "BACKUP_DISK_TYPE": "{{mysql.backup_restore.backup_disk_type}}",
        "BACKUP_NODE_DISK": "{{mysql.backup_restore.backup_disk}}",
        "RESTORE_DISK_TYPE": "{{mysql.backup_restore.restore_disk_type}}",
        "RESTORE_NODE_DISK": "{{mysql.backup_restore.restore_disk}}",
        "BACKUP_FULL_FROM": "yymmddhhmmss-full",
        "RESTORE_TO": "yymmddhhmmss-inc",


        "MYSQL_INNODB_BUFFER_POOL_SIZE_RATIO":
"{{mysql.server.innodb_buffer_pool_size_ratio}}",
        "MYSQL_ROOT_PASSWORD": "{{mysql.server.root_password}}",
        "MYSQL_CHAR_SET_SERVER": "{{mysql.server.character_set_server}}",
        "MYSQL_MAX_CONNECTIONS": "{{mysql.server.max_connections}}",
        "MYSQL_MAX_USER_CONNECTIONS": "{{mysql.server.max_user_connections}}",
        "MYSQL_TIME_HOUR": "{{mysql.server.time_hour}}",
        "MYSQL_EXPIRE_LOGS_DAYS": "{{mysql.server.expire_logs_days}}",
        "MYSQL_TIMEOUT": "{{mysql.server.timeout}}",
        "MYSQL_CONNECT_ERRORS": "{{mysql.server.max_connect_errors}}",


        "EXPORTER_ENABLED": "{{exporter.is_enabled}}",
        "EXPORTER_CPUS": "{{exporter.cpus}}",
        "EXPORTER_MEM": "{{exporter.memory}}",
        "EXPORTER_PLACEMENT": "{{exporter.placement}}"

      },
      "uris": [
```

```
        "{{resource.assets.uris.bootstrap-zip}}",
        "{{resource.assets.uris.jre-tar-gz}}",
        "{{resource.assets.uris.scheduler-zip}}",
        "{{resource.assets.uris.libmesos-bundle-tar-gz}}"
      ],
      "upgradeStrategy":{
        "minimumHealthCapacity": 0,
        "maximumOverCapacity": 0
      },
      "healthChecks": [
        {
          "protocol": "MESOS_HTTP",
          "path": "/v1/health",
          "gracePeriodSeconds": 900,
          "intervalSeconds": 30,
          "portIndex": 0,
          "timeoutSeconds": 30,
          "maxConsecutiveFailures": 0
        }
      ],
        "portDefinitions": [
        {
          "port": 0,
          "protocol": "tcp",
          "name": "api",
          "labels": { "VIP_0": "/api.{{service.name}}:3306" }
        }
      ]
    }
```

package.json 文件的内容如下：

```
    {
      "assets": {
        "uris": {
          "jre-tar-gz":
"https://example-external-project.oss-cn-hangzhou.aliyuncs.com/dcos-frameworks/
dcos-commons/openjdk-jre-8u212b03-hotspot-linux-x64.tar.gz",
          "libmesos-bundle-tar-gz":
"https://example-external-project.oss-cn-hangzhou.aliyuncs.com/dcos-frameworks/
dcos-commons/libmesos-bundle-1.12.0.tar.gz",
          "bootstrap-zip":
```

```
"https://example-external-project.oss-cn-hangzhou.aliyuncs.com/dcos-frameworks/dcos-commons/bootstrap.zip",
          "scheduler-zip":
"{{artifact-dir}}/dcos-http-mysql-innodb-cluster/operator-scheduler.zip",

"mysql-shell-tar-gz":"https://example-external-project.oss-cn-hangzhou.aliyuncs.com/dcos-frameworks/mysql-innodb-cluster/mysql-shell-8.0.22-linux-glibc2.12-x86-64bit.tar.gz",

"mysql-init-uri":"https://example-external-project.oss-cn-hangzhou.aliyuncs.com/dcos-frameworks/mysql-innodb-cluster/mysql-init.sh",
          "router-init-uri":
"https://example-external-project.oss-cn-hangzhou.aliyuncs.com/dcos-frameworks/mysql-innodb-cluster/router-init.sh",
          "demo-sql-uri":
"https://example-external-project.oss-cn-hangzhou.aliyuncs.com/dcos-frameworks/mysql-innodb-cluster/demo.sql",

"mysql-router-uri":"https://example-external-project.oss-cn-hangzhou.aliyuncs.com/dcos-frameworks/mysql-innodb-cluster/mysql-router-8.0.21-el7-x86_64.tar.gz",
          "mysql-shell-uri":
"https://example-external-project.oss-cn-hangzhou.aliyuncs.com/dcos-frameworks/mysql-innodb-cluster/mysql-shell-8.0.21-1.el7.x86_64.rpm",
          "mysql-cnf-uri":
"https://example-external-project.oss-cn-hangzhou.aliyuncs.com/dcos-frameworks/mysql-innodb-cluster/my.cnf",
          "mysql-xtrabackup-tar-gz":
"https://example-external-project.oss-cn-hangzhou.aliyuncs.com/dcos-frameworks/mysql-innodb-cluster/percona-xtrabackup-2.4.21-Linux-x86_64.glibc2.12.tar.gz",
          "mysql-xtrabackup-shell-uri":
"https://example-external-project.oss-cn-hangzhou.aliyuncs.com/dcos-frameworks/mysql-innodb-cluster/backupself.sh",
          "rejoin-shell-uri": "https:
//example-external-project.oss-cn-hangzhou.aliyuncs.com/dcos-frameworks/mysql-innodb-cluster/rejoin_instance.sh"
        },
         "container": {
          "docker": {
            "mysql_cluster": "mysql/mysql-server:5.7.33-1.1.19",
```

```
            "mysql_exporter": "prom/mysqld-exporter"
          }
        }
      },
      "images": {
       "icon-small":
"https://downloads.mesosphere.com/assets/universe/000/mysql-icon-small.png",
       "icon-medium":
"https://downloads.mesosphere.com/assets/universe/000/mysql-icon-medium.png",
       "icon-large":
"https://downloads.mesosphere.com/assets/universe/000/mysql-icon-large.png"
      },
      "cli": {
        "binaries": {
          "darwin": {
            "x86-64": {
              "contentHash": [
                {
                  "algo": "sha256",
                  "value": "{{sha256:dcos-service-cli-darwin}}"
                }
              ],
              "kind": "executable",
              "url": "{{artifact-dir}}/dcos-service-cli-darwin"
            }
          },
          "linux": {
            "x86-64": {
              "contentHash": [
                {
                  "algo": "sha256",
                  "value": "{{sha256:dcos-service-cli-linux}}"
                }
              ],
              "kind": "executable",
              "url": "{{artifact-dir}}/dcos-service-cli-linux"
            }
          },
          "windows": {
            "x86-64": {
              "contentHash": [
```

```
            {
              "algo": "sha256",
              "value": "{{sha256:dcos-service-cli.exe}}"
            }
          ],
          "kind": "executable",
          "url": "{{artifact-dir}}/dcos-service-cli.exe"
        }
      }
    }
  }
}
```

resource.json 文件的内容如下：

```
{
  "assets": {
    "uris": {
      "jre-tar-gz":
"https://example-external-project.oss-cn-hangzhou.aliyuncs.com/dcos-frameworks/
dcos-commons/openjdk-jre-8u212b03-hotspot-linux-x64.tar.gz",
      "libmesos-bundle-tar-gz":
"https://example-external-project.oss-cn-hangzhou.aliyuncs.com/dcos-
frameworks/dcos-commons/libmesos-bundle-1.12.0.tar.gz",
      "bootstrap-zip":
"https://example-external-project.oss-cn-hangzhou.aliyuncs.com/dcos-
frameworks/dcos-commons/bootstrap.zip",

      "scheduler-zip":
"{{artifact-dir}}/dcos-http-mysql-innodb-cluster/operator-scheduler.zip",


"mysql-shell-tar-gz":"https://example-external-project.oss-cn-hangzhou.aliyuncs.
com/dcos-frameworks/mysql-innodb-cluster/mysql-shell-8.0.22-linux-glibc2.12-
x86-64bit.tar.gz",


"mysql-init-uri":"https://example-external-project.oss-cn-hangzhou.aliyuncs.
com/dcos-frameworks/mysql-innodb-cluster/mysql-init.sh",
      "router-init-uri":
"https://example-external-project.oss-cn-hangzhou.aliyuncs.com/dcos-frameworks/
mysql-innodb-cluster/router-init.sh",
```

```
          "demo-sql-uri":
"https://example-external-project.oss-cn-hangzhou.aliyuncs.com/dcos-
frameworks/mysql-innodb-cluster/demo.sql",

"mysql-router-uri":"https://example-external-project.oss-cn-hangzhou.
aliyuncs.com/dcos-frameworks/mysql-innodb-cluster/mysql-router-8.0.21-
el7-x86_64.tar.gz",

          "mysql-shell-uri":
"https://example-external-project.oss-cn-hangzhou.aliyuncs.com/dcos-
frameworks/mysql-innodb-cluster/mysql-shell-8.0.21-1.el7.x86_64.rpm",

          "mysql-cnf-uri":
"https://example-external-project.oss-cn-hangzhou.aliyuncs.com/dcos-
frameworks/mysql-innodb-cluster/my.cnf",

          "mysql-xtrabackup-tar-gz":
"https://example-external-project.oss-cn-hangzhou.aliyuncs.com/dcos-
frameworks/mysql-innodb-cluster/percona-xtrabackup-2.4.21-Linux-x86_64.
glibc2.12.tar.gz",

          "mysql-xtrabackup-shell-uri":
"https://example-external-project.oss-cn-hangzhou.aliyuncs.com/dcos-
frameworks/mysql-innodb-cluster/backupself.sh",

          "rejoin-shell-uri":
"https://example-external-project.oss-cn-hangzhou.aliyuncs.com/dcos-
frameworks/mysql-innodb-cluster/rejoin_instance.sh"
        },
         "container": {
          "docker": {
            "mysql_cluster": "mysql/mysql-server:5.7.33-1.1.19",
            "mysql_exporter": "prom/mysqld-exporter"
          }
        }
      },
      "images": {
        "icon-small":
"https://downloads.mesosphere.com/assets/universe/000/mysql-icon-small.png",

        "icon-medium":
```

```
"https://downloads.mesosphere.com/assets/universe/000/mysql-icon-medium.png",

    "icon-large":
"https://downloads.mesosphere.com/assets/universe/000/mysql-icon-large.png"
    },
    "cli": {
      "binaries": {
        "darwin": {
          "x86-64": {
            "contentHash": [
              {
                "algo": "sha256",
                "value": "{{sha256:dcos-service-cli-darwin}}"
              }
            ],
            "kind": "executable",
            "url": "{{artifact-dir}}/dcos-service-cli-darwin"
          }
        },
        "linux": {
          "x86-64": {
            "contentHash": [
              {
                "algo": "sha256",
                "value": "{{sha256:dcos-service-cli-linux}}"
              }
            ],
            "kind": "executable",
            "url": "{{artifact-dir}}/dcos-service-cli-linux"
          }
        },
        "windows": {
          "x86-64": {
            "contentHash": [
              {
                "algo": "sha256",
                "value": "{{sha256:dcos-service-cli.exe}}"
              }
            ],
            "kind": "executable",
            "url": "{{artifact-dir}}/dcos-service-cli.exe"
```

```
            }
          }
        }
      }
    }
```

另外，与编译和版本相关的文件还有 version.sh 和 build.sh。其中，version.sh 文件的内容如下：

```
#!/usr/bin/env bash
export TEMPLATE_MYSQL_VERSION="5.7.25"
export TEMPLATE_DCOS_SDK_VERSION="0.56.2"
```

version.sh 文件主要是将 TEMPLATE_MYSQL_VERSION 和 TEMPLATE_DCOS_SDK_VERSION 变量作为 MIC 框架的环境变量输入。

build.sh 文件的内容如下：

```
#!/usr/bin/env bash
set -e

FRAMEWORK_DIR="$( cd "$( dirname "${BASH_SOURCE[0]}" )" && pwd )"
REPO_ROOT_DIR=$(dirname $(dirname $FRAMEWORK_DIR))

# grab TEMPLATE_x vars for use in universe template:
source $FRAMEWORK_DIR/versions.sh

# Build/test scheduler.zip/CLIs/setup-helper.zip
${REPO_ROOT_DIR}/gradlew -p ${FRAMEWORK_DIR} check distZip
$FRAMEWORK_DIR/cli/build.sh

# Build package with our scheduler.zip/CLIs/setup-helper.zip and the SDK
artifacts we built:
$REPO_ROOT_DIR/tools/build_package.sh \
    mysql-innodb-cluster \
    $FRAMEWORK_DIR \
    -a "$FRAMEWORK_DIR/build/distributions/operator-scheduler.zip" \
    -a "$FRAMEWORK_DIR/cli/dcos-service-cli-linux" \
    -a "$FRAMEWORK_DIR/cli/dcos-service-cli-darwin" \
    -a "$FRAMEWORK_DIR/cli/dcos-service-cli.exe" \
    $@
```

然后，打包和添加本地 Universe 仓库，命令如下：

```
./frameworks/mysql-innodb-cluster/build.sh local
```

最后，通过 DC/OS 的安装服务界面来安装 MIC 框架。安装成功后，其 innodb-cluter 框架服务正常启动界面如图 6.7 所示（整个框架并没有开启 mysql-exporter 组件，是最小化运行）。

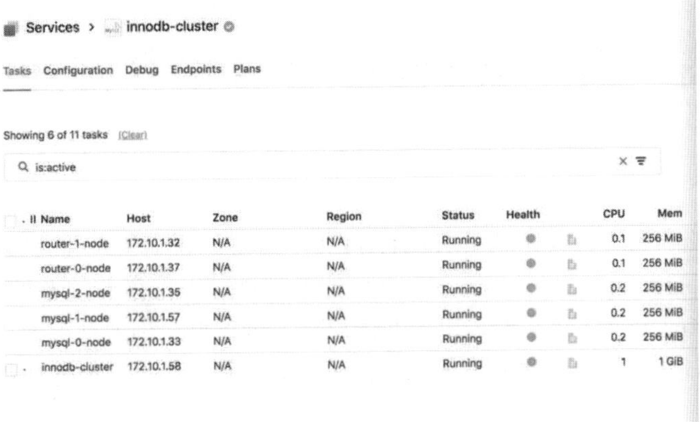

Services > innodb-cluster

Tasks Configuration Debug Endpoints Plans

Showing 6 of 11 tasks (Clear)

is:active

Name	Host	Zone	Region	Status	Health	CPU	Mem
router-1-node	172.10.1.32	N/A	N/A	Running		0.1	256 MiB
router-0-node	172.10.1.37	N/A	N/A	Running		0.1	256 MiB
mysql-2-node	172.10.1.35	N/A	N/A	Running		0.2	256 MiB
mysql-1-node	172.10.1.57	N/A	N/A	Running		0.2	256 MiB
mysql-0-node	172.10.1.33	N/A	N/A	Running		0.2	256 MiB
innodb-cluster	172.10.1.58	N/A	N/A	Running		1	1 GiB

图 6.7 innodb-cluter 框架服务正常启动界面

后续持续优化工作：

（1）系统自带的是开源 mysql-exporter，但它在实际生产环境中并不能直观地反映各 MGR 集群的详细状态，如 MGR 集群是否正常、有没有 MySQL 节点实例失效、MGR 节点实例的数量等。因此后续我们需要用 Go 语言编写基于 mgr-exporter 的监控组件，以便 Pormetheus 能够更有效地监控 MGR 集群的各项指标，并及时发出告警。

（2）基于节约资源的考虑，目前 MIC 框架服务的 MGR 集群是一主两从（即 3 节点）。但从官方文档得知，MGR 集群是支持 9 节点的，所以这个应该配置成可选项，其实例可以选择 3 ~ 9（节点越多，容忍出现 Crash 的节点就更多，集群的高可用性也越高），方便用户根据实际资源情况分配实际节点数量。

6.6 小　　结

本章通过分析 DC/OS SDK 工具的使用，学习其 SDK 的设计和思路（其实后续的 KUDO 工具也沿袭了此设计）；通过阅读源码，包括 Plan 工作流和 Universe 的设计，可以借鉴和学习其中的思想。不过随着 Kubernetes 的流行，云原生生态也向其偏移，在业务环境和云计算环境下，见得最多的应该还是 Kubernetes 集群，后续的工作和学习重心建议放在 Kubernetes 上。

第 7 章　Kubernetes 的基础知识

Kubernetes 系统的 API Server 基于 HTTP/HTTPS 接收并响应客户端的操作请求，它提供了一种“基于资源（resource-based）”的 Restful 风格的编程接口，将集群的各种组件都抽象成标准的 REST 资源。了解 Kubernetes 的基础知识且阅读源码，可以更深刻地理解 Kubernetes 的核心设计。这些基础知识对于日常工作中二次开发 Kubernetes，是非常重要的，希望读者重点掌握。

7.1　利用 kind 搭建 Kubernetes 本地开发环境

演示 client-go 需要一个 Kubernetes 集群环境，作者这里就以本地 Mac 环境演示，通过 kind 来安装一个本地的 Kubernetes 集群。

首先，提前开启 Docker 进程，可以查看 Docker 的版本号，命令如下：

```
docker -version
```

命令显示结果如下：

```
Docker version 20.10.11, build dea9396
```

确定 Docker 已启动后，在此基础上安装 kind，命令如下：

```
brew install kind
```

然后查看 kind 版本，命令如下：

```
kind --version
```

命令显示结果如下：

```
kind version 0.13.0
```

众所周知，Kubernetes 的 master 主要起资源调度作用，node 节点是真正执行任务资源的，所以这里为了模拟完整的 Kubernetes 集群，也安装一个 1 master、3 node 的 Kubernetes 集群，并且安装一个本地 local-registry，以方便推送本地的 Docker Image。可以用一个完整的 Shell 脚本实现，其内容如下：

在 /tmp/ 目录下创建一个 kind.yaml 文件，其内容如下：

```
kind: Cluster
apiVersion: kind.x-k8s.io/v1alpha4
nodes:
  - role: control-plane
```

```
- role: worker
- role: worker
- role: worker
```

这里指定 Kubernetes 版本为 1.20.7，命令如下：

```
kind create cluster --image
kindest/node:v1.20.7@sha256:cbeaf907fc78ac97ce7b625e4bf0de16e3ea725daf6b04f93
0bd14c67c671ff9 --config kind.yaml
```

注：不手动使用 --name 指定集群名称时，集群名称默认为 kind。

前面默认创建了一个名为 kind 的集群，如果创建了多个集群（可以使用 --name 指定集群名称），那么使用 kubectl 命令时要用 --context 指定集群名称，命令如下：

```
kubectl  cluster-info --context kind
```

如果有且只有一个 kind 集群，那么后面的命令可以省略，即

```
kubectl cluster-info
```

命令显示结果如下：

```
Kubernetes control plane is running at https://127.0.0.1:57992
KubeDNS is running at https://127.0.0.1:57992/api/v1/namespaces/kube-
system/services/kube-dns:dns/proxy

To further debug and diagnose cluster problems, use 'kubectl cluster-info
dump'.
```

然后查看名为 kind 集群的 node 分布情况，命令如下：

```
kubectl get node
```

命令显示结果如下：

```
NAME                 STATUS   ROLES                  AGE   VERSION
kind-control-plane   Ready    control-plane,master   13m   v1.20.7
kind-worker          Ready    <none>                 12m   v1.20.7
kind-worker2         Ready    <none>                 12m   v1.20.7
kind-worker3         Ready    <none>                 12m   v1.20.7
```

再查看相应系统自带的 Pod，命令如下：

```
kubectl get pod -n kube-system
```

命令显示结果如下：

```
NAME                          READY   STATUS    RESTARTS   AGE
coredns-74ff55c5b-2blkq        1/1    Running      0       14m
coredns-74ff55c5b-pn7nd        1/1    Running      0       14m
etcd-kind-control-plane        1/1    Running      0       14m
```

```
kindnet-2c7jn                               1/1     Running   0     14m
kindnet-mhmd9                               1/1     Running   0     14m
kindnet-q2r8j                               1/1     Running   0     14m
kindnet-vsqkt                               1/1     Running   0     14m
kube-apiserver-kind-control-plane           1/1     Running   0     14m
kube-controller-manager-kind-control-plane  1/1     Running   0     14m
kube-proxy-8t2tx                            1/1     Running   0     14m
kube-proxy-97xw4                            1/1     Running   0     14m
kube-proxy-gfrd5                            1/1     Running   0     14m
kube-proxy-p4qnk                            1/1     Running   0     14m
kube-scheduler-kind-control-plane           1/1     Running   0     14m
```

一般来说，在 Kubernetes 内部署应用需要先把容器镜像推送到镜像仓库中，这样在本地开发时相对来说会比较麻烦，特别是当镜像比较大时，往返会有两次网络消耗，为了解决这个问题，可以使用 kind load 镜像的功能直接把镜像加载到集群中。

所以这里需要提前导入需要的镜像，命令如下：

```
kind load docker-image nginx:1.14.2 --name kind
```

命令显示结果如下：

```
    Image: "nginx:1.14.2" with ID
"sha256:7bbc8783b8ecfdb6453396805cc0fb5fcdaf1b16cbb907c8ab1b8685732d50a4" not
yet present on node "kind-control-plane", loading...
    Image: "nginx:1.14.2" with ID
"sha256:7bbc8783b8ecfdb6453396805cc0fb5fcdaf1b16cbb907c8ab1b8685732d50a4" not
yet present on node "kind-worker", loading...
    Image: "nginx:1.14.2" with ID
"sha256:7bbc8783b8ecfdb6453396805cc0fb5fcdaf1b16cbb907c8ab1b8685732d50a4" not
yet present on node "kind-worker2", loading...
    Image: "nginx:1.14.2" with ID
"sha256:7bbc8783b8ecfdb6453396805cc0fb5fcdaf1b16cbb907c8ab1b8685732d50a4" not
yet present on node "kind-worker3", loading...
```

从以上显示结果可知，操作是成功的。

Kubernetes 集群搭建成功后，需要写个简单的 yaml 文件来验证 Kubernetes 集群的正常功能。/tmp/test-nginx.yaml 文件内容如下：

```
apiVersion: apps/v1
kind: Deployment
metadata:
  name: nginx-deployment
  labels:
    app: nginx
spec:
```

```
  replicas: 3
  selector:
    matchLabels:
      app: nginx
  template:
    metadata:
      labels:
        app: nginx
    spec:
      containers:
        - name: nginx
          image: nginx:1.14.2
          ports:
            - containerPort: 80
```

用命令查看容器详情：

```
docker ps
```

命令显示结果如下：

```
CONTAINER ID   IMAGE                         COMMAND                    CREATED
STATUS          PORTS                          NAMES
1200f017b5c4   kindest/node:v1.20.7   "/usr/local/bin/entr…"   23 hours ago
Up 23 hours    127.0.0.1:57992->6443/tcp    kind-control-plane
739d5343fd12   kindest/node:v1.20.7   "/usr/local/bin/entr…"   23 hours ago
Up 23 hours                                   kind-worker
dec3d91cdb61   kindest/node:v1.20.7   "/usr/local/bin/entr…"   23 hours ago
Up 23 hours                                   kind-worker2
e848f8ce7a4f   kindest/node:v1.20.7 "/usr/local/bin/entr…"   23 hours ago
Up 23 hours                                   kind-worker3
```

随便进入一个容器，如名为 kind-worker3 的容器，命令如下：

```
docker exec -ti kind-worker3 /bin/bash
```

用 crictl image 命令查看镜像，命令显示结果如下：

```
IMAGE                                       TAG                    IMAGE ID        SIZE
docker.io/kindest/kindnetd          v20210326-1e038dc5   f37b7c809e5dc    54.8MB
docker.io/library/nginx                     1.14.2           7bbc8783b8ecf    107MB
docker.io/rancher/local-path-provisioner v0.0.14             2b703ea309660    12.3MB
k8s.gcr.io/build-image/debian-base          v2.1.0           3cc9c70b44747    22.8MB
k8s.gcr.io/coredns                          1.7.0            db91994f4ee8f    12.8MB
k8s.gcr.io/etcd                             3.4.13-0         05b738aa1bc63    135MB
k8s.gcr.io/kube-apiserver                   v1.20.7          0b0cef4cd861d    114MB
k8s.gcr.io/kube-controller-manager          v1.20.7          2d2dc842de585    109MB
```

```
k8s.gcr.io/kube-proxy              v1.20.7        825a3d5895492    119MB
k8s.gcr.io/kube-scheduler    v1.20.7     ccc0dc44acc58        45.3MB
k8s.gcr.io/pause              3.5         f7ff3c4042631         253kB
```

从以上结果可以看到，下面的 nginx:1.14.2 镜像是真实存在的。

```
docker.io/library/nginx    1.14.2      7bbc8783b8ecf        107MB
```

由于这里没有做服务暴露，因此不能直接访问对应的服务，但可以用 kubectl 提供的端口转发功能将流量从本地转发给 Kubernetes 集群，命令如下：

```
kubectl port-forward nginx-deployment-66b6c48dd5-f9fhk 30080:80
```

命令显示结果如下：

```
Forwarding from 127.0.0.1:30080 -> 80
Forwarding from [::1]:30080 -> 80
Handling connection for 30080
Handling connection for 30080
```

然后可以用浏览器访问 http://127.0.0.1:30080，结果会弹出正常的 nginx 界面。这表示 Kubernetes 集群是正常的，也能正常对外提供服务；后续就用集群来演示。

kind 的其他常规操作如下：

查看集群：

```
kind get clusters
```

删除 kind 集群：

```
kind delete cluster --name kind
```

7.2 Kubernetes 的核心数据结构

Kubernetes 对象模型通常由 TypeMeta 和 MetaData 组成，下面介绍 TypeMeta 和 MetaData 两种对象模型。

7.2.1 TypeMeta

TypeMeta 是 Kubernetes 对象的最基本定义，它通过引入 GKV（Group Kind Version）定义了一个对象类型。

在整个 Kubernetes 庞大而复杂的系统中，只有资源是远远不够的，还要包括以下内容。

（1）Group：资源组，Kubernetes API Server 中也可称其为 API Group。

Kubernetes 系统支持多个 Group，每个 Group 支持多个 Version，每个 Version 支持多个 Resource，其中部分资源同时拥有自己的子资源（SubResource）。例如，Deployment 资源拥有 Status 子资源。资源组、资源版本、资源、子资源的完整表现形式为 {group}/{version}/{resource}/{subsource}。以常用的 Deployment 资源为例，其完整表现形式为 apps/v1/deployment/status。

（2）Version：资源版本，在 Kubernetes API Server 中也可称为 API Versions。

Kubernetes 的资源版本可以分为 3 种，分别为 Alpha、Beta、Stable，它们之间的迭代顺序为 Alpha → Beta → Stable，通常用来表示软件测试中的 3 个阶段。其中，Alpha 是第 1 个阶段，一般用于内部测试；Beta 是第 2 个阶段，该版本已经修复了大部分不完善之处，但仍有可能存在缺陷和漏洞，一般由特定的用户群来测试；Stable 是第 3 个阶段，表示此时产品达到了一定的成熟度，可以稳定运行。常见的多是 Stable 版本，其命名一般为 v1、v2 和 v3。

每个资源都至少有两个版本，分别是外部版本和内部版本。外部版本用于对外暴露给用户请求的接口所使用的资源对象。内部版本不对外暴露，仅在 Kubernetes API Server 内部使用。

（3）Resource：资源，在 Kubernetes API Server 中也可以称为 API Resource，其表现形式为 {group}/{version},Kind={kind}。例如，apps/v1,Kind=Deployment。

Resource 资源在 Kubernetes 中的重要性是不言而喻的，常见的 pod、service、deployment 等都是资源。

每个资源都拥有一定数量的资源操作方法（Verbs），资源操作方法用于 etcd 集群存储中对资源对象的增、删、改、查操作。目前 Kubernetes 系统支持 8 种资源操作方法，分别为 create、delete、deletecollection、get、list、patch、update 和 watch。

Kubernetes 资源也可以分成两种，分别是内置资源和自定义资源，开发者通过 CRD（Custom Resource Definitions）可实现自定义资源，允许用户将自己定义的资源添加到 Kubernetes 资源中，并像使用 Kubernetes 内置资源一样使用。

（4）Kind: 资源种类，用于定义一个对象的基本类型，如 Node、Pod、Deployment 等。

常用的 Kubernetes 资源结构见表 7.1。

表 7.1 常用的 Kubernetes 资源结构

资源结构名称	简　称	说　明
GroupVersionResource	GVR	描述资源组、资源版本及资源
GroupVersion	GV	描述资源组、资源版本
GroupResource	GR	描述资源组及资源
GroupVersionKind	GVK	描述资源组、资源版本及资源种类
GroupKind	GK	描述资源组及资源种类
GroupVersions	GVS	描述资源组内多个资源版本

7.2.2 MetaData

TypeMeta 定义了“我是什么”，MetaData 定义了“我是谁”。为方便管理，Kubernetes 将不同用户或不同业务的对象用不同的 Namespace 隔离。MetaData 中有两个最重要的属性——Namespace 和 Name，分别用于定义对象的命名空间及名字，这两个属性唯一定义了某个对象实例。

前面说过，所有对象都会以 API 的形式发布供用户访问，TypeMeta、Namespace 和 Name 唯一确定了对象所在的 API 访问路径。

MetaData 的对象有以下几种。

1. Label

在传统的面向对象设计系统中，对象组合的方法通常是内嵌或引用，即将对象 A 内嵌到对象 B 中，或者将对象 A 的 ID 内嵌到对象 B 中。这种设计的弊端是对象关系是固化的。一个对象可能与多个其他对象发生关联，如果该对象发生变更，系统需要遍历所有与其关联的对象并做修改。

Kubernetes 采用了更巧妙的方式管理对象之间的松耦合关系，其依赖的就是 Label 和 Selector。Label，顾名思义就是给对象打标签，一个对象可以有任意对标签，其存在形式是键值对。与名字和 UID 不同，标签不需要独一无二，多个对象可以共用一个标签，每个对象可以有多组标签。

Label 定义了这些对象的可识别属性，Kubernetes API 支持以 Label 作为过滤条件查询对象。因此 Label 通常用最简形式定义：

```
apiVersion: apps/v1
kind: Deployment
metadata:
  name: nginx-deployment
  labels:
    app: nginx
spec:
  replicas: 3
  selector:
    matchLabels:
      app: nginx
  template:
    metadata:
      labels:
        app: nginx
```

2. Annotation

Annotation 与 Label 一样用键值对来定义，但其功能与 Label 不同，所以在用法上也有不同的原则，API 也不支持用 Annotation 做条件过滤（Annotation 不能用于标识及选择对象）。虽然 Kubernetes 把对象做了很好的抽象，但在实际运用中特别是生产化落地过程中，总是需要保存一些在对象内置属性中无法保存的信息，Annotation 属性的作用就是为了满足这类需求，事实上 Annotation 是对象的属性扩展。kubernetes 社区在开发新功能，需要对象发生变更之前，往往会先把需要变更的属性放在 Annotation 中，当功能经历完实验阶段再将其移至正式属性中。

Annotation 作为属性扩展，更多是面向系统管理员和开发人员的，因此 Annotation 需要像其他属性一样做合理归类。与 Java 开发中的包名设计类似，通常需要将系统以不同的功能规划为不同的 Annotation Namespace，其键应以如下形式存在：<namespace>/key:value。如 Annotation 用于 Nginx Ingress Controller 蓝绿发布，代码如下：

```
apiVersion: extensions/v1beta1
```

```
kind: Ingress
metadata:
  annotations:
    kubernetes.io/ingress.class: nginx
    nginx.ingress.kubernetes.io/canary: "true"
    nginx.ingress.kubernetes.io/canary-weight: "50"
  labels:
    app: echoserverv2
  name: echoserverv2
  namespace: echoserver
```

需要注意的是，Nginx Ingress Controller 目前不支持 HTTP 头的匹配，也不支持 rewrite 规则和限流，如果想在 Nginx Ingress Controller 中提供这些功能，就需要扩展 Annotation 功能，这使得 Annotation 越来越复杂，越来越难以维护。

3. ResourceVersion

在多线程操作共享资源时，为保证数据的一致性，通常需要在对象进行访问时加锁，以确保在一个线程访问该对象时，其他线程无法修改该对象。排他锁的存在可以确保某一对象在同一时刻只有一个线程在修改，但其排他的特性会让其他线程等待锁，使得系统整体效率显著降低。

可以将 ResourceVersion 看作一种乐观锁，每个对象在任意时刻都有其 ResourceVersion，当 Kubernetes 对象被客户端读取以后，ResourceVersion 信息也被一并读取。客户端更改对象并回写 APIServer 时，ResourceVersion 会被增加，同时 APIServer 需要确保回写的版本比服务器端当前版本高，在回写成功后，服务器端的版本会更新为新的 ResourceVersion。因此，当两个线程同时访问某个对象时，假设它们获取的对象 ResourceVersion 为 1。紧接着第一个线程修改了对象，ResourceVersion 会变为 2，回写至 APIServer 以后，该对象服务器端 ResourceVersion 会被更新为 2。此时如果第二个线程对该对象在 1 的版本基础上做了更改，回写 APIServer 时，所带的新的版本信息也为 2，APIServer 校验会发现第二个线程新写入的对象 ResourceVersion 与服务器端 ResourceVersion 冲突，从而拒绝写入，需要第二个线程读取最新版本重新更新。

此机制确保了在 Kubernetes 分布式系统中，可以任意多线程无锁并发访问对象，极大地提升了系统的整体效率。

4. Spec 和 Status

Spec 和 Status 才是对象的核心。Spec 是用户的期望状态，由创建对象的用户端定义。Status 是对象的实际状态，由对应的控制器收集实际状态并更新。

Kubernetes 对象设计完全遵循互补的原则，鼓励 API 对象尽量实现面向对象设计时的要求，即高内聚，松耦合，对与业务相关的概念有一个合适的分解，提高分解出来的对象的可重用性。高层 API 对象设计一定是从业务出发的，低层 API 对象能够被高层 API 对象所使用，从而实现减少冗余、提高重用性的目的。

5. Finalizer

如果只看社区实现，那么 Finalizer 属性毫无存在感。因为在社区代码中，很少有对

Finalizer 属性的操作。但在企业级应用过程中，它是一个十分重要、值得重点强调的属性。因为 Kubernetes 不是一个独立存在的系统，它最终会跟企业资源和系统整合，这意味着 Kubernetes 会操作这些集群外部资源或系统。试想一个场景，用户创建了一个 Kubernetes 对象，假设对应的控制器需要从外部系统获取资源，当用户删除该对象时，控制器接收到删除事件后，会尝试释放该资源。但是如果此时外部系统无法连通，并且同时控制器发生重启会有何后果？该对象永远泄露了。

Finalizer 本质上是一个资源锁，Kubernetes 在接收到某个对象的删除请求时，会检查 Finalizer 是否为空，如果为空，则只对其做逻辑删除，即只会更新对象中的 metadata.deletionTimestamp 字段。具有 Finalizer 属性的对象不会立刻被删除，需等到 Finalizer 列表中的所有字段被删除后，也就是与该对象相关的所有外部资源已被删除，这个对象才会被最终删除。

因此，如果控制器需要操作集群外部资源，则一定要在操作外部资源之前为对象添加 Finalizer，确保资源不会因对象删除而泄露。同时控制器需要监听对象的更新时间，当对象的 deletionTimestamp 不为空时，则处理对象删除逻辑，回收外部资源，并清空自己之前添加的 Finalizer。

这里用前面的资源对象描述文件 test.yaml 说明一下 TypeMeta 和 MetaData，这样效果更加直观，如图 7.1 所示。

图 7.1 TypeMeta 和 MetaData 的 YAML 样式说明

7.3 client-go 编程式交互

Kubernetes 系统使用 client-go 作为 Go 语言的官方编程交互式客户端，提供对 Kubernetes API Server 服务的交互式访问。client-go 源码在 Kubernetes 官方 github 下载，注意，其版本要与 Kubernetes 版本保持一致。

client-go 源码目录结构见表 7.2。

表 7.2 client-go 源码目录结构

源码目录	说 明
discovery	提供 DiscoveryClient（发现客户端）
dynamic	提供 DynamicClient（动态客户端）
informers	每种 Kubernetes 资源的动态实现
kubernetes	提供 ClientSet 客户端
listers	为每个 Kubernetes 资源提供 Lister 功能，该功能对 Get 和 List 请求提供只读的缓存数据
plugin	提供 OpenStack、GCP 和 Azure 等云服务商授权插件
rest	提供 RESTClient 客户端，对 Kubernetes API Server 执行 RESTful 操作
scale	提供 ScaleClient 客户端，用于扩容或缩容 Deployment、ReplicaSet、Replication Controller 等资源对象
tools	提供常用工具，如 Sharedinformer、Reflector、DealtFIFO 及 Indexers。提供 Client 查询和缓存机制，以减少向 kube-apiserver 发起的请求数等
transport	提供安全的 TCP 连接，支持 Http Stream，某些操作需要在客户端和容器之间传输二进制流，如 exec、attach 等操作。该功能由内部的 spdy 包提供支持
util	提供常用方法，如 WorkQueue（工作队列）、Certificate（证书管理）等

client-go 支持 4 种 Client 客户端对象与 Kubernetes API Server 交互。Client 客户端对象如图 7.2 所示。

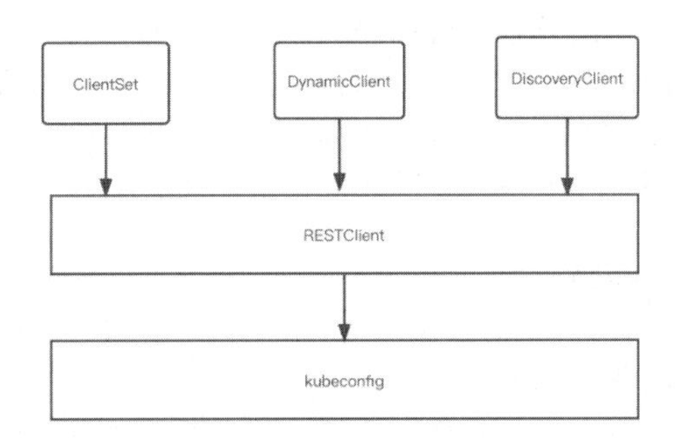

图 7.2 Client 客户端对象

RESTClient 是最基础的客户端。它对 HTTP Request 进行了封装，实现了 RESTful 风格的 API。ClientSet、DynamicClient 及 DiscoveryClient 客户端都是基于 RESTClient 实现的。

一般情况下，开发者对 Kubernetes 进行二次开发时通常使用 ClientSet。下面介绍 client-go 支持的 4 种 Client 客户端对象。

（1）RESTClient：这是最基础的客户端对象，仅对 HTTPRequest 进行了封装，实现 RESTFul 风格的 API，这个对象的使用并不方便，因为很多参数都需要使用者来设置，于是 client-go 基于

RESTClient 又实现了三种新的客户端对象。

（2）ClientSet：把 Resource 和 Version 也封装成方法，用起来更简单直接，一个资源是一个客户端，多个资源就对应了多个客户端，所以 ClientSet 就是多个客户端的集合，这样就好理解了，不过 ClientSet 只能访问内置资源，访问不了自定义资源。

（3）DynamicClient：可以访问内置资源和自定义资源，类似 Java 的集合操作，获取的内容是 Object 类型，按实际情况自己做强制转换，也会有强制转换失败的风险。

（4）DiscoveryClient：用于发现 Kubernetes 的 API Server 支持的 Group、Version、Resources 等信息。

以上 4 种客户端都可以通过 kubeconfig 配置信息连接到指定的 Kubernetes API Server，后面将详解它们的实现。

7.3.1 kubeconfig 配置文件说明

kubeconfig 配置文件的默认地址为 ~/.kube/config，用于管理 Kubernetes 集群。下面以前面创建的名为 kind 的 Kubernetes 集群为例进行说明，文件内容如下：

```
    apiVersion: v1
    clusters:
    - cluster:
        certificate-authority-data:
LS0tLS1CRUdJTiBDRVJUSUZJQ0FURS0tLS0tCk1JSUM1ekNDQWMrZ0F3SUJBZ01CQURBTkJna3Foa
2lHOXcwQkFRc0ZBREFWTVJNd0VRWURWUVFERXdwcmRXSmwKY201bGRHVnpNQjRYRFRJeU1Ea3hNak
UyTWpBeE5Gb1hEVE15TURrd09URTJNakF4TkZvd0ZURVRNQkVHQTFVRQpBeE1LYTNWaVpYSnVaWFJ
sY3pDQ0FTSXdEUVlKS29aSWh2Y05BUUVCQlFBRGdnRVBBRENDQVFvQ2dnRUJBTHFLCnVpWCsySkh5
Y1BLSTYwOVdrK2VBZDZVMWFCTFltQkYwVXZOV08wdXZJUjhFT3R0dTd5S1NBY0I4R1cvNzZJVGQKV
i80NnlBRFUyYjRXQnM5NjVuLytTMGRQOUJDYVdyWHNrMnZvby9ER2tTK2xXV1Z4RkF2VHBsQ0dzak
dRMGxaWgpWL1pWTHhRUlpOanVWRVErcC9BN0cwa3ArQjRuRFFDYTRxL0drUVRobGx4eDNLRHBER3R
Bc3BkKzFUbnhaK1BYClZ4MWp6WTRLdmc0ZklEdWRmM3Z4VkVNenJJZTdEVEMweGd5UDN0TU5mZTRv
ckU1ejc5NmliNXNUKzlTQ2Y2SXQKSUFXUXFtTlFzUUdoMjFkaHh0UEZEVEozVkNTa2NPOCtwd1drM
zcvYVlDT1JUR0pMdkEvTWp4MG1IcnF0eWhpMwowM3o0SmpkZWlxRGxSa1BHblkwQ0F3RUFBYU5DTU
VBd0RnWURWUjBQQVFIL0JBUURBZ0trTUE4R0ExVWRFd0VCCi93UUZNQU1CQWY4d0hRWURWUjBPQkJ
ZRUZNZ1FoM3RpdUV4WFFCSHd2MTRQSWZCbHRUU21NQTBHQ1NxR1NJYjMKRFFFQkN3VUFBNElCQVFD
ZFJRTzdJbUFqd3JFeWJNdTlGVFo0L1hsbWZBcGhvaEdOSUpTT1BwZ2cwZG5hZC9ITwowalJzLzNoY
XhKUUo1elJCMlViREdFTjEzZ3IvNUtWbDFDTG80ZktCT1pHUnRwNjlzNnpWWVpISEJoUTJFZjd3Ck
U2TjBpNStNazhyenV3WU5RT2lzYkVBd0UwWEZSdHBjaktJOEVkdzY4Nnl0UGprd1pncC9pQW5oY2F
KczZUU1AKblk2MzFCdWVPbm55NldLdUZ3aG5WeVcydDZqd2t4ZU5ZUWJ3d2F2NjBhSWltTXVibG5p
T2VHbHp2b3JCVHRTLwplWkVJME1PT1ZxMEo0VHhsNkUwMFJqZWtWYWFXNmpNMFpaTmxlTDlrVG1RM
05uZkE3ZVhTNGFlZWhsSlFUbGpECnVkM1ZBRUpiUzUvblB0WHI2TGZZbElRVFFFTOElpK0xSRkZucQ
otLS0tLUVORCBDRVJUSUZJQ0FURS0tLS0tCg==
        server: https://127.0.0.1:60354
      name: kind-kind
```

```
contexts:
- context:
    cluster: kind-kind
    user: kind-kind
  name: kind-kind
current-context: kind-kind
kind: Config
preferences: {}
users:
- name: kind-kind
  user:
    client-certificate-data:
LS0tLS1CRUdJTiBDRVJUSUZJQ0FURS0tLS0tCk1JSURFekNDQWZ1Z0F3SUJBZ0lJR0I3cWtud0dDW
nN3RFFZSktvWklodmNOQVFFTEJRQXdGVEVUTUJFR0ExVUUKQXhNS2EzVmlaWEp1WlhSbGN6QWVGdz
B5TWpBNU1USXhOakl3TVRSYUZ3MHlNekE1TVRJeE5qSXdNVFZhTURReApGekFWQmdOVkJBb1REbk4
1YzNSbGJUcHRZWE4wWlhKek1Sa3dGd1lEVlFRREV4QnJkV0psY201bGRHVnpMV0ZrCmJXbHVNSUlC
SWpBTkJna3Foa2lHOXcwQkFRRUZBQU9DQVE4QU1JSUJDZ0tDQVFFQTZoWWpXZ1RnOE5vUzhFZVIKZ
CszeWgybkd0aWEyVHJmVTlDY2pVSnNmRnZTZ25qWW1Ecmw1dHM3UlZQajBWS2t2cmVFdHY4d01OdT
lLNEYxYgo0MXlPUWhzTy9WYWh2bHZqaDhXWXhqU1RnWFpUKzlVcWNkSThOZlhBNHcxWmpYdWdDTi9
iVUlDY1BFV2ZNMkNJCi9QNzlvT3REWlJJWTJpbnBQZVoxTzYyd3ZLZmNtSXFlWmJhQnNaMWtZMG9R
UlZPWFlLUitLRm1nY3RGVVVQdjQKRThTSTFPOE1OY3R3MHdydGxGWHZSSXBFMlptUllISncwM0dIN
VQ5eHY4cXYyUlNTVTYwcjkvdVNqOGlNaXBtdgpZaDRYLy9LZHE2VU9LQmtHaXZCTGkvYkdqWlNoST
dWTFJ4THBUMDYzMkxTVzNkYkE1aEMrWVV0Nk0rSWJUVE43CkNRVDA2UUlEQVFBQm8wZ3dSakFPQmd
OVkhROEJBZjhFQkFNQ0JhQXdFd1lEVlIwbEJBd3dDZ1lJS3dZQkJRVUgKQXdJd0h3WURWUjBqQkJn
d0ZvQVV5QkNIZTJLNFRGZEFFZkMvWGc4aDhHVzFOS1l3RFFZSktvWklodmNOQVFFTApCUUFFZ2dFQ
kFEc3djSEI2ajd0MFdNYllvdy9zYjVvcHY1MkZ6MDVNRmpRQW1OaHo3bUs3ajFRTGZYL291cFNhCm
oyMDI4U200SDJZY29VUDBFb3Q4TFRvamU2SEJkNjBJN3BIS29uWWdRTDEvR3lNNDZtekQ5K1AwSDF
4NDZhaVkKaDdha2g5MWdkK2R6c0U0c0tKTmE0bTlmK0NRWkpBenp6T2dXMXprSmt3OHdwUzNvUGFw
bEY5RFVEc0dycXZDTgp5VTdnWTlkT1BBZlRZb2hnL0FlL215SEhoMmlJeU9rTE96aG0vL3MwMTRxZ
2FTVG5FYXdvVjk2eXRLeG1TTW1OCi94emtKenFNaXk5a0NNTlZMRmN3Z3NBYWNUN1NPTjVaUzM5UV
BBYUlPazdINHNpWDFnRUI3QXFTdVlXejJPRVYKZ0RrY0c0UVFneWVRZlNNdG1EUkt5amxSTGd3K2Z
tdz0KLS0tLS1FTkQgQ0VSVElGSUNBVEUtLS0tLQo=
    client-key-data:
LS0tLS1CRUdJTiBSU0EgUFJJVkFURSBLRVktLS0tLQpNSUlFcEFJQkFBS0NBUUVBNmhZaldnVGc4T
m9TOEVlUmQrM3loMm5HTmlhMlRyZlU5Q2NqVUpzZkZ2U2dualltCkRybDV0czdSVlBqMFZLa3ZyZU
V0djh3TU51OUs0RjFiNDF5T1Foc08vVmFodmx2amg4V1l4alNUZ1haVCs5VXEKY2RJOE5mWEE0dzF
aalh1Z0NOL2JVSUNjUEVXZk0yQ0kvUDc5b090RFpSSVkyaW5wUGVaMU82Mnd2S2ZjbUlxZQpaYmFC
c1oxa1kwb1FSVk9YWUtSK0tGbWdjdEZVVVB2NEU4U0kxTzhNTmN0dzB3cnRsRlh2UklwRTJabVJZS
Ep3CjAzR0g1VDl4djhxdjJSU1NVNjByOS91U2o4aU1pcG12WWg0WC8vS2RxNlVPS0JrR2l2QkxpL2
JHalpTaTdWTkwKUnhMcFQwNjMyTFNXM2RiQTVoQytZVXQ2TStJYlRUTjdDUVQwNlFJREFRQUJBb0l
CQUhLdzhoVXkrcnRIUGF3NApoSStNS0I0c3R3N2xITUdvMUUvanRTdlROUUpjVFMzM1BlSk5BZE9O
```

```
dFlhNmk0dkhjVEx5ZHlYWDVjd3NwalVUCnlPZTIrNFJzR1lZS29palFnTVNQRGZGZTVUNnpLU2JrS
kcyc0loQVdIckdnWXRZdFVoWEhDSCtDNUFtcjJ2cWoKTzdzcGZEOFJ4ZTJKVmwvZDFRZWRHS2Rxek
lOQms3RG1GbFN6N01GR2lpNUYrZ1ZyTGhPK2NRVjdUNGpNM2Q4UwozMUgvczhMT2NvV0lGbmJGdnE
waktXVlRhNEtqR1BzMS8xbUxTeGhQRlkzM1JOcGVEa0NqRnRBcVZxOGZucm5TCm9WMjhzR1hmTW92
L3g3SUlDN21yV2hEM0JOTTN5N2c4dEpIb3ZmbFV3NHBCTXltUFNhN29TTkhaSkl3Q2FpUVQKUGZzN
itoa0NnWUVBOG5BaWtOcFlkM0VxWVF2Q1ExSlNmb0lJR1QxK3hhQWFGN08zZHQ1YUpzVEt0YktIbl
IzTQpwUWUycUV5WXdBc2NLZk1PeXlwYTNEa210K2NRNWd0cTJYY3BQVUNnelVBdlhIRjJDcEVETC9
nZjl3TGpEZFlGClBHWDV4ZmlZa1A2bXYvbklhMzNvWUVvdXRETTk3dzV1WEZ2VE4rNFR3YTFlNk9v
a3NoOTVPY2NDZ1lFQTl5NW4KVTFpY25RQnZXWmxScmlYZ05NeFZabTdNZHRCaWMyc2RNbHBOK0h4a
1pTYU9CUjJCMlBsS09LbHNQZ0p2TjBFaQpNQWxZTTBIaVhySU5WYVZFUGZOS1U2L21KQS9BWGMweW
l3ZlFCdFVLYUhLT3VWV0tHdGhpdGtVTmN0RmVMSTFuCmFCQitsVkFzVlRyUThScXFYVEFOK3BQcjZ
nckJGRExrYXdIZTI4OENnWUFLdFpkaFRMUGh1dGNEYUhDUlpyNk0KSmhSVGdScnJnZklGSUFRM3RR
VElkNkw0dWNVeDFKSW9yQm1PZjZ1enVBbnJkSGd2MDFRbUxFUWdsWTF5akFpUgpTS2xDc21zb0JmR
FJwWlZ3VEEyVy9kL3ptbVlMUTJ3QnlnTnBrbHl5OHZ6TXVDejVTcHVNUFZxb3pTYXNPN2YyClRud2
JHV0k0Q1ArTFRHU0lsY2xXR3dLQmdRRFJKUFpRVXozUXFqbU5PbWtlVFBYME5xQmIxNHd4VEFqQ1Y
vU1EKVWxCKzIrR21wSVN1VjFVeTNWdXJyYzdrZVBMWmNjY1hsaENNHN2t3M2k3ZzBxMjJlYldpS2Z0
N2Z6R1FiK2krMQo2NXBjN1kyczZxRlF0OHZ2QXdBaThUSjc0Q01xMTBxVGhzRlVYVVNsbjZxREZ1a
m5pRzJMZy9XdFJnUXdydzhPCmE3ckJpUUtCZ1FDVi9GT2ZwRWJ4d2I1UTNNZjZsemF5SjBVSElIRE
1VdjN5MldMaG1MVTdpbjQyZVB5R1hjdHkKRzF0STR1amlkeDYwRCtNSFhMcm9UMkFiM0FUWjg3NGJ
pK3hZTXNZUzFDKzYrMUhzT29yQ2RFNDNYRDhqME1vKwpKNW4xR2JYeGxMitleFZNRGNnaGV4Y1lVk
RlRlWERZUk1NOVkxZHRmT00xY3htSld0ei9yQUE9PQotLS0tLUVORCBSU0EgUFJJVkFURSBLRVktL
S0tLQo=
```


kubeconfig 配置信息通常分为三个部分，这里分别进行介绍。

（1）clusters：定义 Kubernetes 集群信息，如 kube-apiserver 的服务器地址及集群的证书信息等。

（2）users：定义 Kubernetes 集群用户身份验证的客户端凭证，如 client-certificate-data、client-key-data 及 token 等。

（3）contexts：定义 Kubernetes 集群的用户信息和命名空间等，用于将请求发送到指定的集群。

可以用 git clone 命令在本地下载，方便对照参考源码，最终路径为 /Users/yuhongchun/repo/workspace/Kubernetes/client-go。

这里提前创建名为 k8stest 的目录，然后新建 module，命令如下：

```
go mod init k8stest
```

添加 k8s.io/api 和 k8s.io/client-go 这两个依赖，注意要匹配 Kubernetes 版本，即 1.20.0，命令如下：

```
go get k8s.io/api@v0.20.0
go get k8s.io/client-go@v0.20.0
```

然后创建名为 testconfig.go 的文件，代码如下：

```
package main
```

```
import (
        "context"
        "flag"
        "fmt"
        "path/filepath"
        metav1 "k8s.io/apimachinery/pkg/apis/meta/v1"
        "k8s.io/client-go/kubernetes"
        "k8s.io/client-go/tools/clientcmd"
        "k8s.io/client-go/util/homedir"
)

func main() {
        var kubeconfig *string
        // 默认会从 ~/.kube/config 路径下获取配置文件
        if home := homedir.HomeDir(); home != ""{
                kubeconfig = flag.String("kubeconfig", filepath.Join(home,
".kube", "config"), "(optional) absolute path to the kubeconfig file")
        } else {
                kubeconfig =flag.String("kubeconfig", "", "absolute path to the
kubeconfig file")
        }
        flag.Parse()

        // use the current context in kubeconfig
        config, err := clientcmd.BuildConfigFromFlags("", *kubeconfig)
        if err != nil{
                panic(err.Error())
        }

        // create the clientset
        clientset, err := kubernetes.NewForConfig(config)
        if err != nil{
                panic(err.Error())
        }
        pods, err := clientset.CoreV1().Pods("").List(context.TODO(), metav1.
ListOptions{})
        if err != nil{
                panic(err.Error())
        }

        fmt.Printf("There are %d pods in the cluster\n", len(pods.Items))
}
```

用 go mod tidy 增加相应的依赖，然后运用此文件，命令如下：

```
go run testconfig.go
```

命令显示结果如下：

```
There are 18 pods in the cluster
```

这里需要重点关注以下加载配置文件的代码：

```
	var kubeconfig *string
	// 默认会从 ~/.kube/config 路径下获取配置文件
	if home := homedir.HomeDir(); home != ""{
		kubeconfig = flag.String("kubeconfig", filepath.Join(home, ".kube", 
"config"), "(optional) absolute path to the kubeconfig file")
	} else {
		kubeconfig = flag.String("kubeconfig", "", "absolute path to the 
kubeconfig file")
	}
	flag.Parse()

	// use the current context in kubeconfig
	config, err := clientcmd.BuildConfigFromFlags("", *kubeconfig)
	if err != nil{
		panic(err.Error())
	}
```

调用 clientcmd.BuildConfigFromFlags 加载配置文件后，会生成一个 rest.Config 对象，这个对象中包含如 apiserver 地址、用户名、密码和 token 等信息。

7.3.2 RESTClient 客户端

RESTClient 是最基础的客户端，ClientSet、DynamicClient 及 DiscoveryClient 都是基于 RESTClient 实现的。RESTClient 对 HTTP Request 进行了封装，实现了 RESTful 风格的 API。它具有很高的灵活性，数据不依赖方法和资源，因此 RESTClient 能够处理多种类型的调用，并返回不同的数据格式。

下面用 podlist.go 文件来演示 RESTClient 客户端的功能。代码如下：

```
package main

import (
	"context"
	"flag"
	"fmt"
	"k8s.io/client-go/kubernetes/scheme"
	"k8s.io/client-go/rest"
```

```
        "k8s.io/client-go/tools/clientcmd"
        "k8s.io/client-go/util/homedir"
        corev1 "k8s.io/api/core/v1"
        metav1 "k8s.io/apimachinery/pkg/apis/meta/v1"
        "path/filepath"
    )

    func main() {
        var kubeconfig *string

        // home 是家目录，如果能取得家目录的值，就可以用来做默认值
        if home:=homedir.HomeDir(); home != ""{
                // 如果输入了 kubeconfig 参数，该参数的值就是 kubeconfig 文件的绝对路径
                // 如果没有输入 kubeconfig 参数，就用默认路径 ~/.kube/config
                kubeconfig = flag.String("kubeconfig", filepath.Join(home, ".kube",
"config"), "(optional) absolute path to the kubeconfig file")
        } else {
                // 如果取不到当前用户的家目录，就没办法设置 kubeconfig 的默认目录了，只
                // 能从参数中取
                kubeconfig = flag.String("kubeconfig", "", "absolute path to the
kubeconfig file")
        }

        flag.Parse()

        // 从本机加载 kubeconfig 配置文件，因此第一个参数为空字符串
        config, err := clientcmd.BuildConfigFromFlags("", *kubeconfig)

        // 如果 kubeconfig 加载失败，就直接退出
        if err != nil{
                panic(err.Error())
        }

        // 参考 path : /api/v1/namespaces/{namespace}/pods
        config.APIPath = "api"
        // pod 的 group 是空字符串
        config.GroupVersion = &corev1.SchemeGroupVersion
        // 指定序列化工具
        config.NegotiatedSerializer = scheme.Codecs

        // 根据配置信息构建 restClient 实例
        restClient, err := rest.RESTClientFor(config)
```

```
        if err!=nil {
            panic(err.Error())
        }

        // 保存pod结果的数据结构实例
        result := &corev1.PodList{}


        // 指定namespace
        namespace := "kube-system"
        // 设置请求参数，然后发起请求
        // Get请求
        err = restClient.Get().
                //指定namespace，参考path : /api/v1/namespaces/{namespace}/pods
                Namespace(namespace).
                // 查找多个pod，参考path : /api/v1/namespaces/{namespace}/pods
                Resource("pods").
                // 指定大小限制和序列化工具
                VersionedParams(&metav1.ListOptions{Limit:100},
scheme.ParameterCodec).
                // 请求
                Do(context.TODO()).
                // 结果存入result
                Into(result)

        if err != nil{
                panic(err.Error())
        }

        // 表头
        fmt.Printf("namespace\t status\t\t name\n")

        // 每个pod都打印Namespace、Status.Phase、Name三个字段
        for _, d := range result.Items{
            fmt.Printf("%v\t %v\t %v\n",
                    d.Namespace,
                    d.Status.Phase,
                    d.Name)
        }
}
```

代码显示结果如下：

```
namespace        status         name
kube-system      Running        coredns-74ff55c5b-gkg4q
kube-system      Running        coredns-74ff55c5b-k24kq
kube-system      Running        etcd-kind-control-plane
kube-system      Running        kindnet-hfd6b
kube-system      Running        kindnet-nxmxk
kube-system      Running        kindnet-qnp82
kube-system      Running        kindnet-zgh4w
kube-system      Running        kube-apiserver-kind-control-plane
kube-system      Running        kube-controller-manager-kind-control-plane
kube-system      Running        kube-proxy-6t8ld
kube-system      Running        kube-proxy-8d7v2
kube-system      Running        kube-proxy-hs47m
kube-system      Running        kube-proxy-m7lhh
kube-system      Running        kube-scheduler-kind-control-plane
```

运行以上代码，列出 kube-system 命名空间下的所有 Pod 资源对象的相关信息。首先加载 ~/.kube/config 配置信息，并设计 config.APIPath 请求的 HTTP 路径；然后设置 config.GroupVersion 请求的资源组 / 资源版本；最后设置 config.NegotiatedSerializer 数据的编解码器。

下面看一下 client-go 源码中的部分代码：

```
// /workspace/kubernetes/client-go/rest/client.go
type RESTClient struct {
    // base 是客户端所有调用的根 URL
    base *url.URL
    // versionedAPIPath 是连接基本 URL 到资源根的路径段
    versionedAPIPath string

    // content 描述了 RESTClient 如何编码和解码响应
    content ClientContentConfig

    // 创建传递给请求的 BackoffManager
    createBackoffMgr func() BackoffManager

    // 除非被特别覆盖重写，否则 RrateLimiter 在此客户端创建的所有请求之间共享
    rateLimiter flowcontrol.RateLimiter

  // warningHandler 在此客户端创建的所有请求之间共享。如果未设置，则使用
//defaultWarningHandler
    warningHandler WarningHandler

    // 设置客户端的特定行为。如果没有设置，http.DefaultClient 将被使用
```

```
        // Client *http.Client
    }
    func (c *RESTClient) Verb(verb string) *Request{...}
    func (c *RESTClient) Post() *Request{...}
    func (c *RESTClient) Put() *Request{...}
    func (c *RESTClient) Patch(pt types.PatchType) *Request{...}
    func (c *RESTClient) Get() *Request{...}
    func (c *RESTClient) Delete() *Request{...}
    func (c *RESTClient) APIVersion() schema.GroupVersion{...}
```

以上代码定义了 RESTClient 结构体，下面具体介绍该结构体中封装的对象。

（1）RESTClient 结构体封装了上面介绍的 ClientContentConfig 对象作为属性，用来设置访问某一组下某一个版本的资源，以及完成对这些资源的序列化和反序列化操作。

（2）RESTClient 结构体封装了 http.Client 对象指针，它是 Go 语言中的原生 http 访问组件，底层利用该组件完成 HTTP 请求。

（3）RESTClient 结构体也封装了 HTTP 请求中的其他属性，如 baseurl 等，同时也提供了限速（rateLimiter）和回退（BackoffManager）等高级功能。

（4）RESTClient 结构体生成了 Request 对象，提供相应的 HTTP 方法访问资源。

7.3.3 ClientSet 客户端

RESTClient 是一种较基础的客户端，使用时需要指定 Resource 和 Version 等信息。相比 RESTClient，ClientSet 使用起来更加便捷。一般情况下，开发者对 Kubernetes 进行二次开发时通常会使用 ClientSet。

ClientSet 在 RESTClient 的基础上封装了对 Resource 和 Version 的管理方法。每个 Resource 可以理解成一个客户端，而 ClientSet 则是多个客户端的集合。每个 Resource 和 Version 都以函数的方式暴露给开发者。

ClientSet 多资源集合如图 7.3 所示。

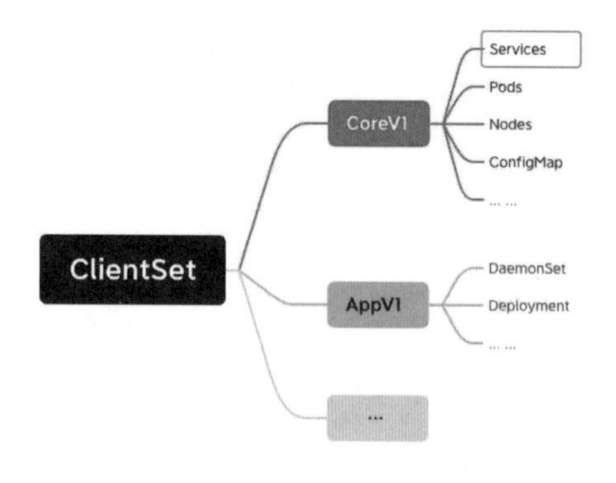

图 7.3 ClientSet 多资源集合

ClientSet 中关于接口的源码片段如下：

```
    // /Users/yuhongchun/repo/workspace/Kubernetes/client-go/Kubernetes/clientset.go
    type Interface interface {
         Discovery() discovery.DiscoveryInterface
         AdmissionregistrationV1()
admissionregistrationv1.AdmissionregistrationV1Interface
         AdmissionregistrationV1beta1()
admissionregistrationv1beta1.AdmissionregistrationV1beta1Interface
         InternalV1alpha1() internalv1alpha1.InternalV1alpha1Interface
         AppsV1() appsv1.AppsV1Interface
         AppsV1beta1() appsv1beta1.AppsV1beta1Interface
         AppsV1beta2() appsv1beta2.AppsV1beta2Interface
         AuthenticationV1() authenticationv1.AuthenticationV1Interface
         AuthenticationV1beta1()
authenticationv1beta1.AuthenticationV1beta1Interface
         AuthorizationV1() authorizationv1.AuthorizationV1Interface
         AuthorizationV1beta1()
authorizationv1beta1.AuthorizationV1beta1Interface
         AutoscalingV1() autoscalingv1.AutoscalingV1Interface
         AutoscalingV2() autoscalingv2.AutoscalingV2Interface
         AutoscalingV2beta1() autoscalingv2beta1.AutoscalingV2beta1Interface
         AutoscalingV2beta2() autoscalingv2beta2.AutoscalingV2beta2Interface
         BatchV1() batchv1.BatchV1Interface
         BatchV1beta1() batchv1beta1.BatchV1beta1Interface
         CertificatesV1() certificatesv1.CertificatesV1Interface
         CertificatesV1beta1()
certificatesv1beta1.CertificatesV1beta1Interface
         CoordinationV1beta1()
coordinationv1beta1.CoordinationV1beta1Interface
         CoordinationV1() coordinationv1.CoordinationV1Interface
    CoreV1() corev1.CoreV1Interface
    …
    }
```

在以上代码中，Interface 接口定义了获得 Kubernetes 中所有资源的方法。例如，获得 CoreV1 资源可以调用该接口的 CoreV1() 方法；同理，获得 AppV1 资源可以调用该接口的 AppsV1() 方法。源码中再往下就是各种具体的 struct 定义及其实现方法了。

下面可以看一下 ClientSet 的简单例子，代码如下：

```
    package main

    import (
         "context"
```

```
        "flag"
        "fmt"
        metav1 "k8s.io/apimachinery/pkg/apis/meta/v1"
        "k8s.io/client-go/kubernetes"
        "k8s.io/client-go/tools/clientcmd"
        "k8s.io/client-go/util/homedir"
        "log"
        "path/filepath"
    )

    func main() {

        var kubeconfig *string

        // home 是家目录，如果能取得家目录的值，就可以用来做默认值
        if home := homedir.HomeDir(); home != ""{
            // 如果输入 kubeconfig 参数，该参数的值就是 kubeconfig 文件的绝对路径
            // 如果没有输入 kubeconfig 参数，就用默认路径 ~/.kube/config
            kubeconfig = flag.String("kubeconfig", filepath.Join(home, ".kube",
"config"), "(optional) absolute path to the kubeconfig file")
        } else {
            // 如果取不到当前用户的家目录，就没有办法设置 kubeconfig 的默认目录，只能
            //从入参中取
            kubeconfig = flag.String("kubeconfig", "", "absolute path to the
kubeconfig file")
        }

        flag.Parse()

        // 从本机加载 kubeconfig 配置文件，因此第一个参数为空字符串
        config, err := clientcmd.BuildConfigFromFlags("", *kubeconfig)

        // 如果 kubeconfig 加载失败、就直接退出
        if err != nil{
            panic(err.Error())
        }

        clientset, err := kubernetes.NewForConfig(config)
        if err != nil{
            panic(err.Error)
        }
```

```
        nodes, err := clientset.CoreV1().Nodes().List(context.TODO(), metav1.
ListOptions{})
        if err != nil{
             log.Fatalln("failed to get nodes:", err)
        }

        for i, node := range nodes.Items{
             fmt.Printf("[%d] %s\n", i, node.GetName())
        }

    }
```

用 go run example.go 执行上面这个文件，则命令显示结果如下：

```
    [0] kind-control-plane
    [1] kind-worker
    [2] kind-worker2
    [3] kind-worker3
```

这是一个非常典型的访问 Kubernetes 集群资源的方式，通过 client-go 提供的 ClientSet 客户端来获取资源数据，主要有以下三个步骤：

（1）使用 kubeconfig 文件或者 ServiceAccount(InCluster 模式）来创建访问 Kubernetes API 的 Restful 配置参数，也就是代码中的 rest.Config 对象。

（2）使用 rest.Config 参数创建 ClientSet 对象，直接调用 kubernetes.NewForConfig(config) 即可初始化。

（3）使用 ClientSet 对象的方法获取各个 Group 下面的对应资源对象进行 CURD 操作。

再看一个用 ClientSet 对象获取 Deployments 的例子，代码如下：

```
package main

import (
     "context"
     "flag"
     "fmt"
     metav1 "k8s.io/apimachinery/pkg/apis/meta/v1"
     "k8s.io/client-go/kubernetes"
     "k8s.io/client-go/tools/clientcmd"
     "k8s.io/client-go/util/homedir"
     "path/filepath"
)

func main() {
```

```
        var kubeconfig *string

        // home 是家目录，如果能取得家目录的值，就可以用来做默认值
        if home := homedir.HomeDir(); home != ""{
             // 如果输入 kubeconfig 参数，该参数的值就是 kubeconfig 文件的绝对路径
             // 如果没有输入 kubeconfig 参数，就用默认路径 ~/.kube/config
              kubeconfig = flag.String("kubeconfig", filepath.Join(home, ".kube",
"config"), "(optional) absolute path to the kubeconfig file")
        } else {
        // 如果取不到当前用户的家目录，就没办法设置 kubeconfig 的默认目录了，只能从入参中取
              kubeconfig = flag.String("kubeconfig", "", "absolute path to the
kubeconfig file")
        }

        flag.Parse()

        // 从本机加载 kubeconfig 配置文件，因此第一个参数为空字符串
        config, err := clientcmd.BuildConfigFromFlags("", *kubeconfig)

        // kubeconfig 加载失败就直接退出
        if err != nil{
                panic(err.Error())
        }

        clientset, err := kubernetes.NewForConfig(config)
        if err != nil{
                panic(err.Error)
        }

        // 使用 clientset 获取 Deployments
          deployments, err := clientset.AppsV1().Deployments("default").
List(context.TODO(),metav1.ListOptions{})
        if err != nil{
                panic(err)
        }
        for idx, deploy := range deployments.Items{
             fmt.Printf("%d -> %s\n", idx+1, deploy.Name)
        }

   }
```

命令显示结果如下：

```
1 -> nginx-deployment
```

上面了解了如何使用 Clientset 对象来获取集群资源，接下来分析 Clientset 对象的实现。上面的源码实现实际上是对各种资源类型的 Clientset 对象的一次封装：

```
// NewForConfig 使用给定的 config 创建一个新的 Clientset
func NewForConfig(c *rest.Config) (*Clientset, error) {
  configShallowCopy := *c
  if configShallowCopy.RateLimiter == nil && configShallowCopy.QPS > 0 {
    configShallowCopy.RateLimiter =
flowcontrol.NewTokenBucketRateLimiter(configShallowCopy.QPS, configShallowCopy.
Burst)
  }
  var cs Clientset
  var err error
  cs.admissionregistrationV1beta1, err =
admissionregistrationv1beta1.NewForConfig(&configShallowCopy)
  if err != nil {
    return nil, err
  }
  // 将其他 Group 和版本资源的 RESTClient 封装到全局的 Clientset 对象中
  cs.appsV1, err = appsv1.NewForConfig(&configShallowCopy)
  if err != nil {
    return nil, err
  }
  ...
  cs.DiscoveryClient, err =
discovery.NewDiscoveryClientForConfig(&configShallowCopy)
  if err != nil {
    return nil, err
  }
  return &cs, nil
}
```

在以上代码中，NewForConfig 函数将其他各种资源的 RESTClient 封装到全局的 Clientset 对象中，这样当需要访问某个资源时，只需使用 Clientset 对象中包装的属性即可。例如，clientset.CoreV1() 就是访问 Core 这个 Group 下面 V1 这个版本的 RESTClient，而 clientset.AppsV1() 就是访问 Apps 这个 Group 下面 V1 这个版本的 RESTClient。下面同样以 Deployment 为例，其源码如下：

```
// /workspace/Kubernetes/client-go/Kubernetes/clientset.go
// NewForConfig 根据 rest.Config 创建一个 AppsV1Client
func NewForConfig(c *rest.Config) (*AppsV1Client, error) {
```

```
    config := *c
    // 为 rest.Config 设置资源对象默认的参数
    if err := setConfigDefaults(&config); err != nil {
      return nil, err
    }
    // 实例化 AppsV1Client 的 RestClient
    client, err := rest.RESTClientFor(&config)
    if err != nil {
      return nil, err
    }
    return &AppsV1Client{client}, nil
  }

  func setConfigDefaults(config *rest.Config) error {
    // 资源对象的 GroupVersion
    gv := v1.SchemeGroupVersion
    config.GroupVersion = &gv
    // 资源对象的 root path
    config.APIPath = "/apis"
    // 使用注册的资源类型 Scheme 对请求和响应进行编解码，Scheme 是资源类型的规范
    config.NegotiatedSerializer = serializer.DirectCodecFactory{CodecFactory:
scheme.Codecs}

    if config.UserAgent == "" {
      config.UserAgent = rest.DefaultKubernetesUserAgent()
    }

    return nil
  }

  func (c *AppsV1Client) Deployments(namespace string) DeploymentInterface {
    return newDeployments(c, namespace)
  }
  // staging/src/k8s.io/client-go/kubernetes/typed/apps/v1/deployment.go
  // deployments 实现了 DeploymentInterface 接口
  type deployments struct {
    client rest.Interface
    ns     string
  }
  // newDeployments 实例化 deployments 对象
  func newDeployments(c *AppsV1Client, namespace string) *deployments {
```

```
    return &deployments{
      client: c.RESTClient(),
      ns:     namespace,
    }
  }
```

clientset.AppsV1().Deployments("default") 可以获取一个 Deployments 对象，该对象下面定义了 Deployments 对象的 CURD 操作，如调用的 List 函数：

```
  func  (c  *deployments)  List(opts  metav1.ListOptions)  (result  *v1.
DeploymentList, err error) {
    var timeout time.Duration
    if opts.TimeoutSeconds != nil {
      timeout = time.Duration(*opts.TimeoutSeconds) * time.Second
    }
    result = &v1.DeploymentList{}
    err = c.client.Get().
      Namespace(c.ns).
      Resource("deployments").
      VersionedParams(&opts, scheme.ParameterCodec).
      Timeout(timeout).
      Do().
      Into(result)
    return
  }
```

从上面代码可以看出，最终是通过 c.client 发起的请求，具体来说，局部的 restClient 是在初始化函数中通过调用 rest.RESTClientFor(&config) 创建的，该函数将 rest.Config 对象转换为一个用于网络操作的 RESTful 客户端对象。代码如下：

```
  // RESTClientFor 返回一个满足客户端 Config 对象上的属性的 RESTClient 对象
  // 注意在初始化客户端时，RESTClient 可能需要一些可选的属性
  func RESTClientFor(config *Config) (*RESTClient, error) {
    if config.GroupVersion == nil {
      return nil, fmt.Errorf("GroupVersion is required when initializing a
RESTClient")
    }
    if config.NegotiatedSerializer == nil {
      return nil, fmt.Errorf("NegotiatedSerializer is required when
initializing a RESTClient")
    }
    qps := config.QPS
    if config.QPS == 0.0 {
      qps = DefaultQPS
```

```
    }
    burst := config.Burst
    if config.Burst == 0 {
      burst = DefaultBurst
    }

    baseURL, versionedAPIPath, err := defaultServerUrlFor(config)
    if err != nil {
      return nil, err
    }

    transport, err := TransportFor(config)
    if err != nil {
      return nil, err
    }
    // 初始化一个 HttpClient 对象
    var httpClient *http.Client
    if transport != http.DefaultTransport {
      httpClient = &http.Client{Transport: transport}
      if config.Timeout > 0 {
        httpClient.Timeout = config.Timeout
      }
    }

    return NewRESTClient(baseURL, versionedAPIPath, config.ContentConfig, qps,
burst, config.RateLimiter, httpClient)
  }
```

综上所述可知，ClientSet 是基于 RESTClient 的，RESTClient 是底层用于网络请求的对象，可以直接通过 RESTClient 提供的 RESTful 方法，如 Get、Put、Post、Delete 等，和 APIServer 进行交互。

ClientSet 的作用：

- 同时支持 JSON 和 protobuf 两种序列化方式。
- 支持所有原生资源。

7.3.4 DynamicClient 客户端

DynamicClient 是一种动态客户端，它可以对任意 Kubernetes 资源进行 RESTful 操作，包括 CRD 自定义资源。DynamicClient 与 ClientSet 操作类似，同样封装了 RESTClient，也提供了 Create、Update、Delete、Get、List、Watch、Patch 等方法。

DynamicClient 与 ClientSet 最大的区别在于，ClientSet 仅能访问 Kubernetes 自带的资源，不能访问 CRD 自定义资源。ClientSet 需要预先实现每种 Resource 和 Version 的操作，其内部的数据都是结构化数据（即已知数据结构），而 DynamicClient 内部实现了 Unstructured，用于处理非结构化数据（即无法提前预知的数据结构），这也是使 DynamicClient 能够处理 CRD 自定义资源的关键。

下面用简单例子进行说明，其代码如下：

```
package main

import (
  "context"
  "flag"
  "fmt"
  apiv1 "k8s.io/api/core/v1"
  metav1 "k8s.io/apimachinery/pkg/apis/meta/v1"
  "k8s.io/apimachinery/pkg/runtime"
  "k8s.io/apimachinery/pkg/runtime/schema"
  "k8s.io/client-go/dynamic"
  "k8s.io/client-go/tools/clientcmd"
  "k8s.io/client-go/util/homedir"
  "path/filepath"
)

func main() {

  var kubeconfig *string

  // home 是家目录，如果能取得家目录的值，就可以用来做默认值
  if home:=homedir.HomeDir(); home != "" {
    // 如果输入 kubeconfig 参数，该参数的值就是 kubeconfig 文件的绝对路径
    // 如果没有输入 kubeconfig 参数，就用默认路径 ~/.kube/config
      kubeconfig = flag.String("kubeconfig", filepath.Join(home, ".kube",
"config"), "(optional) absolute path to the kubeconfig file")
  } else {
    // 如果取不到当前用户的家目录，就没办法设置 kubeconfig 的默认目录，只能从参数中取得
      kubeconfig = flag.String("kubeconfig", "", "absolute path to the
kubeconfig file")
  }

  flag.Parse()
```

```
// 从本机加载 kubeconfig 配置文件，因此第一个参数为空字符串
config, err := clientcmd.BuildConfigFromFlags("", *kubeconfig)

// kubeconfig 加载失败就直接退出
if err != nil {
  panic(err.Error())
}

dynamicClient, err := dynamic.NewForConfig(config)

if err != nil {
  panic(err.Error())
}

// dynamicClient 的唯一关联方法所需的入参
gvr := schema.GroupVersionResource{Version: "v1", Resource: "pods"}

// 使用 dynamicClient 的查询列表方法，查询指定 namespace 下的所有 pod
// 注意此方法返回的数据结构类型是 UnstructuredList
unstructObj, err := dynamicClient.
  Resource(gvr).
  Namespace("kube-system").
  List(context.TODO(), metav1.ListOptions{Limit: 10})

if err != nil {
  panic(err.Error())
}

// 实例化一个 PodList 数据结构，用于接收从 unstructObj 转换后的结果
podList := &apiv1.PodList{}

// 转换
err = runtime.DefaultUnstructuredConverter.FromUnstructured(unstructObj.
UnstructuredContent(), podList)

if err != nil {
  panic(err.Error())
}

// 表头
fmt.Printf("namespace\t status\t\t name\n")
```

```
  // 每个 pod 都打印 Namespace、Status.Phase、Name 三个字段
  for _, d := range podList.Items {
    fmt.Printf("%v\t %v\t %v\n",
      d.Namespace,
      d.Status.Phase,
      d.Name)
  }
}
```

命令显示结果如下：

```
namespace        status        name
kube-system      Running       coredns-74ff55c5b-gkg4q
kube-system      Running       coredns-74ff55c5b-k24kq
kube-system      Running       etcd-kind-control-plane
kube-system      Running       kindnet-hfd6b
kube-system      Running       kindnet-nxmxk
kube-system      Running       kindnet-qnp82
kube-system      Running       kindnet-zgh4w
kube-system      Running       kube-apiserver-kind-control-plane
kube-system      Running       kube-controller-manager-kind-control-plane
kube-system      Running       kube-proxy-6t8ld
```

注：DynamicClient不是类型安全的，因此在访问CRD自定义资源时需要特别注意。例如，在操作指针不当的情况下可能会导致程序崩溃。

7.3.5 DiscoveryClient 客户端

DiscoveryClient 是发现客户端，主要用于发现 Kubernetes API Server 所支持的资源组、资源版本和资源信息。这里举例说明其功能，代码如下：

```
package main

import (
  "flag"
  "fmt"
  "k8s.io/apimachinery/pkg/runtime/schema"
  "k8s.io/client-go/discovery"
  "k8s.io/client-go/tools/clientcmd"
  "k8s.io/client-go/util/homedir"
  "path/filepath"
)
```

```
func main() {

    var kubeconfig *string

    // home 是家目录，如果能取得家目录的值，就可以用来做默认值
    if home:=homedir.HomeDir(); home != "" {
        kubeconfig = flag.String("kubeconfig", filepath.Join(home, ".kube",
"config"), "(optional) absolute path to the kubeconfig file")
    } else {
        // 如果取不到当前用户的家目录，就没办法设置 kubeconfig 的默认目录，只能从参数中获取
        kubeconfig = flag.String("kubeconfig", "", "absolute path to the kubeconfig
file")
    }

    flag.Parse()

    // 从本机加载 kubeconfig 配置文件，因此第一个参数为空字符串
    config, err := clientcmd.BuildConfigFromFlags("", *kubeconfig)

    // kubeconfig 加载失败就直接退出
    if err != nil {
      panic(err.Error())
    }

    // 新建 discoveryClient 实例
    discoveryClient, err := discovery.NewDiscoveryClientForConfig(config)

    if err != nil {
      panic(err.Error())
    }

    // 获取所有分组和资源数据
    APIGroup, APIResourceListSlice, err :=
discoveryClient.ServerGroupsAndResources()

    if err != nil {
      panic(err.Error())
    }

    // 先看 Group 信息
```

```
    fmt.Printf("APIGroup :\n\n %v\n\n\n\n",APIGroup)

    // APIResourceListSlice是个切片，其中的每个元素代表一个GroupVersion及其资源
    for _, singleAPIResourceList := range APIResourceListSlice {

      // GroupVersion是个字符串，如“apps/v1”
      groupVerionStr := singleAPIResourceList.GroupVersion

      // 用ParseGroupVersion方法将字符串转成数据结构
      gv, err := schema.ParseGroupVersion(groupVerionStr)

      if err != nil {
        panic(err.Error())
      }

fmt.Println("*****************************************************************")
        fmt.Printf("GV string [%v]\nGV struct [%#v]\nresources :\n\n",
groupVerionStr, gv)

      // APIResources字段是个切片，其中是当前GroupVersion下的所有资源
      for _, singleAPIResource := range singleAPIResourceList.APIResources {
        fmt.Printf("%v\n", singleAPIResource.Name)
      }
    }
  }
```

部分显示结果如下：

```
NAME: pods/attach, GROUP: , VERSION: v1
NAME: pods/binding, GROUP: , VERSION: v1
NAME: pods/eviction, GROUP: , VERSION: v1
NAME: pods/exec, GROUP: , VERSION: v1
NAME: pods/log, GROUP: , VERSION: v1
NAME: pods/portforward, GROUP: , VERSION: v1
NAME: pods/proxy, GROUP: , VERSION: v1
NAME: pods/status, GROUP: , VERSION: v1
NAME: podtemplates, GROUP: , VERSION: v1
NAME: replicationcontrollers, GROUP: , VERSION: v1
NAME: replicationcontrollers/scale, GROUP: , VERSION: v1
NAME: replicationcontrollers/status, GROUP: , VERSION: v1
NAME: resourcequotas, GROUP: , VERSION: v1
NAME: resourcequotas/status, GROUP: , VERSION: v1
```

```
NAME: secrets, GROUP: , VERSION: v1
NAME: serviceaccounts, GROUP: , VERSION: v1
NAME: serviceaccounts/token, GROUP: , VERSION: v1
NAME: services, GROUP: , VERSION: v1
NAME: services/proxy, GROUP: , VERSION: v1
NAME: services/status, GROUP: , VERSION: v1
NAME: apiservices, GROUP: apiregistration.k8s.io, VERSION: v1
NAME: apiservices/status, GROUP: apiregistration.k8s.io, VERSION: v1
NAME: apiservices, GROUP: apiregistration.k8s.io, VERSION: v1beta1
NAME: apiservices/status, GROUP: apiregistration.k8s.io, VERSION: v1beta1
NAME: controllerrevisions, GROUP: apps, VERSION: v1
NAME: daemonsets, GROUP: apps, VERSION: v1
NAME: daemonsets/status, GROUP: apps, VERSION: v1
NAME: deployments, GROUP: apps, VERSION: v1
NAME: deployments/scale, GROUP: apps, VERSION: v1
NAME: deployments/status, GROUP: apps, VERSION: v1
NAME: replicasets, GROUP: apps, VERSION: v1
NAME: replicasets/scale, GROUP: apps, VERSION: v1
NAME: replicasets/status, GROUP: apps, VERSION: v1
```

DiscoveryClient 还可以将与资源相关的信息存储于本地，本地缓存可以减轻 client-go 对 Kubernetes API Server 的访问压力。在运行 Kubernetes 组件的机器上，缓存信息默认存储于 ~/.kube/cache 和 ~/.kube/http-cache 下。默认每 10 分钟与 Kubernetes API Server 同步一次，同步周期较长，因为资源组、源版本、资源信息一般很少变动。本地缓存的 DiscoveryClient 工作流程如图 7.4 所示。

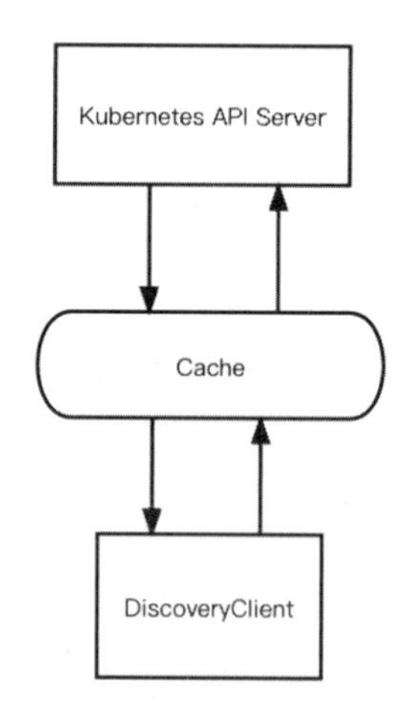

图 7.4　本地缓存的 DiscoveryClient 工作流程

7.4　理解 Kubernetes Informer 机制

在 Kubernetes 中使用 HTTP 进行通信时，如何在不依赖中间件的情况下保证消息的实时性、可靠性和顺序性呢？答案就是利用 Informer 机制。Kubernetes 的其他组件都是通过 client-go 的 Informer 机制与 Kubernetes API Server 进行通信的；Informer 机制，降低了 Kubernetes 各个组件与 Etcd 和 Kubernetes API Server 的通信压力。

Informer 机制的运行原理如图 7.5 所示。

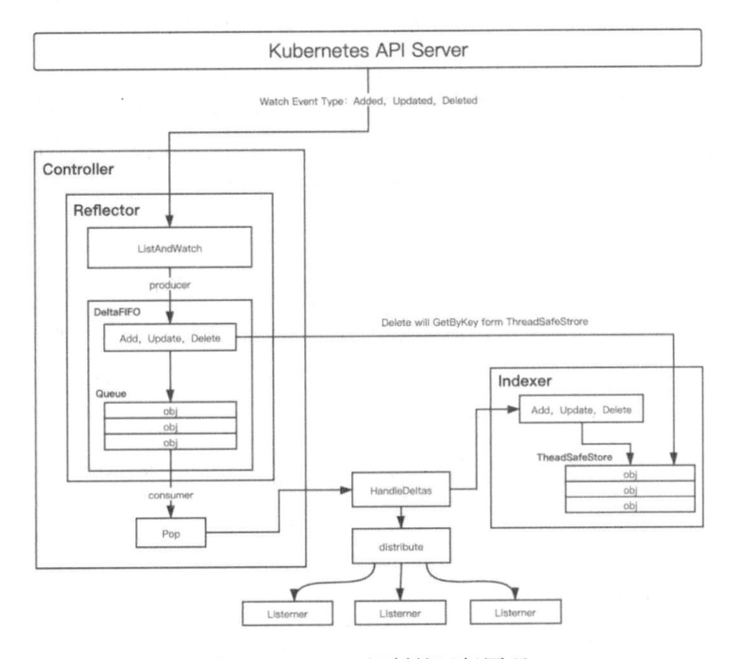

图 7.5　Informer 机制的运行原理

由图 7.5 所示可知，Informer 架构设计中有多个核心组件。

（1）Reflector：用于监控（Watch）指定的 Kubernetes 资源，当监控的资源发生变化时触发相应的变更事件（如 Added 事件、Updated 事件和 Deleted 事件），并将其资源对象存放到本地缓存 DeltaFIFO 中。

（2）DeltaFIFO：DeltaFIFO 是一个生产者 - 消费者的队列，生产者是 Reflector，消费者是 POP 函数。FIFO 是一个先进先出的队列，拥有队列操作的基本方法（Add、Update、Delete、List、Pop、Close 等）；Delta 是一个资源对象存储，可以保存资源对象的操作类型（Added、Updated、Deleted、Sync 等）。

（3）Indexer：Indexer 是 client-go 用来存储资源对象并自带索引功能的本地存储，Reflector 从 DeltaFIFO 中将消费出来的资源对象存储至 Indexer。Indexer 与 Etcd 集群中的数据完全一致。client-go 可以很方便地从本地存储中读取相应的资源对象数据，无须每次从远程 Etcd 集群读取，减轻了 Kubernetes API Server 和 Etcd 集群的压力。

下面用一个实际示例来说明以上核心组件的使用。例如，现在删除一个 Pod，Informer 的执行流程如下。

（1）初始化 Informer，Reflector 通过 List 接口获取所有 Pod 对象。

（2）Reflector 拿到所有 Pod 后，将全部 Pod 放到 Store（本地缓存）中。

（3）如果调用 Lister 的 List/Get 方法获取 Pod，那么 Lister 直接从 Store 中获取数据。

（4）Informer 初始化完成后，Reflector 开始 Watch Pod 相关的事件。

（5）如果删除 Pod1，那么 Reflector 会监听到这个事件，然后将这个事件发送到 DeltaFIFO 中。

（6）DeltaFIFO 首先将这个事件存储在一个队列中，然后操作 Store 中的数据，删除其中的 Pod1。

（7）DeltaFIFO 将 Pop 这个事件放到事件处理器（资源事件处理器）中进行处理。

（8）LocalStore 会周期性地把所有的 Pod 信息重新放回 DeltaFIFO 中。

一个控制器每次需要获取对象的时候都要访问 APIServer，这会给系统带来很高的负载，Informer 的内存缓存就是来解决这个问题的。此外，Informer 还可以实时监控对象的变化，而不需要轮询请求，这样可以保证客户端的缓存数据和服务端的数据一致，从而大大降低 APIServer 的压力。

Informer 的逻辑处理流程如图 7.6 所示。

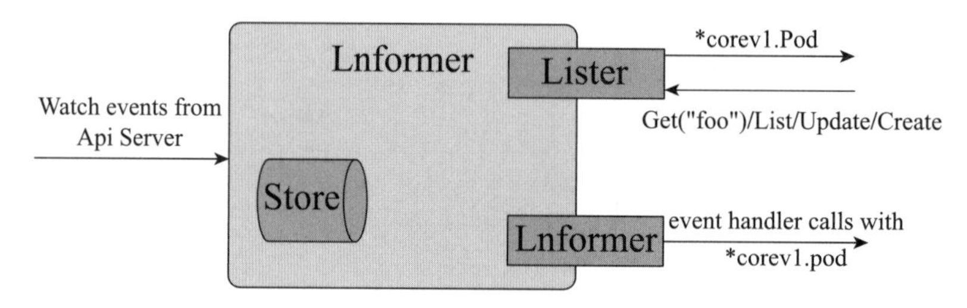

图 7.6　Informer 的逻辑处理流程

Informer 的逻辑处理流程：

（1）以 events 事件类型的方式从 APIServer 获取数据。

（2）提供一个类似客户端的 Lister 接口，从内存缓存中获取 get 和 list 对象。

（3）添加、删除、更新注册事件处理程序。

另外，Informer 也有错误处理方式，当长期运行的 watch 连接中断时，它们会尝试使用另一个 watch 请求来恢复连接，以在不丢失任何事件的情况下恢复事件流。如果中断的时间较长且 APIServer 丢失了事件（Etcd 在新的 watch 请求成功之前从数据库中清除这些事件），那么 Informer 会重新 List 全量数据。而且在重新 List 全量数据时还可以配置一个重新同步的周期参数，用于协调内存缓存数据和业务逻辑数据的一致性，每次过了该周期，注册的事件处理程序就将被所有的对象调用，通常这个周期参数以分钟为单位，如 10 分钟或者 30 分钟。

Informer 的源码比较复杂，这里用 testpod.go 源码来理解 Informer。代码如下：

```
package main

import (
```

```
        "flag"
        "fmt"
        v1 "k8s.io/api/apps/v1"
        "k8s.io/apimachinery/pkg/labels"
        "k8s.io/client-go/informers"
        "k8s.io/client-go/kubernetes"
        "k8s.io/client-go/rest"
        "k8s.io/client-go/tools/cache"
        "k8s.io/client-go/tools/clientcmd"
        "k8s.io/client-go/util/homedir"
        "path/filepath"
        "time"
    )

    func main() {
        // 程序会一直运行，直到按 Ctrl+C 键才能终止
        var err error
        var config *rest.Config
        var kubeconfig *string

        if home := homedir.HomeDir(); home != ""{
                kubeconfig = flag.String("kubeconfig", filepath.Join(home,
".kube", "config"), "[ 可选 ] kubeconfig 绝对路径 ")
        } else {
                kubeconfig = flag.String("kubeconfig", "", "kubeconfig 绝对路径 ")
        }

        // 初始化 rest.Config 对象
        if config, err = rest.InClusterConfig(); err != nil{
                if config, err = clientcmd.BuildConfigFromFlags("", *kubeconfig);
err != nil {
                        panic(err.Error())
                }
        }

        // 创建 clientSet 对象
        clientset, err := kubernetes.NewForConfig(config)
        if err != nil{
            panic(err.Error())
        }
```

```
        // 初始化 informerfactory
        // 为了测试方便，这里设置每 30 秒重新 List 一次
          informerFactory := informers.NewSharedInformerFactory(clientset,
time.Second*30)
        // 对 deployInformer 进行监听
        deployInformer := informerFactory.Apps().V1().Deployments()
        // 创建 Informer
        // 相当于注册到工厂中，这样下面启动时就会执行 list & watch 对应的资源
        informer := deployInformer.Informer()
        // 创建 deployLister
        deployLister := deployInformer.Lister()
        // 注册事件处理程序
        informer.AddEventHandler(cache.ResourceEventHandlerFuncs{
             AddFunc:    onAdd,
             UpdateFunc: onUpdate,
             DeleteFunc: onDelete,
        })

        stopper := make(chan struct{})
        defer close(stopper)

        //启动 informer 的资源监听与缓存功能
        informerFactory.Start(stopper)
        // 等待 Informer 同步
        informerFactory.WaitForCacheSync(stopper)

        // 从本地缓存中获取 default 命名空间中的所有 deployment 列表
        deployments, err :=
deployLister.Deployments("default").List(labels.Everything())
        if err != nil{
                panic(err.Error())
        }

        for idx, deploy := range deployments{
                fmt.Printf("%d -> %s\n", idx+1, deploy.Name)
        }
        <-stopper
    }

    func onAdd(obj interface{}) {
        deploy := obj.(*v1.Deployment)
        fmt.Println("add a deployment:", deploy.Name)
```

```
}

func onUpdate(old, new interface{}) {
	oldDeploy := old.(*v1.Deployment)
	newDeploy := new.(*v1.Deployment)
	fmt.Println("update deployment:", oldDeploy.Name, newDeploy.Name)
}

func onDelete(obj interface{}) {
	deploy := obj.(*v1.Deployment)
	fmt.Println("delete a deployment:", deploy.Name)
}
```

在以上代码中，informer.AddEventHandler 函数可以为 Pod 资源添加以下 3 种资源事件回调函数。

（1）AddFunc：创建 Pod 资源对象时触发的事件回调函数。

（2）UpdateFunc：更新 Pod 资源对象时触发的事件回调函数。

（3）DeleteFunc：删除 Pod 资源对象时触发的事件回调函数。

运行 testpod.go 程序，命令如下：

```
go run testpod.go
```

命令显示结果如下：

```
add a deployment: coredns
add a deployment: local-path-provisioner
update deployment: coredns coredns
update deployment: local-path-provisioner local-path-provisioner
```

下面另外开一个 Terminal，增加名为 nginx-dep 的 Deployment，命令如下：

```
kubectl create deployment nginx-dep --image=nginx:1.17.1
```

命令显示结果如下：

```
add a deployment: nginx-dep
update deployment: coredns coredns
update deployment: local-path-provisioner local-path-provisioner
update deployment: nginx-dep nginx-dep
```

然后尝试删除 nginx-dep 的 Deployment，命令如下：

```
kubectl delete deploy nginx-dep
```

命令显示结果如下：

```
delete a deployment: nginx-dep
update deployment: coredns coredns
update deployment: local-path-provisioner local-path-provisioner
```

因为代码中通过 Informer 注册了事件处理程序，这样当启动 Informer 时首先会将集群的全量 Deployment 数据同步到本地的缓存中，会触发 AddFunc 回调函数，然后又使用 Lister() 来获取 default 命名空间下面的所有 Deployment 数据，此时的数据是从本地的缓存中获取的，所以就看到了上面的结果。由于还配置了每 30 秒重新全量 List 一次，所以正常每 30 秒也可以看到所有的 Deployment 数据会出现在 UpdateFunc 回调函数下面，也可以尝试删除一个 Deployment，同样也会出现对应的 DeleteFunc 回调函数下面的事件。

7.5 小　　结

本章介绍了 Kubernetes 的基础知识和 client-go 的部分源码及基于 client-go 的 Informer 机制，希望能帮助读者提高对 client-go 的熟悉程度。因为整个 Kubernetes（包括 client-go）源码博大精深，流程设计精巧，所以希望读者能熟悉源码和深入理解其工作机制，这样才能有助于云原生 DevOps 开发。

第 8 章　Kubernetes 下的 Operator 脚手架开发工具 KUDO

8.1　Operator 和 CRD 的基本概念

Operator 可以被视为另一种 Controller 的形式。当前的 Controller Manager 主要管理基础的、通用的资源概念，如 RS/Deployment，而对于特定的应用或者服务（如 etcd cluster，都可以认为是一种资源），则放权给了第三方，即 CRD。

那么，什么是 CRD（Custom Resource Definition）呢？

（1）在 Kubernetes 系统中，所有对象都可以视为资源，以便于管理和调度。Kubernetes 1.7 之后增加了对 CRD 自定义资源进行二次开发的功能，以扩展 Kubernetes API。通过 CRD 可以向 Kubernetes API 中增加新的资源类型，而不需要修改 Kubernetes 源码来创建自定义的 API Server，该功能大大提高了 Kubernetes 的扩展能力。

（2）当创建一个新的 CRD 时，Kubernetes API 服务器将为指定的每个版本创建一个新的 RESTful 资源路径，可以根据该 API 路径来创建一些自定义的资源类型。CRD 可以是命名空间的，也可以是集群范围的，由 CRD 的作用域（scope）字段指定。与现有的内置对象一样，删除命名空间将删除该命名空间中的所有自定义对象。CRD 本身没有命名空间，所有命名空间都可以使用。

用户可以通过自定义的资源描述，以及自研的 Controller/Operator 进行接入。因此 Controller 和 Operator 的关系类似于标准库和第三方库的关系。

一般来说，不同的应用需要不同的 Operator 进行处理。以 ReplicaSet Controller 为例，ReplicaSet 的主要功能是保持副本数。当有 Pod 因某种原因挂掉 / 删除时，对于无状态的应用来说，恢复的方式就是再增加对应的 Pod 数量。从这个角度来说，ReplicaSet Controller 其实就是无状态应用的 Operator。

云原生 DevOps，其实就是运维和开发的结合，可以提升开发交付的效率和质量。这也带来了一个趋势，就是运维和开发一体化。例如，原来的开发人员，通过 Docker 镜像的制作，将应用的部署启动等固化在了 Dockerfile 镜像中，分担了运维的许多部署工作。但是实际上，运维的工作内容和范围非常广，特别是现在的分布式系统，包括集群化部署、高可用、故障恢复、系统升级等工作。而这些是无法仅用 Docker 镜像进行固化的。

Operator 提供了一种可能，或者说是一个很好的框架，就是把运维的经验沉淀为代码，实现运维的代码化、自动化、智能化。以往的高可用、扩展收缩，以及故障恢复等运维操作，都通过

Operator 进行沉淀形成代码。从长期来看，将会推进 Dev、Ops、DevOps 的深度一体化。将运维经验、应用的各种方案和功能通过代码的方式进行固化和传承，减少人为故障的概率，提升整个运维的效率。

传统架构中，无论是 MySQL 集群还是 MySQL 主从复制，数据备份、迁移和回滚等操作通常都无法通过 Kubernetes 的原生组件和资源来完成。而 Operator 的作用就是将这些日常的运维经验和操作进行代码化，将复杂的手动操作自动化，从而实现自动化运维的目标。

Operator 的许多理念并不是现在才有的。Yarn 中的 Application Manager 中，Mesos 中的 Framework 中，都可以找到 Operator 的应用。之所以说 Operator 将可能成为 Docker 之后的又一项重大变革，其另外一个重要的因素就是 Operator 是基于 Kubernetes 的应用。

Kubernetes 强化了基础资源的封装，并保持了灵活性和可定制性。它从传统资源（CPU/MEM）的交付，转为了 Pod/SVC/PV/PVC 等资源的交付，扩展了资源的概念，将域名、负载均衡、存储等必要或相关的概念也进行了封装。而 Operator 在这些公共资源的基础上，将应用集群也视为了一种资源，可以向用户提供，并且借助 Kubernetes 已有的工作机制和框架，从而更为便捷灵活地实现。

构建 Operator 是一项重要且复杂的工作，通常需要数千行 Go 代码，这个需要 DevOps 开发人员对 Kubernetes 内部有深入了解以及对应用程序本身运维有专门的领域特定知识。对于许多 IT 工作人员而言，这些要求都过于复杂了，因此难以利用 Operator 的强大功能。

近几年来，D2iQ（原名 Mesosphere，是一家位于加利福尼亚州旧金山的美国技术公司），其开发的软件 KUDO 可简化 Kubernetes 生命周期管理，像混合、多云和边缘环境的部署，并支持高级应用程序用例），在 20 多个国家 / 地区举行了 60 多次会议和聚会演讲。KUDO 现在正式成为 CNCF 项目并成为 CNCF 生态系统的一部分，这使得该项目能够在社区建立的治理模型下继续发展，从而推动项目本身。

建筑工地在建房子时，最开始都要搭建一个脚手架，以便于更快更安全地施工。同样，Operator 工程的构建也要搭建一个脚手架，以方便后续快速地开发和迭代，这就是 Operator 脚手架工具。Kubernetes 社区有很多成熟的构建脚手架的工具供选择，如 Kuberbuilder/Operator SDK，但是它们均要求用户熟练掌握 Go 程序开发和 Kubernetes 控制器模式。KUDO（Kubernetes Universal Declarative Operator，Kubernetes 通用声明性 Operator）是 Operator 的开发工具和运行时，它为 Kubernetes 开发了一个 SDK，该 SDK 提供了一组用于任务调度和编排的基本用语。KUDO 通过抽象层简化了开发过程，该抽象层将通用动作编码为计划、阶段、步骤和任务的概念，KUDO 使 Operator 的编写变得简单高效，允许 Operator 开发者和最终用户使用他们已知的工具来管理有状态服务，这便是 KUDO 的由来。

KUDO 的作用可以简单总结如下：

（1）使用 KUDO 可以方便地开发基于 Kubernetes 有状态的 Operator 复杂应用。

（2）KUDO 是一个低代码化的 Operator 脚手架开发工具，实际上，通常只需编写 YAML 和 Shell 脚本就能完成 Kubernetes Operator 的开发工作。

（3）KUDO 非常适合强依赖性的组件的 Operator 开发工作。例如，开发名为 F 的 Operator，此 Operator 中包含 A ~ E 个应用，并且启动 A 应用需要依赖 C 应用，启动 C 应用又需要依赖 E 应用。

8.2　KUDO 源码简介

KUDO 与 Kubernetes API 进行交互的工作流如图 8.1 所示。

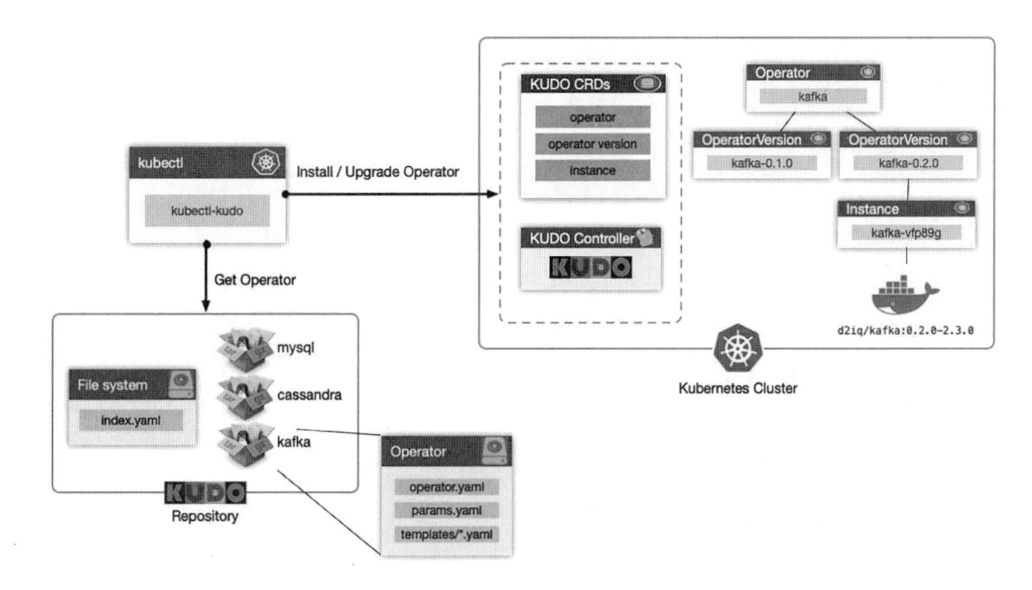

图 8.1　KUDO 与 Kubernetes API 进行交互的工作流

kubectl-kudo 是命令行 CLI 工具客户端，它会调用本地 Repository 或本地文件，与 Kubernetes API Server 进行交互，从而生成相应的 Operator 实例。该客户端可以帮助开发人员创建 KUDO Operator，并且用来管理 Kubernetes 集群中的 Operator 实例。kubectl-kudo 会在 Kubernetes 集群中创建 kudo-system 命名空间，并在其中创建 KUDO Controller 和与 KUDO 相关的 CRD 资源。

KUDO CRD 能扩展 Kubernetes 的 API 资源以支持 KUDO。

KUDO Controller 提供相关 KUDO CRD 定义的服务，并管理 KUDO Operator 实例。KUDO Controller 负责：

- 观察 Kubernetes KUDO 对象并确保期望的状态。
- 创建 KUDO Operator 并调用 Operator 计划。

KUDO 源码的下载命令如下：

```
git clone https://github.com/kudobuilder/kudo.git
```

KUDO 源码最终下载工作目录为 /Users/yuhongchun/repo/workspace/cloudnative/kudo，用 tree 命令可以查看这个名为 kudo 工程的层级结构。命令如下：

```
tree -L 2 /Users/yuhongchun/repo/workspace/cloudnative/kudo
```

命令显示结果如下：

```
./kudo/
├── CONTRIBUTING.md
├── Dispatchfile
├── Dockerfile
├── Dockerfile.goreleaser
├── LICENSE
├── Makefile
├── OWNERS
├── PROJECT
├── README.md
├── RELEASE.md
├── SECURITY_CONTACTS
├── STYLEGUIDE.md
├── build
│   ├── Dockerfile
│   └── test-images.sh
├── cmd
│   ├── kubectl-kudo
│   └── manager
├── code-of-conduct.md
├── config
│   └── crds
├── development.md
├── go.mod
├── go.sum
├── hack
│   ├── boilerplate.go.txt
│   ├── check-commit-signed-off.sh
│   ├── deploy-dev-prereqs.sh
│   ├── generate_krew.sh
│   ├── gh-md-toc.sh
│   ├── install-golangcilint.sh
│   ├── run-e2e-tests.sh
│   ├── run-integration-tests.sh
│   ├── run-kuttl-tests.sh
│   ├── run-operator-tests.sh
│   ├── run-upgrade-tests.sh
│   ├── update-manifests.sh
```

```
|       ├── update-webhook-config.sh
|       ├── update_codegen.sh
|       ├── update_kep_overview.sh
|       ├── verify-generate.sh
|       └── verify-go-clean.sh
├── keps
|       ├── 0000-kep-template.md
|       ├── 0001-kep-process.md
|       ├── 0002-dynamic-instances.md
|       ├── 0003-kep-cli.md
|       ├── 0004-add-testing-infrastructure.md
|       ├── 0005-cluster-resources-for-crds.md
|       ├── 0006-stable-kafka-example.md
|       ├── 0007-cli-generation.md
|       ├── 0008-operator-testing.md
|       ├── 0009-operator-toolkit.md
|       ├── 0010-package-manager.md
|       ├── 0012-operator-extensions.md
|       ├── 0013-external-specs.md
|       ├── 0014-pull-request-process.md
|       ├── 0015-repository-management.md
|       ├── 0017-pipe-tasks.md
|       ├── 0018-controller-overhaul.md
|       ├── 0019-package-api-versioning.md
|       ├── 0020-manual-plan-execution.md
|       ├── 0021-kudo-upgrade.md
|       ├── 0022-diagnostics-bundle.md
|       ├── 0023-enable-disable-features.md
|       ├── 0024-parameter-enhancement.md
|       ├── 0025-template-to-yaml-function.md
|       ├── 0026-reading-parameter-values-from-a-file.md
|       ├── 0027-pod-restart-controls.md
|       ├── 0029-operator-dependencies.md
|       ├── 0030-immutable-parameters.md
|       ├── 0031-namespace-management.md
|       ├── 0032-community-repository-management.md
|       ├── 0033-structured-parameters.md
|       ├── 0034-instance-health.md
|       ├── 0035-json-schema-export.md
|       └── README.md
├── logo
```

```
│    ├── kudo_horizontal_color.png
│    ├── kudo_horizontal_color@2x.png
│    ├── kudo_horizontal_dark.png
│    ├── kudo_horizontal_dark@2x.png
│    ├── kudo_horizontal_light.png
│    ├── kudo_horizontal_light@2x.png
│    ├── kudo_stacked_color.png
│    ├── kudo_stacked_color@2x.png
│    ├── kudo_stacked_dark.png
│    ├── kudo_stacked_dark@2x.png
│    ├── kudo_stacked_light.png
│    └── kudo_stacked_light@2x.png
├── pkg
│    ├── apis
│    ├── client
│    ├── controller
│    ├── engine
│    ├── feature
│    ├── kubernetes
│    ├── kudoctl
│    ├── test
│    ├── util
│    ├── version
│    └── webhook
├── test
│    ├── Dockerfile
│    ├── README.md
│    ├── cert
│    ├── e2e
│    ├── integration
│    ├── kudo-e2e-test.yaml.tmpl
│    ├── kudo-integration-test.yaml
│    ├── kudo-upgrade-test.yaml.tmpl
│    ├── manifests
│    ├── run_tests.sh
│    └── upgrade
└── tools.go

27 directories, 88 files
```

由于工作目录路径太长，因此后面用 workspace 别名代指 /Users/yuhongchun/repo/workspace。

KUDO 是基于 Go 1.16 版本开发的（Go 高版本是往下兼容的）。在阅读源码之前，建议多熟悉下第 7 章中与 Kubernetes 有关的 Group、Version、Resource、Kind 等核心数据结构，因为 KUDO 源码很多都与之有关联。下面简单介绍源码，重点放在 CRD 资源及相应的 CRD Controller 上。

KUDO 会生成相应的 KUDO CRD 第三方资源，分别是 Instances、Operator 及 OperatorVersion CRD，可以关注 workspace/config/crds 下的 yaml 文件，它们会依次在 kudo-system 命名空间下生成 KUDO 相对应的名为 instances.kudo.dev、operators.kudo.dev 及 operatorversions.kudo.dev 的 CRD 资源，如图 8.2 所示。

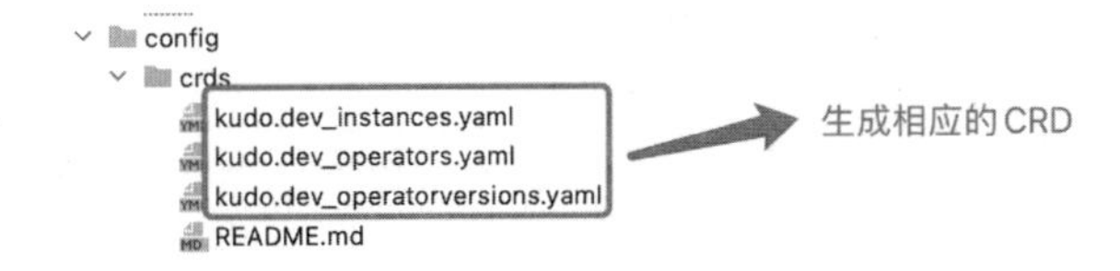

图 8.2 config 目录下的 crds 中的 yaml 文件

下面重点介绍以上这些 CRD 资源对应的 controller 方法。这里首先解释这些 CRD 资源的作用，具体如下：

Operator 包含要在 Kubernetes 集群中运行的应用程序的高级描述（不要与整体 Operatortp 管理器混淆）。它只包含有关应用程序的元数据，但没有特定的计划或资源。例如，可以在当前 Kubernetes 集群中安装多个版本的应用程序，所有这些版本都属于同一个 Operator。

OperatorVersion 是应用程序的具体版本。它包含运营商使用的所有 Kubernetes 资源（部署、服务）、计划和参数。将 OperatorVersion 视为一个类，实例化后，它将成为正在运行的应用程序。

Instance 是应用程序的部署版本。如果把 OperatorVersion 比作类，则 Instance 是该类的实例化对象。它必须提供缺少的参数值，并且可以覆盖 OperatorVersion 中定义的默认值。在大多数情况下，创建实例意味着执行部署计划，该部署计划将根据计划、阶段和步骤呈现和应用 Kubernetes 资源。

下面介绍控制器文件的核心代码。

第一个控制器文件是 operator_controller，其源码文件为 workspace/pkg/controller/instance/operator_controller.go，主要核心代码如下：

```
// Reconciler 结构体协调一个 Operator 对象
type Reconciler struct {
	client.Client
}

// SetupWithManager 函数向控制器管理器注册此协调器
func (r *Reconciler) SetupWithManager(
	mgr ctrl.Manager) error {
	return ctrl.NewControllerManagedBy(mgr).
```

```
                For(&kudoapi.Operator{}).
                Complete(r)
    }

    /*
    Reconcile方法读取Operator对象的集群状态，并根据读取的集群状态和Operator.Spec中的
内容进行更改，自动生成RBAC规则，让Controller读写Deployments
    */
    func (r *Reconciler) Reconcile(request ctrl.Request) (ctrl.Result, error) {
        // Fetch the operator
        operator := &kudoapi.Operator{}
        err := r.Get(context.TODO(), request.NamespacedName, operator)
        if err != nil{
                if errors.IsNotFound(err){
                        return reconcile.Result{}, nil
                }
                return reconcile.Result{}, err
        }

          log.Printf("OperatorController: Received Reconcile request for an
operator named: %v", request.Name)

        return reconcile.Result{}, nil
    }
```

第二个控制器文件是operatorversion_controller，其源码文件为workspace/pkg/controller/instance/operatorversion_controller.go，主要核心代码如下：

```
    // Reconciler结构体协调一个Operator对象
    type Reconciler struct {
        client.Client
    }

    // SetupWithManager函数向控制器管理器注册此协调器
    func (r *Reconciler) SetupWithManager(
        mgr ctrl.Manager) error {
        return ctrl.NewControllerManagedBy(mgr).
                For(&kudoapi.OperatorVersion{}).
                Complete(r)
    }

    /* Reconcile方法读取 OperatorVersion 对象的集群状态，并根据读取的集群状态和
```

```
OperatorVersion.Spec 中的内容进行更改
    // 自动生成 RBAC 规则，让 Controller 读写 Deployments
    */
    func (r *Reconciler) Reconcile(request ctrl.Request) (ctrl.Result, error) {
        // Fetch the operator version
        operatorVersion := &kudoapi.OperatorVersion{}
        err := r.Get(context.TODO(), request.NamespacedName, operatorVersion)
        if err != nil{
            if errors.IsNotFound(err){
                return reconcile.Result{}, nil
            }
            return reconcile.Result{}, err
        }

        log.Printf("OperatorVersionController: Received Reconcile request for
an operatorVersion named: %v", request.Name)

        // 待办事项：验证 OperatorVersion 是否合适
        return reconcile.Result{}, nil
    }
```

第三个控制器文件是 instances_controller，其源码文件为 /workspace/pkg/controller/instance/instances_controller.go，这里主要介绍 instance_controller 的 Reconcile 方法，源码内容如下：

```
    // Reconciler 结构体协调一个 Operator 对象
    type Reconciler struct {
        client.Client
        Discovery discovery.CachedDiscoveryInterface
        Config     *rest.Config
        Recorder   record.EventRecorder
        Scheme     *runtime.Scheme
    }
    /* Reconcile 方法是主要的控制器方法，每次实例发生变化时都会调用该方法
    自动生成 RBAC 规则，让 Controller 读写 Deployments
    */
    func (r *Reconciler) Reconcile(request ctrl.Request) (ctrl.Result, error) {
        //---------- 1. 查询当前状态 ----------

        log.Printf("InstanceController: Received Reconcile request for
instance %s", request.NamespacedName)
        instance, err := r.getInstance(request)
        if err != nil {
```

```
            if apierrors.IsNotFound(err){
                  log.Printf("Instance %s was deleted, nothing to reconcile.",
request.NamespacedName)
                  return reconcile.Result{}, nil
            }
            return reconcile.Result{}, err
      }
      oldInstance := instance.DeepCopy()

      ov, err := instance.GetOperatorVersion(r.Client)
      if err != nil{
              err = fmt.Errorf("InstanceController: Error getting
operatorVersion %s for instance %s/%s: %v",
                        instance.Spec.OperatorVersion.Name,
instance.Namespace, instance.Name, err)
            log.Print(err)
            r.Recorder.Event(instance, "Warning",
"InvalidOperatorVersion", err.Error())
            return reconcile.Result{}, err
      }

      //---------- 2.如果计划存在，则获取当前计划 ----------

      // 得到预定的计划
      plan, uid := scheduledPlan(instance, ov)
      if plan == ""{
            // 如果没有计划在运行，仍然需要确保 readiness 属性是最新的
            err := setReadinessOnInstance(instance, r.Client)
            if err != nil{
                  log.Printf("InstanceController: Error when computing
readiness for %s/%s: %v", instance.Namespace, instance.Name, err)
                  return reconcile.Result{}, err
            }
            if readinessChanged(oldInstance, instance){
                  err = updateInstance(instance, oldInstance, r.Client)
            } else {
                  log.Printf("InstanceController: Readiness did not
change for %s/%s. Not updating.", instance.Namespace, instance.Name)
            }
            return reconcile.Result{}, err
```

```
        }

        ensureReadinessInitialized(instance)
        ensurePlanStatusInitialized(instance, ov)

        // 如果计划是新的，以日志的方式记录它并重置其状态
        planStatus, err := resetPlanStatusIfPlanIsNew(instance, plan, uid)
        if err != nil{
                log.Printf("InstanceController: Error resetting instance %s/%s
status. %v", instance.Namespace, instance.Name, err)
              return reconcile.Result{}, err
        }

        if planStatus.Status == kudoapi.ExecutionPending{
                log.Printf("InstanceController: Going to start execution of
    plan '%s' on instance %s/%s", plan, instance.Namespace, instance.Name)
                r.Recorder.Event(instance, "Normal", "PlanStarted",
fmt.Sprintf("Execution of plan %s started", plan))
        }

        // 如果有必要，检查是否可以解决所有依赖项
        err = r.resolveDependencies(instance, ov)
        if err != nil{
             planStatus.SetWithMessage(kudoapi.ExecutionFatalError, err.Error())
             instance.UpdateInstanceStatus(planStatus, &metav1.Time{Time:
time.Now()})
             err = r.handleError(err, instance, oldInstance)
             return reconcile.Result{}, err
        }

        //---------- 3. 执行预定的计划 ----------

        metadata := &engine.Metadata{
             OperatorVersionName: ov.Name,
             OperatorVersion:     ov.Spec.Version,
             AppVersion:          ov.Spec.AppVersion,
             ResourcesOwner:      instance,
             OperatorName:        ov.Spec.Operator.Name,
             InstanceNamespace:   instance.Namespace,
             InstanceName:        instance.Name,
```

```
        }

        activePlan, err := preparePlanExecution(instance, ov, planStatus, metadata)
        if err != nil{
                err = r.handleError(err, instance, oldInstance)
                return reconcile.Result{}, err
        }
        log.Printf("InstanceController: Going to proceed with execution of the scheduled plan '%s' on instance %s/%s", activePlan.Name, instance.Namespace, instance.Name)
        newStatus, err := workflow.Execute(activePlan, metadata, r.Client, r.Discovery, r.Config, r.Scheme)

        //---------- 4.执行后更新实例及其状态 ----------

        if newStatus != nil{
                instance.UpdateInstanceStatus(newStatus, &metav1.Time{Time: time.Now()})
        }
        if err != nil{
                err = r.handleError(err, instance, oldInstance)
                return reconcile.Result{}, err
        }

        err = updateInstance(instance, oldInstance, r.Client)
        if err != nil{
                log.Printf("InstanceController: Error when updating instance %s/%s. %v", instance.Namespace, instance.Name, err)
            return reconcile.Result{}, err
        }

        // 在实例及其状态成功更新后发布PlanFinished事件
        if instance.Spec.PlanExecution.Status.IsTerminal(){
              r.Recorder.Event(instance, "Normal", "PlanFinished", fmt.Sprintf("Execution of plan %s finished with status %s", newStatus.Name, newStatus.Status))
        }

        return computeTheReconcileResult(instance, time.Now), nil
    }
```

Reconcile方法是instances_controller的主要方法，每次实例发生变化时都会调用该方法。它主要完成以下工作：

（1）查询 instance 和 OperatorVersion 的状态。

（2）如果存在预定的计划，则执行。

（3）使用新的执行状态更新 instance。

（4）即使没有计划运行，也更新准备情况。

KUDO 管理器在日常的运维管理工作中会大量涉及版本更新和升级的需求，所以会有版本比对的强需求。

workspace/pkg/version/version.go 主要用于 KUDO 的版本管理和比对。这里穿插个小知识点，软件版本的制订规则通常为 major.minor(.build)。其中，major 是最大的版本编号；minor 其次；某些软件可能再细分为 build，即更小的版本编号。

version.go 文件中的 CompareMajorMinor 函数仅针对主要元素和次要元素提供比较结果 –1（小于）、0（等于）、1（大于），从而达到版本比对的效果。version.go 源码中关于版本比较的代码片段如下：

```
    /* CompareMajorMinor 函数仅针对主要元素和次要元素提供比较结果 -1、0、1;semver 包忽
略补丁或预发布元素，假如正在寻找 minVersion，这个会很有用，如 1.15.6 是 1.15 或更高版本 */
    func (v *Version) CompareMajorMinor(o *Version) int {
        if d := compareSegment(v.Major(), o.Major()); d != 0 {
                return d
        }
        if d := compareSegment(v.Minor(), o.Minor()); d != 0 {
                return d
        }
        return 0
    }

    // 将 v1 与 v2 进行比较，结果为 -1、0、1，分别表示小于、等于、大于
    func compareSegment(v1, v2 uint64) int {
        if v1 < v2{
                return -1
        }
        if v1 > v2{
                return 1
        }

        return 0
    }

    // New 函数从 semver 字符串提供 Version 的实例
    func New(v string) (*Version, error) {
```

```
    ver, err := semver.NewVersion(v)
    if err != nil{
        return nil, err
    }
    return FromSemVer(ver), nil
}

// FromGithubVersion 函数会解析以 v 开头的版本。例如，会将 v1.5.2 自动转成 1.5.2，注意
// 会回调下面的 New 函数
func FromGithubVersion(v string) (*Version, error) {
    return New(Clean(v))
}

// FromSemVer 函数将 semver.Version 转换为参数指定的版本
func FromSemVer(v *semver.Version) *Version {
    return &Version{v}
}

// MustParse 函数解析给定的版本并在出错时抛出 Panic 级错误
func MustParse(v string) *Version {
    return FromSemVer(semver.MustParse(v))
}

// Clean 函数返回不带前缀 v 的版本（如果存在）
func Clean(ver string) string {
    if strings.HasPrefix(ver, "v"){
        return ver[1:]
    }
    return ver
}
```

以上代码主要调用了外部的第三方开源 semver 包，其地址为 https://github.com/Masterminds/semver/，主要功能如下：

（1）解析语义版本。

（2）排序语义版本。

（3）检查版本是否合法。

（4）可选择使用 v 前缀，即类似 v2.3.4 这种带 v 的版本也认为是合法的。

（5）比较两个版本的大小。

以上功能的具体实现可以看一下 version.go 文件，主要是 Compare 方法。该方法会按照 X.Y.Z 的方式将当前版本与另一个版本进行比较，如果版本小于、等于或大于另一个版本，则分别返

回 -1、0 或 0。其主要实现逻辑如下：

```
/* Compare 方法会将此版本与另一个版本进行比较，返回 -1、0 或 1，如果版本小于、等于或大于另一个版本，则版本将按 X.Y.Z 的方式进行比较
*/
func (v *Version) Compare(o *Version) int {
    // Compare the major, minor, and patch version for differences. If a
    // difference is found return the comparison
    if d := compareSegment(v.Major(), o.Major()); d != 0 {
            return d
    }
    if d := compareSegment(v.Minor(), o.Minor()); d != 0 {
            return d
    }
    if d := compareSegment(v.Patch(), o.Patch()); d != 0 {
            return d
    }

    // 此时 major、minor、patch 版本是一样的
    ps := v.pre
    po := o.Prerelease()

    if ps == "" && po == ""{
            return 0
    }
    if ps == ""{
            return 1
    }
    if po == ""{
            return -1
    }

    return comparePrerelease(ps, po)
}
```

再关注下 test_version.go 的单元测试文件，其 TestCompare 函数代码如下：

```
func TestCompare(t *testing.T) {
    tests := []struct{
        v1       string
        v2       string
        expected int
    }{
```

```
        {"1.2.3", "1.5.1", -1},
        {"2.2.3", "1.5.1", 1},
        {"2.2.3", "2.2.2", 1},
        {"3.2-beta", "3.2-beta", 0},
        {"1.3", "1.1.4", 1},
        {"4.2", "4.2-beta", 1},
        {"4.2-beta", "4.2", -1},
        {"4.2-alpha", "4.2-beta", -1},
        {"4.2-alpha", "4.2-alpha", 0},
        {"4.2-beta.2", "4.2-beta.1", 1},
        {"4.2-beta2", "4.2-beta1", 1},
        {"4.2-beta", "4.2-beta.2", -1},
        {"4.2-beta", "4.2-beta.foo", -1},
        {"4.2-beta.2", "4.2-beta", 1},
        {"4.2-beta.foo", "4.2-beta", 1},
        {"1.2+bar", "1.2+baz", 0},
        {"1.0.0-beta.4", "1.0.0-beta.-2", -1},
        {"1.0.0-beta.-2", "1.0.0-beta.-3", -1},
        {"1.0.0-beta.-3", "1.0.0-beta.5", 1},
    }
```

测试是讲究覆盖率的，这种方式是表组测试，还可以修改 {} 中的数据内容，最后可以运行以下命令：

```
go test -run 'TestCompare' -v
```

从测试结果可以发现，测试达到了预期，即通过了单元测试。命令显示结果如下：

```
=== RUN   TestCompare
1.2.3
1.5.1
2.2.3
1.5.1
2.2.3
2.2.2
3.2.0-beta
3.2.0-beta
1.3.0
1.1.4
4.2.0
4.2.0-beta
4.2.0-beta
4.2.0
4.2.0-alpha
```

```
4.2.0-beta
4.2.0-alpha
4.2.0-alpha
4.2.0-beta.2
4.2.0-beta.1
4.2.0-beta2
4.2.0-beta1
4.2.0-beta
4.2.0-beta.2
4.2.0-beta
4.2.0-beta.foo
4.2.0-beta.2
4.2.0-beta
4.2.0-beta.foo
4.2.0-beta
1.2.0+bar
1.2.0+baz
1.0.0-beta.4
1.0.0-beta.-2
1.0.0-beta.-2
1.0.0-beta.-3
1.0.0-beta.-3
1.0.0-beta.5
--- PASS: TestCompare (0.00s)
PASS
ok      github.com/Masterminds/semver/v3 0.196s
```

因为整个KUDO源码阅读起来比较晦涩，所以这里只介绍了KUDO CRD及CRD相对应的Controller文件内容；后面可以结合具体的KUDO安装、更新配置、升级版本和KUDO Operator实例来熟悉KUDO，如果在这个过程中有针对性地进行代码审阅，效果可能会更好，这里可以简单地熟悉和了解一下。

8.3 KUDO的安装和使用

KUDO是依附在Kubernetes集群上的，因此需要一个存在的Kubernetes集群，底层容器运行时为Docker或containerd均可。

8.3.1 KUDO安装的预置条件

KUDO安装的预置条件有：

（1）需要一个Kubernetes集群（基本条件）。

（2）Kubernetes 集群需要 cert-manager。

（3）kubectl 版本大于或等于 1.13.0。

（4）Kubernetes 版本大于或等于 1.16。

为什么需要 cert-manager 呢？

随着 HTTPS 的不断普及，大多数网站开始由 HTTP 升级到 HTTPS。使用 HTTPS 需要向权威机构申请证书，并且需要付出一定的成本，如果需求数量多，则开支也相对增加。cert-manager 是 Kubernetes 上的全能证书管理工具，支持基于 ACME 协议与 Let's Encrypt 签发免费证书并为证书自动续期，实现永久免费使用证书。cert-manager 部署到 Kubernetes 集群后会查阅其所支持的自定义资源 CRD，可通过创建 CRD 资源来指示 cert-manager 签发证书并为证书自动续期。如果要了解更详细的资料，可以直接参考官方文档。

下面参考第 7 章内容，在一台华为云主机上提前部署名为 kind 的 Kubernetes 集群，然后通过 kubectl get nodes 查看节点分配。命令显示结果如下：

```
      Welcome to Huawei Cloud Service

    # kubectl get nodes
    NAME                 STATUS   ROLES                  AGE   VERSION
    kind-control-plane   Ready    control-plane,master   18d   v1.20.7
    kind-worker          Ready    <none>                 18d   v1.20.7
    kind-worker2         Ready    <none>                 18d   v1.20.7
    kind-worker3         Ready    <none>                 18d   v1.20.7
```

再用 kubectl version 查看 Server 和 Client 端的版本，命令显示结果如下：

```
    Client Version: version.Info{Major:"1", Minor:"26", GitVersion:"v1.26.0",
GitCommit:"b46a3f887ca979b1a5d14fd39cb1af43e7e5d12d", GitTreeState:"clean",
BuildDate:"2022-12-08T19:58:30Z", GoVersion:"go1.19.4", Compiler:"gc",
Platform:"linux/amd64"}
    Kustomize Version: v4.5.7
    Server Version: version.Info{Major:"1", Minor:"20", GitVersion:"v1.20.7",
GitCommit:"132a687512d7fb058d0f5890f07d4121b3f0a2e2", GitTreeState:"clean",
BuildDate:"2021-05-27T23:27:49Z", GoVersion:"go1.15.12", Compiler:"gc",
Platform:"linux/amd64"}
```

Server Version 显示的是 Kubernetes 服务端的版本，Client Version 显示的是客户端的版本，这里发现是 1.26，满足 KUDO 安装需要的 kubectl 版本需求。下面介绍安装 cert-manager 的步骤。

（1）下载 cert-manager.yaml 文件。命令如下：

```
    wget https://github.com/cert-manager/cert-manager/releases/download/
v1.11.0/cert-manager.yaml
```

（2）用 kubectl 命令来安装。安装命令如下：

```
kubectl apply -f cert-manager.yaml
```

命令显示结果如下：

```
namespace/cert-manager created
customresourcedefinition.apiextensions.k8s.io/clusterissuers.cert-manager.
io created
customresourcedefinition.apiextensions.k8s.io/challenges.acme.cert-manager.
io created
customresourcedefinition.apiextensions.k8s.io/certificaterequests.cert-
manager.io created
customresourcedefinition.apiextensions.k8s.io/issuers.cert-manager.io
created
customresourcedefinition.apiextensions.k8s.io/certificates.cert-manager.io
created
customresourcedefinition.apiextensions.k8s.io/orders.acme.cert-manager.io
created
serviceaccount/cert-manager-cainjector created
serviceaccount/cert-manager created
serviceaccount/cert-manager-webhook created
configmap/cert-manager-webhook created
clusterrole.rbac.authorization.k8s.io/cert-manager-cainjector created
clusterrole.rbac.authorization.k8s.io/cert-manager-controller-issuers
created
clusterrole.rbac.authorization.k8s.io/cert-manager-controller-
clusterissuers created
clusterrole.rbac.authorization.k8s.io/cert-manager-controller-certificates
created
clusterrole.rbac.authorization.k8s.io/cert-manager-controller-orders
created
clusterrole.rbac.authorization.k8s.io/cert-manager-controller-challenges
created
clusterrole.rbac.authorization.k8s.io/cert-manager-controller-ingress-shim
created
clusterrole.rbac.authorization.k8s.io/cert-manager-view created
clusterrole.rbac.authorization.k8s.io/cert-manager-edit created
clusterrole.rbac.authorization.k8s.io/cert-manager-controller-
approve:cert-manager-io created
clusterrole.rbac.authorization.k8s.io/cert-manager-controller-
certificatesigningrequests created
clusterrole.rbac.authorization.k8s.io/cert-manager-webhook:subjectaccessreviews
created
```

```
    clusterrolebinding.rbac.authorization.k8s.io/cert-manager-cainjector
created
    clusterrolebinding.rbac.authorization.k8s.io/cert-manager-controller-
issuers created
    clusterrolebinding.rbac.authorization.k8s.io/cert-manager-controller-
clusterissuers created
    clusterrolebinding.rbac.authorization.k8s.io/cert-manager-controller-
certificates created
    clusterrolebinding.rbac.authorization.k8s.io/cert-manager-controller-
orders created
    clusterrolebinding.rbac.authorization.k8s.io/cert-manager-controller-
challenges created
    clusterrolebinding.rbac.authorization.k8s.io/cert-manager-controller-
ingress-shim created
    clusterrolebinding.rbac.authorization.k8s.io/cert-manager-controller-
approve:cert-manager-io created
    clusterrolebinding.rbac.authorization.k8s.io/cert-manager-controller-
certificatesigningrequests created
    clusterrolebinding.rbac.authorization.k8s.io/cert-manager-
webhook:subjectaccessreviews created
    role.rbac.authorization.k8s.io/cert-manager-cainjector:leaderelection
created
    role.rbac.authorization.k8s.io/cert-manager:leaderelection created
    role.rbac.authorization.k8s.io/cert-manager-webhook:dynamic-serving
created
    rolebinding.rbac.authorization.k8s.io/cert-manager-
cainjector:leaderelection created
    rolebinding.rbac.authorization.k8s.io/cert-manager:leaderelection created
    rolebinding.rbac.authorization.k8s.io/cert-manager-webhook:dynamic-serving
created
    service/cert-manager created
    service/cert-manager-webhook created
    deployment.apps/cert-manager-cainjector created
    deployment.apps/cert-manager created
    deployment.apps/cert-manager-webhook created
    mutatingwebhookconfiguration.admissionregistration.k8s.io/cert-manager-
webhook created
    validatingwebhookconfiguration.admissionregistration.k8s.io/cert-manager-
webhook created
```

从以上命令显示结果可以看出，名为 cert-manager 的新命名空间已经建立了，可以用 kubectl 命令来验证：

```
kubectl get ns | grep cert-manager
```

命令显示结果如下：

```
cert-manager          Active   4m3s
```

这里再查看一下 cert-manager 命名空间的 Pod 运行情况，命令如下：

```
kubectl get pod -n cert-manager
```

命令显示结果如下：

```
NAME                                       READY   STATUS    RESTARTS   AGE
cert-manager-5d495db6fc-klk59              1/1     Running   0          19m
cert-manager-cainjector-5f9c9d977f-dks46   1/1     Running   0          19m
cert-manager-webhook-57bd45f9c-qvdklz      1/1     Running   0          19m
```

从以上输出结果可以判断，Kubernetes 的 cert-manager 管理工具已经安装成功，可以进行 KUDO 的部署安装工作了。

8.3.2　利用 KUDO 的 CLI 命令行工具安装 KUDO

KUDO 的 CLI 命令行工具是基于 Go 安装的二进制命令工具，所以将其下载后只要给其执行权限，就能直接运行了。其运行步骤如下：

```
cd /usr/local/src
wget
https://github.com/kudobuilder/kudo/releases/download/v0.19.0/kubectl-kudo_0.19.0_linux_x86_64
mv kubectl-kudo_0.19.0_linux_x86_64 /usr/local/bin/kubectl-kudo
chmod +x /usr/local/bin/kubectl-kudo
```

这里还可以查看 kubectl-kudo 的版本，命令如下：

```
kubectl-kudo version
```

命令显示结果如下：

```
KUDO Version: version.Info{GitVersion:"0.19.0", GitCommit:"4173395f", BuildDate:"2021-04-27T15:30:48Z", GoVersion:"go1.16.3", Compiler:"gc", Platform:"linux/amd64", KubernetesClientVersion:"v0.19.2"}
```

然后用 kubectl-kudo init 来初始化 KUDO 工作时需要的 KUDO Controller 控制器、webhook 及其他所有组件。命令如下：

```
kubectl-kudo init
```

命令显示结果如下：

```
$KUDO_HOME has been configured at /root/.kudo
✅ installed crds
✅ installed namespace
✅ installed service account
```

```
    Warning: admissionregistration.k8s.io/v1beta1 MutatingWebhookConfiguration
is deprecated in v1.16+, unavailable in v1.22+; use admissionregistration.
k8s.io/v1 MutatingWebhookConfiguration
    ✅ installed webhook
    ✅ installed kudo controller
```

再用 kubectl get ns 查看新的命名空间，发现在系统命名空间 kube-system 下新生成了名为 kudo-system 的命名空间。其结果如下：

```
NAME                   STATUS    AGE
cert-manager           Active    21h
default                Active    20d
kube-node-lease        Active    20d
kube-public            Active    20d
kube-system            Active    20d
kudo-system            Active    95s
local-path-storage     Active    20d
```

查看 kudo-system 命名空间对应的 Pod 及其运行状态是否正常。

```
kubectl get pod -n kudo-system
```

命令显示结果如下：

```
NAME                        READY   STATUS    RESTARTS   AGE
kudo-controller-manager-0   1/1     Running   0          5m20s
```

此结果显示名为 kudo-controller-manager-0 的 Pod 运行也是正常的；还可以查看 KUDO 生成的第三方 CRD，命令如下：

```
kubectl get crd | grep kudo
```

命令显示结果如下：

```
instances.kudo.dev                  2023-01-18T08:24:09Z
operators.kudo.dev                  2023-01-18T08:24:09Z
operatorversions.kudo.dev           2023-01-18T08:24:09Z
```

此处的显示结果验证了前面 KUDO 源码的程序结果，是一一对应的关系。

另外，如果不想保留 KUDO 的环境，kubectl kudo 也提供了相应命令：

```
kubectl kudo init --upgrade --dry-run --output yaml | kubectl delete -f -
```

执行以上命令，会从 Kubernetes 集群中删除相应的 KUDO CRD 资源、CRD deployments 资源及其他资源。

8.3.3 KUDO 的基本概念

在使用 KUDO 脚手架工具安装 Kubernetes Operator 实例之前，首先需要熟悉 KUDO 的基本概念及相关语法，这样才能在实际工作中得心应手地使用 KUDO。

下面使用 KUDO 的官方 Operator 库中的 first-operator 安装第一个 Operator 实例，并在安装过程中介绍 KUDO 的基本概念。

首先用 git 命令下载相应代码。命令如下：

```
git clone https://github.com/kudobuilder/operators.git
```

下载成功后，最终工作区如下：

```
/opt/kudo/operators
```

可以查看 first-operator 的目录分配，命令如下：

```
tree -L 3 /opt/kudo/operators/repository/first-operator/
```

命令显示结果如下：

```
/opt/kudo/operators/repository/first-operator/
├── docs
│   └── README.md
└── operator
    ├── operator.yaml       //operator 实例的主要工作控制文件
    ├── params.yaml         // 主要是相应变量的 values 值
    └── templates
        └── deployment.yaml  // 资源模板列表，其中的变量会用 params.yaml 渲染

3 directories, 4 files
```

operator 实例的目录一般分为两级，即 docs 和 operator。下面是其中几个文件的详细说明。

（1）README.md：相关 operator 实例项目的说明文档，可以重点关注。

（2）operator.yaml：定义运算符元数据以及运算符的整个生命周期的主要 yaml 文件。此文件中定义了任务和计划。这里还设置了提供 Operator 的名称、版本和维护者的元数据。

（3）params.yaml：定义运算符的参数。在安装过程中，可以覆盖这些参数以允许自定义。这里需要注意的是，params.yaml 文件中定义的都是 string 格式的字符串，所以要用双引号引起来。

（4）templates：该文件夹包含所有模板化的 Kubernetes 对象，这些对象将在安装后基于 operator.yaml 中定义的工作流应用于 Kubernetes 集群。

然后用 kubectl-kudo install 命令安装名为 first-operator 的实例，安装命令如下：

```
kubectl-kudo install ./operator/
```

命令显示结果如下：

```
operatorversion default/first-operator-1.0 already installed
instance default/first-operator-instance created
```

查看当前 default 命名空间中有哪些 instance，命令如下：

```
kubectl-kudo get instances
```

命令显示结果如下：

```
List of current installed instances in namespace "default":
.
```

```
    └── first-operator-instance
```

用 kubectl-kudo plan 命令查看 Plan 的最终结果，命令如下：

```
kubectl-kudo plan status --instance first-operator-instance
```

从命令显示结果可以发现，命名空间为 default，名为 first-operator-instance 的 Plan 已经安装成功了。

```
Plan(s) for "first-operator-instance" in namespace "default":
.
└ — — first-operator-instance (Operator-Version: "first-operator-1.0"
Active-Plan: "deploy")
      └── Plan deploy (serial strategy) [COMPLETE], last updated 2023-01-
21 13:03:29
          └── Phase main (serial strategy) [COMPLETE]
              └── Step everything [COMPLETE]
```

然后查看相应的 Pod，可以看到有对应的 Pod 生成，命令如下：

```
kubectl get pod
```

命令显示结果如下：

```
NAME                                  READY   STATUS    RESTARTS   AGE
nginx-deployment-5f9dd84b59-nmmsm     1/1     Running   0          4m3s
nginx-deployment-5f9dd84b59-xxxnk     1/1     Running   0          4m3s
```

这里需要重点关注 operator.yaml 文件，文件内容如下：

```
apiVersion: kudo.dev/v1beta1 # 指定 API 版本，此值必须在 kubectl apiversion 中
name: "first-operator"         # operator 实例的名字
operatorVersion: "1.0"         # operator 实例的版本号
kubernetesVersion: 1.20.7      # Kubernetes 集群的版本号
maintainers:
  - name: yhc
    email: yuhongchun027@gmail.com
url: https://kudo.dev
tasks:
  - name: app # 注意此处定义的任务名
    kind: Apply
    spec:
      resources:
        - deployment.yaml # template 目录下定义的资源名
plans:
  deploy:
    strategy: serial
    phases:
      - name: main
```

```
        strategy: parallel
        steps:
          - name: everything
            tasks:
              - app # 要与前面定义的tasks名字对应，如果对应不上，则会报错
```

再查看 params.yaml 文件，文件内容如下：

```
apiVersion: kudo.dev/v1beta1
parameters:
  - name: replicas # 名字要与deployment.yaml文件的模板一一对应，方便渲染
      description: Number of replicas that should be run as part of the
deployment #此处是相关parameters的注释说明
      default: 2 # 建议此处写成string类型格式
```

Operator 实例通常由几个计划组成。可以把它们想象成由 KUDO 执行的结构化方式编写的 Runbook（手册）。计划由阶段组成，阶段有一个或多个步骤。每个 Operator 实例必须至少包含一个部署计划，这是将应用程序部署到集群的默认计划。对于更复杂的应用程序，需要定义备份和恢复或升级计划。

阶段和步骤可以串行或并行执行。默认执行策略（如果未指定）是串行的。

这里参考官方的 operator.yaml 示例文件。文件内容如下：

```
tasks:
  - name: deploy-master
    kind: Apply
    spec:
      resources:
        - master-service.yaml
        - master.yaml
  - name: deploy-agent
    kind: Apply
    spec:
      resources:
        - agent-service.yaml
        - agent.yaml
plans:
  deploy:
    strategy: parallel
    phases:
      - name: deploy-master
        strategy: serial
        steps:
          - name: deploy-master
            tasks:
```

```
            - deploy-master
    - name: deploy-agent
      strategy: serial
      steps:
        - name: deploy-agent
          tasks:
            - deploy-agent
```

以上示例文件表示并行执行 deploy 的计划任务（如果有其他 Plan，则会并行执行），然后是串行执行 deploy-master 和 deploy-agent。

下面依次说明 operator.yaml 文件中涉及的重点关键字。

（1）Plan：计划记录整个任务的每个步骤。计划将这些任务分为阶段和步骤。每个步骤都引用要为此步骤运行的任务。通过将步骤组织成阶段，可以捕获服务的复杂行为。

默认情况下，KUDO 保留三个计划名称用于特殊用途。

1）Deploy Plan：部署计划，这个是必需的，而且也是当前 Runbook 的默认计划，这个计划将应用程序部署到当前 Kubernetes 集群。

2）Upgrade Plan：升级计划，用于将当前的 Operator 实例升级为一个新的版本。如果未定义升级版，则使用部署计划。

3）Cleanup Plan：清除计划，主要是清除当前的 Operator 实例。

（2）Phases：KUDO 的 Runbook 中的执行单元，跟计划一样，也可以并行或串行。

（3）Step：KUDO 的 Runbook 中的执行单元，受阶段调度执行。

（4）Task：在整个 KUDO 的 Runbook 中，最小的执行单元就是 Task，这个需要提前写好，通过步骤来调度执行。

（5）Instance：KUDO 最终部署的结果以 KUDO Operator 实例产出。

（6）Repository：KUDO 仓库为整个 KUDO 实例的安装提供了很大的便利性，这里可以直接引用 D2IQ 官方的 KUDO Repository，也可以自建 KUDO Repository。通常为了便利性，可以直接用本地目录来提供 Operator 的安装部署文件。

KUDO 的 Runbook 中出现的计划、阶段、步骤及作业的逻辑关系（简单理解就是一种“套娃”的关系）如图 8.3 所示。

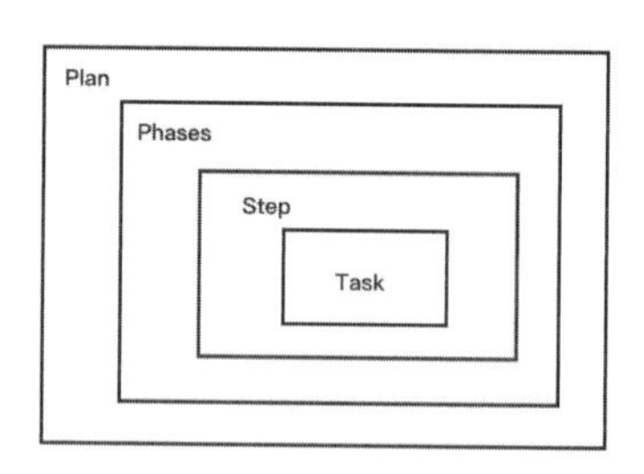

图 8.3　计划、阶段、步骤及作业的逻辑关系

如果两个计划同时运行会发生什么？答案是：视情况而定。如果彼此完全独立（例如部署服务和部署监控服务吊舱），则两者可以并行执行。如果这两个计划是备份和恢复或者是部署和迁

移，则会导致有些计划与其他计划是不兼容的，少数计划甚至可能不能执行。虽然重新启动正在运行的部署计划可能可以执行，但由于数据损坏，恢复计划可能无法执行。

综上所述，KUDO要确保计划正确执行，可以参见Kubernetes准入控制器（Admission Controllers）。Kubernetes准入控制器是管理和强制集群使用方式的插件，可以被认为是一个拦截（认证）API请求的守门人，可以更改请求对象或完全拒绝请求。

KUDO管理器使用一个实例控制器来控制对实例的更改，以确保计划不会受到干扰。KUDO实例更新的生命周期如图8.4所示。

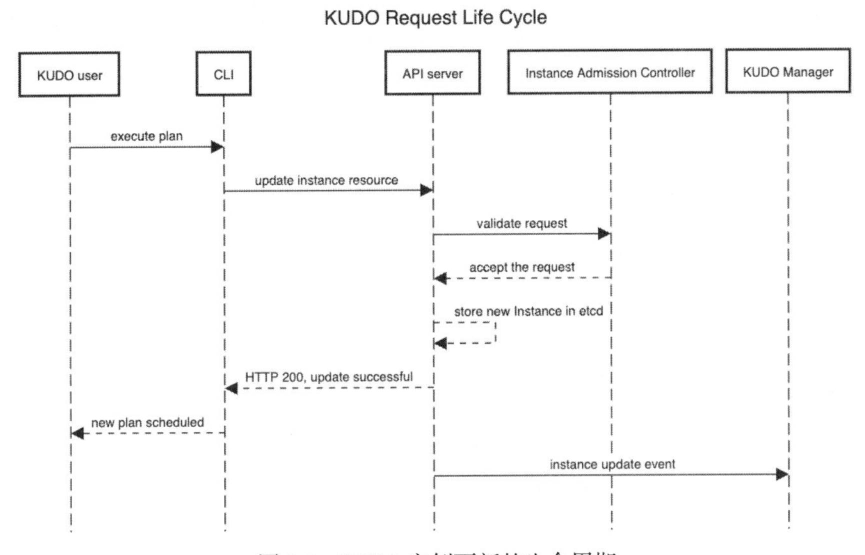

图8.4 KUDO实例更新的生命周期

实例控制器通过手动计划执行或实例参数更新来控制对实例的更新。一般经验法则如下：在允许更新计划开始之前，所有计划都应终止（要么成功完成，状态为COMPLETE，要么失败，状态为FATAL_ERROR）。虽然这种情形并非适用所有计划，但可以认为涵盖了大部分的场景。例如，部署计划被卡住（如从内部镜像仓库拉取失败导致部署计划卡住了），要求必须在每个节点内存较少的情况下重新启动。在请求被拒绝的情况下，实例控制器返回一个错误，解释请求被拒绝的确切原因。

通过阅读workspace/pkg/apis/kudo/v1beta1/instance_types.go处的源码内容来熟悉其流程，代码如下：

```
    // InstanceSpec 结构体定义了 Instance 的期望状态
    type InstanceSpec struct {
        // OperatorVersion 字段指定对特定 OperatorVersion 对象的引用
        OperatorVersion corev1.ObjectReference
`json:"operatorVersion,omitempty"`
        Parameters map[string]string `json:"parameters,omitempty"`
        PlanExecution PlanExecution `json:"planExecution,omitempty"`
```

```
}

/*
可以通过两种方式触发计划执行：
i) 间接通过更新 InstanceSpec.Parameters 中的相应参数
ii) 直接通过 InstanceSpec.PlanExecution.PlanName 字段的设置
虽然每次用户更改参数时都会发生间接 i) 触发，但直接 ii) 触发计划是为参数不更改的情况保留的，
如触发定期备份覆盖现有备份文件。此外，这为将来取消和覆盖当前运行的计划打开了空间。
注意：PlanExecution 结构体定义了当前执行的计划名称和相应的参数。一旦实例控制器 (IC) 完
成执行，该结构体将被清除。每个计划执行都有一个唯一的 UID，因此如果重新触发相同的计划，将有一
个新的 UID
*/
type PlanExecution struct {
	PlanName string                    `json:"planName,omitempty"`
	UID      apimachinerytypes.UID `json:"uid,omitempty"`
	Status   ExecutionStatus           `json:"status,omitempty"`
}

// InstanceStatus 结构体定义了 Instance 的观察状态
type InstanceStatus struct {
	PlanStatus map[string]PlanStatus `json:"planStatus,omitempty"`
	Conditions []metav1.Condition    `json:"conditions,omitempty"`
}


// PlanStatus 表示计划的状态，下面是有效的状态和转换
//
//                           +----------------+
//                           | Never executed |
//                           +-------+--------+
//                                   |
//                                   v
// +-------------+           +-------+--------+
// |    Error    |<------->|    Pending     |
// +------+------+           +-------+--------+
//        ^                          |
//        |                          v
//        |                  +-------+--------+
//        +--------------->|  In progress   |
//        |                  +-------+--------+
//        |                          |
```

```
//        v                        v
// +------+------+        +-------+--------+
// | Fatal error |        |    Complete    |
// +-------------+        +----------------+
//

type PlanStatus struct {
     Name    string          `json:"name,omitempty"`
     Status  ExecutionStatus `json:"status,omitempty"`
     Message string          `json:"message,omitempty"`
     LastUpdatedTimestamp *metav1.Time          `json:"lastUpdatedTimestam
p,omitempty"`
     Phases               []PhaseStatus         `json:"phases,omitempty"`
     UID                  apimachinerytypes.UID `json:"uid,omitempty"`
}

// PhaseStatus结构体表示一个阶段的状态
type PhaseStatus struct {
     Name    string          `json:"name,omitempty"`
     Status  ExecutionStatus `json:"status,omitempty"`
     Message string          `json:"message,omitempty"`
     Steps   []StepStatus    `json:"steps,omitempty"`
}

// StepStatus结构体表示一个步骤的状态
type StepStatus struct {
     Name    string          `json:"name,omitempty"`
     Message string          `json:"message,omitempty"`
     Status  ExecutionStatus `json:"status,omitempty"`
}

func (s *StepStatus) Set(status ExecutionStatus) {
      s.Status = status
      s.Message = ""
}

func (s *StepStatus) SetWithMessage(status ExecutionStatus, message string)
{
      s.Status = status
     s.Message = message
}
```

```
// Step 方法返回具有给定名称的 StepStatus，如果不存在这样的 StepStatus，则返回 nil
func (s *PhaseStatus) Step(name string) *StepStatus {
    for i, stepStatus := range s.Steps{
        if stepStatus.Name == name{
            return &s.Steps[i]
        }
    }
    return nil
}

func (s *PhaseStatus) Set(status ExecutionStatus) {
    s.Status = status
    s.Message = ""
}

func (s *PhaseStatus) SetWithMessage(status ExecutionStatus, message
string) {
    s.Status = status
    s.Message = message
}

// Phase 方法返回具有给定名称的 PhaseStatus，如果不存在这样的 PhaseStatus，则返回 nil
func (s *PlanStatus) Phase(name string) *PhaseStatus {
    for i, phaseStatus := range s.Phases {
        if phaseStatus.Name == name{
            return &s.Phases[i]
        }
    }
    return nil
}

func (s *PlanStatus) Set(status ExecutionStatus) {
    if s.Status != status{
        s.LastUpdatedTimestamp = &metav1.Time{Time: time.Now()}
        s.Status = status
        s.Message = ""
    }
}

func (s *PlanStatus) SetWithMessage(status ExecutionStatus, message string)
```

```
{
        if s.Status != status || s.Message != message{
                s.LastUpdatedTimestamp = &metav1.Time{Time: time.Now()}
                s.Status = status
                s.Message = message
        }
    }

    // ExecutionStatus 捕获退出的状态
    type ExecutionStatus string

    const (
        // ExecutionInProgress 状态表示正在积极部署，但还不健康
        ExecutionInProgress ExecutionStatus = "IN_PROGRESS"

        // ExecutionPending 状态表示未准备好部署，因为依赖阶段 / 步骤不健康
        ExecutionPending ExecutionStatus = "PENDING"

        // ExecutionComplete 状态表示已部署且运行状况良好
        ExecutionComplete ExecutionStatus = "COMPLETE"

        // ErrorStatus 状态表示部署应用程序时出错
        ErrorStatus ExecutionStatus = "ERROR"

        // ExecutionFatalError 状态表示部署应用程序时出现致命错误
        ExecutionFatalError ExecutionStatus = "FATAL_ERROR"

        // ExecutionNeverRun 状态表示当此计划 / 阶段 / 步骤到目前为止从未运行时使用
        ExecutionNeverRun ExecutionStatus = "NEVER_RUN"

        // DeployPlanName 是部署计划的名称
        DeployPlanName = "deploy"

        // UpgradePlanName 是升级计划的名称
        UpgradePlanName = "upgrade"

        // UpdatePlanName 是更新计划的名称
        UpdatePlanName = "update"

        // CleanupPlanName 是清理计划的名称
        CleanupPlanName = "cleanup"
```

```
)


var (
	SpecialPlanNames = []string{
		DeployPlanName,
		UpgradePlanName,
		UpdatePlanName,
		CleanupPlanName,
	}
)

// IsTerminal 是个布尔值类型，它表明如果状态为终端（完成或发生致命错误），则返回 true
func (s ExecutionStatus) IsTerminal() bool {
	return s == ExecutionComplete || s == ExecutionFatalError
}

// IsFinished 是布尔值类型，它表明如果状态成功完成（不处于 FATAL_ERROR 状态），则
// 返回 true
func (s ExecutionStatus) IsFinished() bool {
	return s == ExecutionComplete
}

// IsRunning 是布尔值类型，它表明如果计划当前正在执行，则 IsRunning 返回 true
func (s ExecutionStatus) IsRunning() bool {
	return s == ExecutionInProgress || s == ExecutionPending || s ==
ErrorStatus
}

type Instance struct {
	metav1.TypeMeta   `json:",inline"`
	metav1.ObjectMeta `json:"metadata,omitempty"`

	Spec   InstanceSpec   `json:"spec,omitempty"`
	Status InstanceStatus `json:"status,omitempty"`
}

type InstanceList struct {
	metav1.TypeMeta `json:",inline"`
	metav1.ListMeta `json:"metadata,omitempty"`
	Items           []Instance `json:"items"`
```

```
}

func init() {
    SchemeBuilder.Register(&Instance{}, &InstanceList{})
}
```

在实例控制器升级的过程中，如果 PlanStatus 的状态为 Complete，则表明计划顺利，达到了预期目标；如果 PlanStatus 状态为 Fatal error，则在升级时会由实例控制器返回错误并拒绝。具体实现细节可以参阅 workspace/pkg/webhook/instance_admission.go 源码。

8.3.4　将本地 HTTP Server 作为本地仓库

前面介绍了如何使用本地目录安装名为 first-operator 的 Operator 应用实例，本地目录在调试阶段比较方便，但如果要做业务交付，则带上版本交付才是较好的选择。下面介绍如何利用 kubectl-kudo 命令行工具来互动式交付一个 operator 压缩包，以及启动本地 HTTP 服务来提供本地的 KUDO repository。工作目录为 /opt/kudo/working。

（1）校验并打包 KUDO operator。命令如下：

```
mkdir -p /opt/kudo/working/repo
[root@test-yuhc-tantian working]# kubectl kudo package create /tmp/first-
operator/operator/  --destination=/opt/kudo/working/repo
```

命令显示结果如下：

```
package is valid
Package created: /opt/kudo/working/repo/first-operator-0.1.9.tgz
```

以上结果表示相应文件已经过校验并且在 repo 下面生成了 first-operator-0.1.9.tgz 压缩包，这里需要注意的是，压缩包的名字和版本跟 operator.yaml 文件是一一对应关系，内容如下：

```
name: "first-operator"
operatorVersion: "0.1.9"
```

（2）为本地 repo 建立 index 文件，命令如下：

```
kubectl-kudo repo index ./repo/
```

查看 ./repo/index.yaml 文件，文件结果如下：

```
apiVersion: v1
entries:
  first-operator:
  - digest:
4e2458929df371fd1bb8faaaa3e6db56c2f0a6baf934f7fc4ebb4c97f86f2b99
    maintainers:
    - email: andrew.yu@linktime.cloud>
      name: andrew.yu
    name: first-operator
```

```
    operatorVersion: 0.1.9
    urls:
    - http://localhost/first-operator-0.1.9.tgz
generated: "2023-01-26T11:45:35.837483781+08:00"
```

（3）启动 HTTP Server，这里为了简便，直接用 Go 程序启动 gohttp.go 文件的代码如下：

```
package main

import (
        "fmt"
        "net/http"
)

func main() {
        fmt.Println("Port:80")
        fmt.Println("Directory:/opt/kudo/working/repo")
        http.Handle("/",
http.FileServer(http.Dir("/opt/kudo/working/repo")))
        http.ListenAndServe(":80", nil)

}
```

启动命令如下：

```
go run gohttp.go
```

命令显示结果如下：

```
Port:80
Directory:/opt/kudo/working/repo
```

（4）将本地存储添加到 KUDO 客户端，命令如下：

```
kubectl-kudo repo add local http://localhost
```

命令显示结果如下：

```
"local" has been added to your repositories
```

以上结果表明本地存储已经成功添加了，可以用 kubectl-kudo repo list 命令验证，结果显示如下：

```
NAME            URL
*community      https://kudo-repository.storage.googleapis.com/v1
local           http://localhost
```

（5）设置 local repository 为默认的 KUDO context，其命令如下：

```
kubectl-kudo repo context local
```

下面用 kubectl-kudo repo list 命令来验证，其结果显示如下：

```
NAME            URL
community       https://kudo-repository.storage.googleapis.com/v1
*local          http://localhost
```

> 注：跟上面的结果有明显差异，这表明本地仓库已配置成默认的KUDO context。

（6）验证是否可以通过本地仓库安装 KUDO 实例，命令如下：

```
kubectl kudo install first-operator -v 9
```

这里的 -v 9 代表日志级别，可以用 kubectl kudo install –help 命令查看 kubectl-kudo 的相应子命令 install 的帮助文档。

命令显示结果如下：

```
repo configs: { name:community, url:https://kudo-repository.storage.
googleapis.com/v1 },{ name:local, url:http://localhost }

repository used { name:local, url:http://localhost }
acquiring kudo client
getting operator package
determining package type of first-operator
no local operator discovered, looking for http
no http discovered, looking for repository
getting package reader for first-operator, _
repository using: { name:local, url:http://localhost }
attempt to retrieve package from url:
http://localhost/first-operator-0.1.9.tgz
first-operator is a repository package from { name:local,
url:http://localhost }
folder walking through directory
folder walking through directory templates
Preparing default/first-operator:0.1.9 for installation
parameters in use: map[]
operator.kudo.dev default/first-operator unchanged
operatorversion default/first-operator-0.1.9 already installed
instance first-operator-instance created in namespace default
instance default/first-operator-instance created
```

需要重点关注以下几行命令显示结果，其表明此次安装名为 first-operator 的 Operator 实例是通过本地仓库安装的：

```
repository using: { name:local, url:http://localhost }
attempt to retrieve package from url:
http://localhost/first-operator-0.1.9.tgz
```

```
    first-operator is a repository package from { name:local, url:http://
localhost }
```

日志输出主要是由 /workspace/cloudnative/kudo/pkg/kudoctl/ 下的 clog 包提供，其主要程序由 clog.go 来实现，代码内容如下：

```
package clog

import (
      "fmt"
      "io"
      "os"
      "strconv"
      "github.com/spf13/pflag"
)

// 此包提供（C)lient 日志或 CLI 日志，clog 为 CLI 输出提供详细程度控制
// 5 种日志级别使用指南
//  0-1 正常标准输出
//  2-4 作用调试级日志
//  5-6 逻辑选择
//  7-8 输入 / 输出详细信息
//  9-10 作为跟踪级别（HTTP 详细信息）

// Level specifies a level of verbosity for V logs
type Level int8
```

8.3.5 使用 KUDO upgrade 升级版本

在安装和调试 KUDO Operator 的包文件时，有时会出现虽然更新了其中相应的文件，但 Plan 没有发现改变的情况，这是什么原因呢？

其实 KUDO 的源码包也提供了很详细的单元测试案例，具体可以先参考 workspace/kudo/test/e2e 下的单元测试案例。这里以 helloworld-operator 为例进行演示，其 operator.yaml 文件内容如下：

```
apiVersion: kudo.dev/v1beta1
name: "helloworld-operator"
operatorVersion: "0.1.18"
kubernetesVersion: 1.20.0
maintainers:
  - name: yuhongchun
    email: yuhongchun027@gmail.com
url: https://kudo-operator.dev
tasks:
```

```
    - name: app
      kind: Apply
      spec:
        resources:
          - deployment.yaml
plans:
  deploy:
    strategy: serial
    phases:
      - name: deploy-nginx
        strategy: parallel
        steps:
          - name: everything
            tasks:
              - app
```

这里注意 operatorVersion 版本，即 0.1.18；不改动这个，但是要改动其他配置文件，例如，改动 template/deployment.yaml 文件中的内容，然后重新部署，执行以下命令：

```
kubectl-kudo install ./operator/
```

命令显示结果已经报错了，提示命名空间 default 下的名为 helloworld-operator、版本为 0.1.18 的实例已经安装了，报错如下：

```
operatorversion default/helloworld-operator-0.1.18 already installed
Error: cannot install instance 'helloworld-operator-instance' because an
instance of that name already exists in namespace default
```

判断版本逻辑由下面的函数来处理，即 kudo/pkg/kudoctl/resources/install/operator.go，其核心代码如下：

```
func OperatorAndOperatorVersion(
        client *kudo.Client,
        operator *kudoapi.Operator,
        operatorVersion *kudoapi.OperatorVersion,
        dependencies []deps.Dependency) error {
        if !client.OperatorExistsInCluster(operator.Name,
operator.Namespace){
                if _, err := client.InstallOperatorObjToCluster(operator,
operator.Namespace); err != nil{
                        return fmt.Errorf(
                                "failed to install operator %s/%s: %v",
                                operator.Namespace,
                                operator.Name,
```

```
                                err)
                }
                clog.Printf("operator %s/%s created", operator.Namespace,
operator.Name)
        }
        // OperatorVersionInstalled 函数的作用是将当前 Kubernetes 的指定命名空间的
        // 所有 Operator 实例版本罗列出来并返回所有版本的字符串切片
        versionsInstalled, err := client.OperatorVersionsInstalled(operator.
Name, operator.Namespace)
        if err != nil{
                return fmt.Errorf(
                        "failed to retrieve existing operator versions of
   operator %s/%s: %v",
                        operator.Namespace,
                        operator.Name,
                        err)
        }
   // 利用 go-funk 库的 funk.ContainsString 函数将需要安装的 OperatorVersion 版本与
   // versionsInstalled 切片字符串比较，判断是否在此切片中
        if !funk.ContainsString(versionsInstalled,
operatorVersion.Spec.Version){
                if err := installDependencies(client, operatorVersion,
dependencies); err != nil {
                        return fmt.Errorf(
                                "failed to install dependencies of
operatorversion %s/%s: %v",
                                operatorVersion.Namespace,
                                operatorVersion.Name,
                                err)
                }
                //
                if _, err := client.InstallOperatorVersionObjToCluster(
                        operatorVersion,
                        operatorVersion.Namespace); err != nil {
                        return fmt.Errorf(
                                "failed to install
operatorversion %s/%s: %v",
                                operatorVersion.Namespace,
                                operatorVersion.Name,
                                err)
                }
```

```
            // 前面判断的是不存在切片字符串中的逻辑，下面是存在的逻辑
            clog.Printf("operatorversion %s/%s created",
operatorVersion.Namespace, operatorVersion.Name)
        } else {
                clog.Printf("operatorversion %s/%s already installed",
operatorVersion.Namespace, operatorVersion.Name)
        }

        return nil
    }
```

当然了，KUDO也提供了相应命令来提供支撑，可以用kukbectl-kudo get operatorversions命令查看当前helloworld-operator已部署的实例版本，命令如下：

```
kukbectl-kudo get operatorversions
```

命令显示结果如下：

```
List of current installed operatorversions in namespace "default":
.
├── first-operator-0.1.0-0.1.0
├── first-operator-0.1.9
├── first-operator-0.2.0
├── first-operator-1.0
├── helloworld-operator-0.1.0
├── helloworld-operator-0.1.1
├── helloworld-operator-0.1.14.7
├── helloworld-operator-0.1.15
├── helloworld-operator-0.1.15.1
├── helloworld-operator-0.1.15.2
├── helloworld-operator-0.1.16
├── helloworld-operator-0.1.17
├── helloworld-operator-0.1.18
├── helloworld-operator-0.1.2
├── helloworld-operator-0.1.3
├── helloworld-operator-0.1.4.1
├── helloworld-operator-0.1.4.2
├── helloworld-operator-0.1.4.3
├── helloworld-operator-0.1.4.4
├── helloworld-operator-0.1.4.5
└── helloworld-operator-0.1.5
```

更详细的命令推荐用kubectl-kudo get all，命令显示结果如下：

```
List of current installed operators including versions and instances in
```

```
namespace "default":
    .
    ├── first-operator
    │   ├── first-operator-0.1.0-0.1.0
    │   ├── first-operator-0.1.9
    │   ├── first-operator-0.2.0
    │   └── first-operator-1.0
    └── helloworld-operator
        ├── helloworld-operator-0.1.0
        ├── helloworld-operator-0.1.1
        ├── helloworld-operator-0.1.14.7
        ├── helloworld-operator-0.1.15
        ├── helloworld-operator-0.1.15.1
        ├── helloworld-operator-0.1.15.2
        ├── helloworld-operator-0.1.16
        ├── helloworld-operator-0.1.17
        ├── helloworld-operator-0.1.18
        ├── helloworld-operator-0.1.2
        ├── helloworld-operator-0.1.3
        ├── helloworld-operator-0.1.4.1
        ├── helloworld-operator-0.1.4.2
        ├── helloworld-operator-0.1.4.3
        ├── helloworld-operator-0.1.4.4
        ├── helloworld-operator-0.1.4.5
        └── helloworld-operator-0.1.5
```

以上结果说明，default/helloworld-operator-0.1.18 已经在当前 Kubernetes 集群中部署完成，所以如果还是以此版本来部署，则系统会报错如下：

```
operatorversion default/helloworld-operator-0.1.18 already installed
```

那么究竟怎么才能升级 helloworld-operator 的版本呢？这时可以使用前面提到的本地 repository，这里首先生成新的版本压缩包，如名为 helloworld-operator-0.1.19，按照前面的流程操作如下：

```
kubectl kudo package create /tmp/helloworld-operator/operator
--destination=/opt/kudo/working/repo
```

命令显示结果如下：

```
package is valid
Package created: /opt/kudo/working/repo/helloworld-operator-0.1.19.tgz
```

从结果可以看出，0.1.19 的版本压缩包已经在本地 repository 生成；另外，repo 下的 index.yaml 文件不会自动更新，这里需要手动操作，步骤如下：

```
rm -f /opt/kudo/working/repo/index.yaml
kubectl-kudo repo index /opt/kudo/working/repo/
```

然后查看 index.yaml 文件，这里可以看到，确定是有新的文件内容添加进来了，文件内容如下：

```
apiVersion: v1
entries:
  first-operator:
  - digest:
4e2458929df371fd1bb8faaaa3e6db56c2f0a6baf934f7fc4ebb4c97f86f2b99
    maintainers:
    - email: andrew.yu@linktime.cloud>
      name: andrew.yu
    name: first-operator
    operatorVersion: 0.1.9
    urls:
    - http://localhost/first-operator-0.1.9.tgz
  helloworld-operator:
  - digest:
b4fdce0544d1da06eb3d0517c54b53c866bae94566e6c572cff388f6b683813d
    maintainers:
    - email: yuhongchun027@gmail.com
      name: yhc
    name: helloworld-operator
    operatorVersion: 0.1.16
    urls:
    - http://localhost/helloworld-operator-0.1.16.tgz
  - digest:
9158e715ac0bfdcb1fdd6620cbdb9e68cb335a289b47183f33d790834d25bc1b
    maintainers:
    - email: andrew.yu@linktime.cloud>
      name: andrew.yu
    name: helloworld-operator
    operatorVersion: 0.1.17
    urls:
    - http://localhost/helloworld-operator-0.1.17.tgz
  - digest:
d863afbe854d28706b90a0878e84f11c2788ca742e6453c3a6e05fcdb9352170
    maintainers:
    - email: yuhongchun027@163.com
      name: yhc
```

```
      name: helloworld-operator
      operatorVersion: 0.1.19
      urls:
      - http://localhost/helloworld-operator-0.1.19.tgz
  generated: "2023-02-12T15:22:19.965168987+08:00"
```

接下来，执行以下升级命令来更新版本：

```
kubectl-kudo upgrade helloworld-operator --operator-version=0.1.19
--instance helloworld-operator-instance
```

命令执行后，最终版本将成功更新。这里可以自行用命令来验证：

```
kubectl kudo plan status
--instance=helloworld-operator-instance
```

命令显示结果如下：

```
Plan(s) for "helloworld-operator-instance" in namespace "default":
.
└ — — helloworld-operator-instance (Operator-Version: "helloworld-
operator-0.1.19" Active-Plan: "deploy")
       └──── Plan deploy (serial strategy) [COMPLETE], last updated 2023-02-
12 15:23:50
           └──── Phase deploy-nginx (serial strategy) [COMPLETE]
               └──── Step everything [COMPLETE]
```

如果升级的版本低于当前 instance 的 operationversion，则会报错。例如，执行以下命令：

```
kubectl-kudo upgrade helloworld-operator --operator-version=0.1.19
--instance helloworld-operator-instance
```

报错信息如下：

```
Error: upgraded version 0.1.19 is the same or smaller as current version
0.1.20 -> not upgrading
```

此处逻辑受以下文件中的 compareVersions 文件控制，文件目录如下：

```
/Users/yuhongchun/repo/workspace/cloudnative/kudo/pkg/kudoctl/resources/
upgrade/operatorversion.go
```

其主要功能处理函数如下：

```
func compareVersions(old string, new string) error {
    oldVersion, err := semver.NewVersion(old)
    if err != nil{
        return fmt.Errorf("failed to parse %s as semver: %v", old, err)
    }

    newVersion, err := semver.NewVersion(new)
```

```
        if err != nil{
                return fmt.Errorf("failed to parse %s as semver: %v", new, err)
        }

        if !oldVersion.LessThan(newVersion){
            return fmt.Errorf("upgraded version %s is the same or smaller as
current version %s -> not upgrading", new, old)
        }

        return nil
    }
```

8.3.6 使用 KUDO 触发 update 更新

如果在实际操作 instance 中不想用 upgrade 升级版本，而只想做一些常规操作，如只备份 MySQL 数据，可以使用 update 操作。

这里以官方的 KUDO MySQL Operator 为例进行说明，可以先执行 git clone 命令（工作目录还是在 /opt/kudo/working 下）：

```
cd /opt/kudo/working
git clone https://github.com/kudobuilder/operators
```

然后查看 operator 的目录结构，命令如下：

```
tree -L 2 ./operator/
```

命令显示结果如下：

```
./operator/
├── operator.yaml
├── params.yaml
└── templates
    ├── backup-pv.yaml
    ├── backup.yaml
    ├── init.yaml
    ├── mysql.yaml
    └── restore.yaml
```

这里为了更好地展示结果，可以修改 backup.yaml 文件，修改后的文件如下：

```
apiVersion: batch/v1
kind: Job
metadata:
  name: {{ .PlanName }}-job
  namespace: {{ .Namespace }}
spec:
```

```
    template:
      metadata:
        name: {{ .PlanName }}-job
      spec:
        restartPolicy: OnFailure
        containers:
        - name: {{ .PlanName }}
          image: mysql:5.7
          imagePullPolicy: IfNotPresent
          command:
          - /bin/sh
          - -c
          - "mysqldump -u root -h {{ .Name }}-svc -p{{ .Params.PASSWORD }} kudo >
/backups/{{ .Params.BACKUP_FILE }} && sleep 600"
          volumeMounts:
          - name: backup-pv
            mountPath: /backups
        volumes:
        - name: backup-pv
          persistentVolumeClaim:
            claimName: {{ .Name }}-backup-pv
```

然后修改 restore.yaml 文件，修改后的文件如下：

```
apiVersion: batch/v1
kind: Job
metadata:
  name: {{ .PlanName }}-job
  namespace: {{ .Namespace }}
spec:
  template:
    metadata:
      name: {{ .PlanName }}-job
    spec:
      restartPolicy: OnFailure
      containers:
      - name: {{ .PlanName }}
        image: mysql:5.7
        imagePullPolicy: IfNotPresent
        command:
        - /bin/sh
        - -c
          - "mysql -u root -h {{ .Name }}-svc -p{{ .Params.PASSWORD }}
```

```
--database=kudo < /backups/{{ .Params.RESTORE_FILE }} && sleep 600"
            volumeMounts:
            - name: backup-pv
              mountPath: /backups
          volumes:
          - name: backup-pv
            persistentVolumeClaim:
              claimName: {{ .Name }}-backup-pv
```

主要改动如下：

（1）修改了备份 Job 的执行命令，在其后增加了休眠语句，如 sleep 600，目的是在执行以后查看 Pod 中相应的备份数据是否存在。

（2）修改了恢复 Job 的执行命令，在其后增加了休眠语句，如 sleep 600，目的是在执行以后查看 Pod 的执行状态。

首先，利用 kubectl-kudo 命令安装名为 mysql-instance 的 Operator 实例，安装命令如下：

```
kubectl-kudo install ./operator/
```

可以查看 mysql-instance 的 plan 执行情况，命令如下：

```
kubectl-kudo plan status --instance mysql-instance
```

命令执行结果如下：

```
Plan(s) for "mysql-instance" in namespace "default":
.
└ — — mysql-instance (Operator-Version: "mysql-5.7-0.3.0" Active-Plan:
"deploy")
    ├──── Plan backup (serial strategy) [NOT ACTIVE]
    │   └──── Phase backup (serial strategy) [NOT ACTIVE]
    │       ├──── Step pv [NOT ACTIVE]
    │       ├──── Step backup [NOT ACTIVE]
    │       └──── Step cleanup [NOT ACTIVE]
    ├──── Plan deploy (serial strategy) [COMPLETE], last updated 2023-03-05
17:35:43
    │   └──── Phase deploy (serial strategy) [COMPLETE]
    │       ├──── Step deploy [COMPLETE]
    │       ├──── Step init [COMPLETE]
    │       └──── Step cleanup [COMPLETE]
    └──── Plan restore (serial strategy) [NOT ACTIVE]
        └──── Phase restore (serial strategy) [NOT ACTIVE]
            ├──── Step restore [NOT ACTIVE]
            └──── Step cleanup [NOT ACTIVE]
```

到了这里，假如想备份名为 kudo 的数据库，可以执行 KUDO 的与实例相关的 update 命令：

```
kubectl-kudo plan trigger --name backup  --instance mysql-instance --v 9
```

命令执行结果如下：

```
Triggered backup plan for default/mysql-instance instance
```

然后发现名为 backup-job-rxg22 的 Pod 也已经正常运行，这里执行 kubectl get pod，命令执行结果如下：

```
NAME                               READY    STATUS     RESTARTS    AGE
backup-job-rxg22                   1/1      Running    0           7s
mysql-instance-56c9fddc6d-zlrvh    1/1      Running    0           5m33s
```

进到名为 backup-job-rxg22 的 Pod 中，查看 /backups 目录，用 ls 命令查看是否有相应数据存在，发现存在 backup.sql 备份文件，即

```
ls -lsart backup.sql
```

命令显示结果如下：

```
4 -rw-r--r-- 1 root root 1851 Feb 19 12:30 backup.sql
```

这里还可以用 docker 命令查看：

```
docker exec -ti kind-worker3 /bin/bash && ls -lstart /data/mysql/data/
backup.sql
```

查看 backup.sql 文件，命令如下：

```
bash-4.2# ls -lsart backup.sql
```

命令显示结果如下：

```
4 -rw-r--r-- 1 root root 1851 Mar  5 09:41 backup.sql
```

如果这个时候触发了一个不存在的 Plan，命令如下：

```
kubectl-kudo plan trigger --name backup  --instance mysql-instance --v 9
```

则会有如下报错信息：

```
Error: admission webhook "instance-admission.kudo.dev" denied the request:
plan backp does not exist
```

这块逻辑主要是由 workspace/pkg/apis/kudo/v1beta1/instance_types_helpers.go 来判断，其主要代码实现如下：

```
// 如果输入的计划名字不存在，SelectPlan 方法返回 nil 空值，否则返回列表中第一个存在的计划
func SelectPlan(possiblePlans []string, ov *OperatorVersion) *string {
    for _, plan := range possiblePlans{
            if _, ok := ov.Spec.Plans[plan]; ok{
                    return &plan
            }
    }
    return nil
```

```
}
```

执行完 backup 的 Plan 以后，可能还要验证 restore 的 Plan，这里可以删除 kudo 数据库中的 example 表，具体过程这里略过。

接下来，测试 backup 的 job，命令如下：

```
kubectl-kudo plan trigger --name restore  --instance mysql-instance --v 9
```

命令显示结果如下：

```
Triggered restore plan for default/mysql-instance instance
```

这里用 kubectl get pod 命令查看 pod，命令如下：

```
[root@test-yuhc-tantian templates]# kubectl get pod
```

命令显示结果如下：

```
NAME                               READY   STATUS    RESTARTS   AGE
mysql-instance-56c9fddc6d-7dbwj    1/1     Running   0          36m
restore-job-p9mpc                  1/1     Running   0          2s
```

最后，从输出结果可以发现，kudo 的 example 表恢复了。

8.3.7　关于 KUDO 的实践思考

Kubernetes 调度程序的一个基本特质就是能够将 Pod 转移至它认为合适的位置，但有状态应用服务在运维过程中不适合改变其磁盘或网络的接入路径，保障数据的一致性和服务的可用性需要一系列的复杂步骤。不只如此，在不同的服务中，部署、扩展等标准任务的步骤也完全不同。例如，Kafka 就与 HDFS 部署不一样。为了运行有状态应用服务，Kubernetes 引入了 StatefulSets。StatefulSets 的功能包括一致且特有的网络和存储识别，以及对独特性和特定部署和拓展顺序的约束。

不过，StatefulSets 不是万能的。因为其控制器的通用性，StatefulSets 对 Pod 没有任何洞察，也不可观测，所以这样的运维是很粗糙的，即便顺序可被定义，但它不能处理特殊状况，好比没法识别复制状态、集群状态等，更不用提在这些状态的基础上执行什么行动了。然而，有状态应用几乎都是独一无二的，让这一问题更复杂。

Kubernetes Operator 是由定制化的控制器和一套 CRD 组成的。其中，控制器包含管理特定运维所需要的知识；CRD 则提供了 API 原语，这是在与集群中运行的应用程序互动时所需要的。 也就是说，应用管理员不用与 COE 纠缠，就能够将复杂的运维工作委托给已在 Operator 控制器上编译好的逻辑。

但事实上，要让 Kubernetes Operator 达到生产就绪状态是很复杂的。首先，必须处理边缘用例并进行综合测试，这就要完成大量的软件工程工作。例如，官方 Elastic Operator 的代码已经超过 5 万行，很多企业无法承受这样的工作量。其次，Kubernetes Operator 缺少一致性。每个 Operator 都是独特的，管理员与它们各自的 CRD 互动的方式各不相同。最后，Kubernetes 项目的开发和发布速度之快使人惊叹。这意味着 Operator 开发人员必须时刻关注最新的开发成果，这进一步加重了在公司内部推进这项工作的工程负担。

KUDO 是低代码 Operator 脚手架开发工具，只要熟悉运维特性和 Kubernetes 基础，研发工程师、DevOps 工程师甚至高级运维工程师就能较轻松地写出适合自己业务场景的 Operator。如果 Kubernetes 集群需要管理有状态分布式应用，如数据库、缓存、监控等，可以用 KUDO 尝试编写相关的 Operator。

8.4 小　　结

本章主要介绍了 Kubernetes 下的 Operator 脚手架开发工具 KUDO 的基础概念、核心源码及基础用法。KUDO 是纯命令形式的持续部署工具。通过前面的章节内容可以发现，KUDO 也有自己的版本管理理念，这个作为持续部署工具而言也是非常重要的功能之一。通过对源码的阅读和理解可以发现，KUDO 也是款优秀的工程，其设计精良，也适合在生产环境下作为持续部署工具来使用，以提高交付效率。

第 9 章　基于云原生监控 DevOps 生产实践

监控系统对于业务部门的重要性不言而喻，其搭建过程既复杂又有挑战性，涉及系统运维人员、网络工程师、业务运营人员，重要研发环境还会涉及研发团队，所以要考虑的东西很多。例如，至少要从网络、系统和业务等各方面进行监控，另外还要实现预警 + 告警功能，方便通知技术运维人员和业务运营人员。

9.1　监控系统的选择

在实际运维过程中通常会遇到以下情况：

（1）网络请求突增，超出网关承受的压力。

（2）某个业务模块出现问题时，运维人员不知道，发现时问题已经很严重了。

（3）系统出现瓶颈，CPU 占用持续升高，内存不足，磁盘被写满。

（4）某关键性容器出现 OOM，已经影响到核心业务。

以上这些问题一旦发生，将对业务产生巨大的影响。因此，每个公司或者 IT 团队都会针对此类情况建立自己的 IT 监控系统。

一般将监控系统分为三层，具体如下。

（1）网络层监控：用于监控网络状况参数，如网关流量情况、丢包率、错包率、连接数、客户 VPN 连接失败等情况。

（2）系统层监控：用于监控物理主机、虚拟主机以及操作系统的参数，如 CPU 利用率、内存利用率、磁盘空间利用率等情况。

（3）业务层监控：用于监控核心业务流程，如登录、注册、下单、支付等。

完善的监控体系需要达到以下效果。

（1）趋势分析：长期收集并统计监控样本数据，对监控指标进行趋势分析。例如，通过分析磁盘的使用空间增长率，可以预测何时需要对磁盘进行扩容。

（2）故障定位：机器故障时，运维人员需要对故障进行调查和处理。通过分析监控系统记录的各种历史数据，可以迅速定位到故障根源并解决问题。

（3）预警或告警：当系统即将出现故障或已经出现故障时，监控可以迅速反应并发出告警。这样，运维人员可以提前预防问题发生或快速处理已产生的问题，从而保证业务服务的正常运行。

（4）数据可视化：通过监控面板可以生成可视化仪表盘，使运维人员能够直观地了解系统运行状态、资源使用情况、服务运行状态等，还可以使运营人员直观地了解业务数据。

综合以上因素，这里选择 Prometheus 作为监控系统。

9.2 Prometheus 监控系统

Prometheus 不仅是监控系统，还是时序数据库（自带 TSDB），更是一套完备的监控生态解决方案。

9.2.1 Prometheus 监控系统的基础架构

Prometheus 是目前最为流行的云原生监控系统，所以这里重点介绍 Prometheus 系统的架构及其特点，其架构如图 9.1 所示。

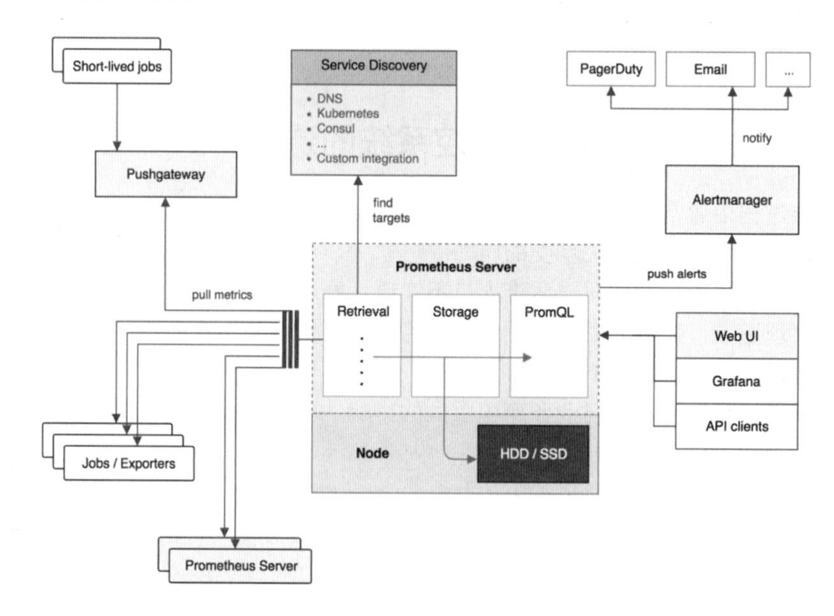

图 9.1　Prometheus 系统架构

这里简单介绍 Prometheus 系统架构相应的重要组件。

1. Job/Exporter

Job/Exporter 属于 Prometheus target，是 Prometheus 监控的对象。

Job 分为长时间执行和短时间执行两种。对于长时间执行的 Job，可以使用 Prometheus Client 集成进行监控；对于短时间执行的 Job，可以将监控数据推送到 PushGateway 中缓存。

Prometheus 社区提供了丰富的 Exporter 实现，涵盖了从基础设施、中间件以及网络等各个方面的监控实现。当然社区中也出现了很多其他 Exporter，如果有必要，也可以完全根据自己的需求开发一个 Exporter，但是最好以官方 Exporter 开发的最佳实践文档作为参考实现方式，后续内容中将介绍如何开发一个合格的 Exporter。官方提供的主要 Exporter 如下。

（1）数据库：Consul exporter、Memcached exporter、MySQL server exporter。

（2）硬件相关：Node metrics exporter。

（3）HTTP：HAProxy exporter。

（4）其他监控系统：AWS CloudWatch exporter、Collectd exporter、Graphite exporter、InfluxDB exporter、JMX exporter、SNMP exporter、StatsD exporter、Blackbox exporter。

除了开源的 Exporter 以外，对于很多业务级别的监控，就需要用 Go 或 Java 程序来开发相应的 Exporter 业务监控组件，以实现其相应的监控告警功能。

2. PushGateway

PushGateway 作为 Prometheus 生态中的一个重要成员，允许任何客户端 push 符合规范地自定义监控指标，再结合 Prometheus 统一收集监控。

Prometheus 采用定时 pull 模式，若由于子网络或者防火墙，不能直接 pull 各个 Target 的指标数据，则可以采用每个 Target 往 PushGateway 上 push 数据，然后 Prometheus 去 PushGateway 上定时 pull。

另外，在监控各个业务数据时，需要将各个不同的业务数据进行统一汇总，此时也可以采用 PushGateway 来统一收集，然后 Prometheus 来统一 pull。

Prometheus 的服务发现有以下几种：

（1）服务发现。Prometheus 通过服务发现机制对云及容器环境下的监控场景提供完善的支持，除了支持文件的服务发现（Prometheus 会周期性地从文件中读取最新的 target 信息）外，Prometheus 还支持多种常见的服务发现组件，如 Kubernetes、DNS、Consul 及 Zookeeper 等。例如，Prometheus 可以使用 Kubernetes 的 API 获取容器信息的变化（如容器的创建和删除）来动态更新监控对象。

（2）基于文件的服务发现。若要实现文件服务发现，就需要额外建立一个文件，Prometheus 执行时会读取文件，从文件中抓取 target。

文件可以使用 JSON、YAML 格式，其中含有定义的 target 列表，以及可选的标签信息。示例如下：

```
- targets:
 - localhost:9090
  labels:
    app: prometheus
    job: prometheus

- targets:
 - localhost:9100
  labels:
    app: node_exporter
    job: node
```

这些文件通常也是由另一个系统生成的，如 ansiblea 或 saltstack 等配置管理系统，也有脚本基于 CMDB 定期查询生成。可以配合 ansible 使用，设置 Prometheus 周期性地从 ansible 获取所有能被监控到的target，来生成对应的YAML或JSON文件。

```
static_configs:
  file_sd_configs:
   - refresh_interval: 2m       # 抓取周期为2分钟
     files:
     - targets/prometheus-*.yaml # 在targets目录下，所有prometheus-后缀的yaml
                                 # 文件，统统会被prometheus当成target识别
```

通过服务发现的方式，管理员可以在不重启Prometheus服务的情况下动态发现需要监控的target实例信息。

3. Prometheus服务

Prometheus服务（Prometheus Server）是Prometheus最核心的模块，主要包括抓取、存储和查询3个功能。

（1）抓取：Prometheus Server通过服务发现组件，周期性地从job、Exporter、PushGateway这3个组件中通过HTTP轮询的形式拉取监控指标数据。

（2）存储：通过一定规则清理和整理抓取到的监控数据，把得到的结果存储到新的时间序列中进行持久化。Prometheus的存储分为本地存储和远程存储。

1）本地存储：即Prometheus自带的TSDB时序数据库，它会直接保存到本地磁盘，性能上建议使用SSD存储，默认是15天。

2）远程存储：适用于存储大量的监控数据。Prometheus支持的远程存储包括OpenTSDB、influxDB、ElasticSerarch及TiKV等。远程存储需要和中间层的适配器进行转换，主要涉及Prometheus中的remote_write和remote_read接口。

（3）查询：Prometheus持久化数据以后，客户端就可以通过PromQL语句对数据进行查询了。

4. Dashboard

Prometheus服务除了内置查询语言PromQL以外，还支持表达式浏览器及表达式浏览器中的数据图形界面，实际工作中经常使用Grafana等作为前端展示界面。

5. AlertManager

AlertManager是独立于Prometheus的一个告警组件，需要单独安装部署。AlertManager接收Prometheus推送过来的告警，用于管理、整合和分发告警到不同的目的地。AlertManager提供多种内置的第三方告警通知方式，同时还提供对webhook通知的支持，通过webhook可以完成对告警的更多个性化的扩展。AlertManager除了提供基本的告警通知功能外，还提供了如分组、抑制以及静默等告警特性。

Prometheus的优点如下：

（1）通过PromQL语句实现多维度数据模型的灵活查询。以这种查询方式使监控指标可以关联到多个标签，并对时间序列进行切片和切块，以支持多种图形、表格和告警场景，这是Prometheus的优势所在，也是很多监控系统做不到的，包括Zabbix。

（2）Go语言编写。二进制文件直接启动，基本上不需要什么依赖；基于云原生与Kubernetes

可以无缝衔接。

（3）采用 pull 模式为主、push 模式为辅的方式采集数据。

（4）定义了开放指标数据的标准，自定义探针（如 Exporter 等），编写简单方便。

（5）高效的存储，不管是本地 TSDB 时序数据库，还是远程的 InfluxDB 时序数据库，平均采样数据占 3.5Byte；这里假设有 300 万个时间序列，每 30 秒采样一次，如果持续运行 60 天，占用磁盘空间大约 200GB。

（6）精确告警。Prometheus 基于灵活的 PromQL 语句可以进行告警设置、预测等。

Prometheus 也存在一定的局限性，具体如下：

（1）Prometheus 主要针对性能和可用性监控，不适于针对日志（Log）、事件（Event）、调用链（Tracing）等的监控。

（2）Prometheus 关注的是近期的事情，因为大多数监控查询及告警针对的都是最近（通常不到一天）的数据，监控数据默认保留 15 天。

（3）Prometheus 对数据的统计无法做到 100% 准确，如订单、支付和计量计费等精确数据监控场景。

（4）本地存储有限，存储大量的历史数据需要对接第三方远程存储，如 InfluxDB。

（5）Prometheus 绝大多数的操作和维护都通过配置文件进行。对大批量监控对象的维护，必须要依赖第三方的配置管理工具，因此运维复杂度比 Zabbix 高。

9.2.2　Prometheus 的基础概念

Prometheus 有 4 大指标类型，分别是 Counter（计算器）、Gauge（仪表盘）、Histogram（直方图）和 Summary（摘要）。

1. Counter：只增不减的计数器

Counter 类型的监控指标的工作方式和计数器一样，只增不减（除非系统发生重置）。常见的监控指标，如 http_requests_total、node_cpu 都是 Counter 类型的监控指标。一般在定义 Counter 类型指标的名称时推荐使用 _total 作为后缀。

Counter 是一个简单又强大的工具。例如，在应用程序中记录某些事件发生的次数，通过以时序的形式存储这些数据，可以轻松了解该事件产生速率的变化。PromQL 内置的聚合操作和函数可以让用户对这些数据进行进一步的分析。

例如，通过 rate() 函数获取 HTTP 请求量的增长率，函数语法如下：

```
rate(http_requests_total[5m])
```

2. Gauge：可增可减的仪表盘

与 Counter 不同，Gauge 类型的监控指标侧重于反应系统的当前状态。因此这类指标的样本数据可增可减。常见指标如 node_memory_MemFree（主机当前空闲的内容大小）、node_memory_MemAvailable（可用内存大小）都是 Gauge 类型的监控指标。

3. Histogram（直方图）

大多数情况下，人们都倾向于使用某些量化指标的平均值，如 CPU 的平均使用率、页面的平均响应时间。这种方式的问题很明显，以系统 API 调用的平均响应时间为例，如果大多数 API 请求都维持在 100 毫秒的响应时间范围，而个别请求的响应时间需要 5 秒，就会导致某些 Web 页面的响应时间落到中位数的情况，而这种现象称为长尾问题。

为了区分是平均的慢还是长尾的慢，最简单的方式就是按照请求延迟的范围进行分组。例如，统计延迟在 0~10 毫秒的请求数有多少而 10~20 毫秒的请求数又有多少。通过这种方式可以快速分析系统慢的原因。Histogram 和 Summary 都是为了能够解决这些问题而存在的，通过 Histogram 和 Summary 类型的监控指标，可以快速了解监控样本的分布情况。

Histogram 在一段时间范围内对数据进行采样（通常是请求持续时间或响应大小等），并将其计入可配置的存储桶（bucket）中，后续可通过指定区间筛选样本，也可以统计样本总数，最后一般将数据展示为直方图。

4. Summary（摘要）

与 Histogram 类型类似，用于表示一段时间内的数据采样结果（通常是请求持续时间或响应大小等），但它直接存储了分位数（通过客户端计算，然后展示出来），而不是通过区间来计算。

9.2.3 Prometheus API 相关开发

实际上，与云原生监控相关的很多开发工作都会涉及利用 Prometheus API 调用 PromQL 语句来查询数据，这里用实际例子来说明。

1. Prometheus API 响应格式

Prometheus API 使用了 JSON 格式的响应内容。 当 API 调用成功后将会返回 2xx 的 HTTP 状态码；反之，当 API 调用失败时可能返回以下几种不同的 HTTP 状态码。

（1）404 Bad Request：当参数错误或者缺失时。

（2）422 Unprocessable Entity：当表达式无法执行时。

（3）503 Service Unavailable：当请求超时或者被中断时。

所有的 API 请求返回的格式均使用以下 JSON 格式：

```
{
  "status": "success" | "error",
  "data": <data>,
     // 错误信息
  "errorType": "<string>",
  "error": "<string>"
 // 如果存在不会阻止请求执行的错误，可能会返回一系列告警
 "warnnings": "[<string>]"
}
```

输入时间戳可以由 RFC3339 格式或者 UNIX 时间戳提供，后面可选的小数位可以精确到亚秒级别，输出时间戳以 UNIX 时间戳的方式呈现。

查询参数名称可以用中括号 [] 重复次数。

（1）<series_selector> 占位符提供类似 http_requests_total 或者 http_requests_total{method=~"(GET|POST)"} 的 Prometheus 时间序列选择器，并且需要在 URL 中编码传输。

（2）<duration> 占位符是指 [0-9]+[smhdwy] 形式的 Prometheus 持续时间字符串。例如，5m 表示 5 分钟的持续时间。

（3）<bool> 提供布尔值（字符串 true 和 false）。

2. 表达式查询

HTTP API 可以分别通过 /api/v1/query 和 /api/v1/query_range 查询 PromQL 表达式当前或者一定时间范围内的计算结果。

（1）瞬时数据查询：使用 QUERY API 可以查询 PromQL 语句在特定时间点下的计算结果。其 API 路径如下：

```
GET /api/v1/query
```

URL 请求参数如下。

1）query=<string>：PromQL 表达式。

2）time=<rfc3339 | unix_timestamp>：用于指定计算 PromQL 的时间戳。可选参数，默认情况下使用当前系统时间。

3）timeout=<duration>：超时设置。可选参数，默认情况下使用全局设置的参数 -query.timeout。

当 API 调用成功后，Prometheus 会返回 JSON 格式的响应内容，格式如上小节所示，并且在 data 部分返回查询结果。data 部分格式如下：

```
{
  "resultType": "matrix" | "vector" | "scalar" | "string",
  "result": <value>
}
```

下面例子执行了时间为 2020-03-01T20:10:51.781Z 的 UP 表达式，其命令如下：

```
curl 'http://localhost:9090/api/v1/query?query=up&time=2020-03-01T20:10:51.781Z'
```

结果如下：

```
{
 "status": "success",
 "data":{
    "resultType": "vector",
    "result" : [
          {
             "metric" : {
                "__name__" : "up",
                "job" : "prometheus",
```

```
                "instance" : "localhost:9090"
            },
            "value": [ 1435781451.781, "1" ]
        },
        {
            "metric" : {
                "__name__" : "up",
                "job" : "node",
                "instance" : "localhost:9100"
            },
            "value" : [ 1435781451.781, "0" ]
        }
    ]
  }
}
```

（2）区间数据查询。使用 QUERY_RANGE API 可以直接查询 PromQL 表达式在一段时间内返回的计算结果。其表达式如下：

```
GET /api/v1/query_range
```

URL 请求参数如下。

1）query=<string>：PromQL 表达式。

2）start=<rfc3339|unix_timestamp>：起始时间戳。

3）end=<rfc3339|unix_timestamp>：结束时间戳。

4）step=<duration|float>：查询步长，时间区间内每 step 秒执行一次。

5）timeout=<duration>：超时设置。可选参数，默认情况下使用全局设置的参数 -query.timeout。

当使用 QUERY_RANGE API 查询 PromQL 表达式时，返回结果一定是一个区间向量：

```
{
  "resultType": "matrix",
  "result": <value>
}
```

以下示例在 30 秒范围内评估表达式，查询步长为 15 秒，其命令如下：

```
curl
'http://localhost:9090/api/v1/query_range?query=up&start=2020-03-01T20:10:30.781Z&end=2020-03-01T20:11:00.781Z&step=15s'
```

命令显示结果如下：

```
{
   "status" : "success",
   "data" : {
```

```
        "resultType" : "matrix",
        "result" : [
            {
                "metric" : {
                    "__name__" : "up",
                    "job" : "prometheus",
                    "instance" : "localhost:9090"
                },
                "values" : [
                    [ 1435781430.781, "1" ],
                    [ 1435781445.781, "1" ],
                    [ 1435781460.781, "1" ]
                ]
            },
            {
                "metric" : {
                    "__name__" : "up",
                    "job" : "node",
                    "instance" : "localhost:9091"
                },
                "values" : [
                [ 1435781430.781, "0" ],
                    [ 1435781445.781, "0" ],
                    [ 1435781460.781, "1" ]
                ]
            }
        ]
    }
}
```

3. 实际工作中用到的示例用法

应用场景：有3台主机d0601、d0603、d0604，均已提前部署了node-exporter监控组件并建立了与Prometheus相应的关联（采集频率是60秒/次），Prometheus主机地址为10.1.0.200:30900，现在想通过Prometheus API查询CPU利用率。其完整Go代码如下：

```
package main

import (
    "encoding/json"
    "fmt"
    log "github.com/sirupsen/logrus"
    "io/ioutil"
```

```
		"net/http"
		"net/url"
		"os"
		"os/exec"
		"regexp"
		"strconv"
		"strings"
	)
	// 定义 Prometheus API 地址变量
	var baseUrl = "http://10.1.0.200:30900/api/v1/query"

	// 定义 Prometheus API 的返回结果结构体
	type monitoring struct {
		Status string `json:"status"`
		Data   struct{
			ResultType string `json:"resultType"`
			Result     []struct{
				Metric struct{
					Instance string `json:"instance"`
				} `json:"metric"`
				Value []interface{} `json:"value"`
			} `json:"result"`
		} `json:"data"`
	}

	func main() {
		nodeslice := []string{"d0601", "d0602", "d0603"}
		for _, vnode := range nodeslice{
			// 利用 fmt.Sprintf 函数将 node 节点分别格式化成 PromQL 语句
		queryUrl :=
fmt.Sprintf("(1-avg(irate(node_cpu_seconds_total{instance=~\"%s\",mode=\"id
le\"}[5m]))by(instance))*100", vnode)
			// url 参数需要用 url.QueryEscape 函数进行转码
			reqUrl := fmt.Sprintf("%s?query=%s", baseUrl,
url.QueryEscape(queryUrl))
			req, err := http.NewRequest("GET", reqUrl, nil)
			if err != nil{
				log.Error(err)
			}
			nodeRsp, err := http.DefaultClient.Do(req)
			if err != nil{
```

```
                log.Error(err)
        }
        defer nodeRsp.Body.Close()
        if nodeRsp.StatusCode != http.StatusOK{
                log.Error("get host monitoring data failed")
        }
        rspResult, _ := ioutil.ReadAll(nodeRsp.Body)
        tmpData := &monitoring{}
        err = json.Unmarshal([]byte(rspResult), &tmpData)
        if err != nil {
                log.Error("get host monitoring data")
        }
        if tmpData != nil{
            for _, result := range tmpData.Data.Result{
                    // 抓取 node 节点的 CPU 利用率
                    cpuUtil := result.Value[1].(string)
                    // 取小数点后 4 位数字，方便后续做 AlertManager 告警处理
                    CPUUtil := cpuUtil[0:6]
                    log.Info("CpuUtils:", CPUUtil)
            }
        }
    }
}
```

最后运行结果如下：

```
INFO[0000] CpuUtils:0.0346
INFO[0000] CpuUtils:0.0364
INFO[0000] CpuUtils:0.0408
```

> 注：net/url包是Go代码的标准库，用于解析URL并实现查询的编码，其中的关键函数是QueryEscape。其函数语法如下：
>
> func QueryEscape(s string) string

QueryEscape 函数对 string 字符串进行转码，使之可以安全地用于 URL 查询。

9.3　Prometheus 的安装部署

Prometheus 有多种部署方式，如将二进制配置成自启动服务的方式或 Docker、Kubernetes 部署方式，考虑到各种研发场景和客户私有化场景（有网络限制或无公网环境），这里推荐使用 Docker 化部署方式。

9.3.1 Docker 化部署安装 Prometheus

考虑到各种研发场景和客户私有化场景，Docker 化部署安装 Prometheus 是其中较方便的一种方式（不受环境和网络的限制，Docker 镜像可以提前导出，但 Kubernetes 的二进制安装和配置还是较复杂），只需提前部署 Docker 服务并启动即可。

1. 部署预置条件

需要开启 Docker 程序，并且提前关闭 iptables 和 SELinux。

本地机器需要 Prometheus 和 InfluxDB 镜像（镜像可以提前 docker load）。

2. 机器基础环境

Docker 版本：20.10.20。

机器 IP: 10.1.0.214。

机器系统：CentOS Linux release 7.9.2009 (Core)。

系统内核版本：3.10.0-1160.el7.x86_64。

3. 安装 InfluxDB

这里以内网机器 10.1.0.214 为例来说明安装 InfluxDB 的流程，因为 Prometheus 部署需要时序数据库保存永久数据，所以这里先要部署 InfluxDB，版本为 1.8，用 docker load 命令载入。

（1）建立 InfluxDB 的数据目录、配置文件目录和备份目录（备用，防止后面需要备份数据），命令如下：

```
mkdir -p /data/influxdb/{data,backup,conf}
```

（2）利用 docker –rm 命令生成 influxdb 的配置文件，利用镜像生成 influxdb.conf 文件，注意 --rm 参数。命令如下：

```
docker run --rm influxdb:1.8 influxd config | sudo tee /data/influxdb/conf/
influxdb.conf
```

（3）用下面的脚本生成名为 influxdb、开放端口为 8086 的容器，其脚本内容如下：

```
docker run -p 8086:8086 \
        --name influxdb \
        --restart unless-stopped \
    -v /data/influxdb/data:/var/lib/influxdb \
    -v /data/influxdb/backup:/backup \
    -v /data/influxdb/conf/influxdb.conf:/etc/influxdb/influxdb.conf \
    -v /etc/localtime:/etc/localtime \
    -d influxdb:1.8
```

（4）创建名为 Prometheus 的数据库之前，进入 influxdb 容器中，命令如下：

```
docker exec -ti influxdb /bin/bash
```

创建 Prometheus 数据库，命令如下：

```
$ influx
```

```
# 登录 influxdb 容器
> create database prometheus;
# 创建 prometheus 数据库
> show databases;
```

正确返回结果应该如下：

```
name: databases
name
----
_internal
prometheus
```

（5）调整 Prometheus 数据库的保存周期（如果不调整，后期数据可能会占满磁盘，这里默认配置为三个月周期）。

查看当前 Prometheus 的默认数据保存策略为 autogen（默认是永久保存），命令如下：

```
show retention policies on "prometheus"
```

命令显示结果如下：

```
name    duration shardGroupDuration replicaN default
----    -------- ------------------ -------- -------
autogen 0s       168h0m0s           1        true
```

这里调整默认策略为 90d，命令如下：

```
use prometheus
alter retention policy "autogen" on "prometheus " duration 90d default
```

还可以用以下命令来验证，发现 autogen 策略确实发生变化了。

```
show retention policies on "prometheus"
```

命令显示结果如下：

```
ame     duration  shardGroupDuration replicaN default
----    --------  ------------------ -------- -------
autogen 2160h0m0s 168h0m0s           1        true
```

> **注**：如果需要备份，则需进入influxdb容器中，执行以下命令备份prometheus数据库：
> influxd backup -portable -database prometheus /backup/

4. 安装 Prometheus

Prometheus 的安装与 InfluxDB 类似，需要提前导入镜像，提前分配好相应的配置文件等，命令如下：

```
mkdir -p /data/prometheus/conf/
```

这里提前写好 Prometheus 的配置文件（默认抓取频率为 60 秒 / 次），prometheus.yml 文件如下（样例文件，真正监控时需要修改对应的 targets 和 jobname 等目标）：

```
global:
  scrape_interval: 60s
  evaluation_interval: 60s
scrape_configs:
  - job_name: 'prometheus'
    static_configs:
    - targets: ['10.1.0.214:9090']
  - job_name: 'node1'
    static_configs:
    - targets: ["10.1.0.214:9100"]
  - job_name: 'node2'
    static_configs:
    - targets: ["10.1.0.215:9100"]
  - job_name: 'node3'
    static_configs:
    - targets: ["10.0.3.11:9182"]
  - job_name: 'node4'
    static_configs:
    - targets: ["10.2.10.91:9100"]
remote_write:
  - url: "http://10.1.0.214:8086/api/v1/prom/write?db=prometheus"
remote_read:
  - url: http://10.1.0.214:8086/api/v1/prom/read?db=prometheus
```

因为已经提前预置了 InfluxDB 的读写 API，所以用下面的命令就可以启动 Prometheus 容器，脚本如下：

```
docker run -d --name prometheus --restart=unless-stopped -p 9090:9090 -u
root -v /data/prometheus/conf/prometheus.yml:/etc/prometheus/prometheus.yml
prom/prometheus:latest \
  --config.file=/etc/prometheus/prometheus.yml --web.enable-lifecycle
```

后续就是要修改 prometheus.yml 文件的监控目标了，修改以后可以用以下命令来尝试热加载 Prometheus 配置，命令如下：

```
curl -X POST http://10.1.0.214:9090/-/reload
```

如果没有生效，可以用 docker restart prometheus 命令重启容器，以达到重载 Prometheus 配置的目的，然后就可以用 http://10.1.0.214:9090 来访问 Prometheus 界面了。

9.3.2 Prometheus 的 YAML 配置详解

在 Prometheus 监控系统中，Prometheus 用于采集、查询、存储和推送报警到 Alertmanager。Prometheus 的配置文件内容如下：

```
global:
  scrape_interval:60s             # 采集目标数据指标的频率，默认为 60s
  evaluation_interval:60s         # 评估规则的频率，默认为 60s
                                  #  采集请求的超时时间，默认为 10s
                                  # scrape_timeout is set to the global default(10s)

# Alertmanager configuration
alerting:
  alertmanagers:
  - static_configs:
    - targets:
      # - alertmanager:9093

# Load rules once and periodically evaluate them according to the global
# 'evaluation_interval'.
rule_files:
  # - "first_rules.yml"
  # - "second_rules.yml"

# A scrape configuration containing exactly one endpoint to scrape:
# Here it's Prometheus itself.
scrape_configs:
  # The job name is added as a label `job=<job_name>` to any timeseries scraped
  # from this config.
  - job_name: 'prometheus'

    # metrics_path defaults to '/metrics'
    # scheme defaults to 'http'.

    static_configs:
    - targets: ['localhost:9090']

remote_write:
  - url: "http://192.168.1.1:8086/api/v1/prom/write?db=prometheus"
remote_read:
  - url: http://192.168.1.1:8086/api/v1/prom/read?db=prometheus
```

下面介绍 Prometheus 配置文件中的主要代码片段。

（1）global：此片段指定的是 Prometheus 的全局配置，如采集间隔、抓取超时时间等。

（2）rule_files：此片段指定报警规则文件，Prometheus 根据这些规则信息，会推送报警信息到 AlertManager 中。

（3）scrape_configs：此片段指定抓取配置，Prometheus 的数据采集通过此片段配置。

（4）alerting：此片段指定报警配置，这里主要是指定 Prometheus 将报警规则推送到指定的 AlertManager 实例地址。

（5）remote_write：指定后端的存储写入 API 地址。

（6）remote_read：指定后端的存储读取 API 地址。

global 片段主要参数，其默认 YAML 配置文件如下：

```
# 抓取间隔
  [ scrape_interval: <duration> | default = 1m ]

# 抓取超时时间
  [ scrape_timeout: <duration> | default = 10s ]

# 评估规则间隔
  [ evaluation_interval: <duration> | default = 1m ]

# 外部一些标签设置
  external_labels:
    [ <labelname>: <labelvalue> ... ]
```

一个 scrape_configs 片段指定一组目标和参数，目标就是实例，指定采集的端点，参数描述如何采集这些实例。主要参数如下。

（1）scrape_interval：抓取间隔，默认继承 global 值。

（2）scrape_timeout：抓取超时时间，默认继承 global 值。

（3）metric_path：抓取路径，默认是 /metrics。

（4）scheme：指定采集使用的协议，http 或 https。

（5）params：指定 url 参数。

（6）basic_auth：指定认证信息。

（7）file_sd_configs：指定服务发现配置。

（8）static_configs：静态指定服务 job。

（9）relabel_config：relabel 设置。

scrape_configs 配置文件如下：

```
scrape_configs:
  # 作业名称作为标签 'job=<job_name>' 添加到从此配置中抓取的任何时间序列中
  - job_name: 'prometheus'

    # '/metrics' 作为默认的 metrics_path 目录
    # 方法默认为 'http'

    static_configs:
```

```
    - targets: ['localhost:9090']
  - job_name: "node"
    static_configs:
    - targets:
    - "192.168.1.10:20001"
    - "192.168.1.11:20001
    - "192.168.1.12:20001"
```

file_sd_configs 配置文件如下：

```
static_configs:
 file_sd_configs:
    - refresh_interval: 2m           // 抓取周期为 2 分钟
      files:
      - targets/promethues-*.yaml    // 在 targets 目录下，所有以 prometheus- 为前缀
                                     // 的 yaml 文件
```

targets/prometheus-node.yaml 文件内容如下：

```
- targets:
  - "192.168.1.10:20001"
  labels:
    hostname: node00
- targets:
  - "192.168.1.11:20001"
  labels:
    hostname: node01
- targets:
  - "192.168.1.12:20001"
  labels:
    hostname: node02
```

9.3.3 InfluxDB 时序数据库的优化

在生产环境下，InfluxDB 在短时间运行后，磁盘 I/O 及内存使用都会居高不下，所以还要针对其配置文件适当进行优化处理，其配置文件建议优化值如下：

```
[data]
  #wal 日志落盘周期，官方建议 0-100ms
  wal-fsync-delay = "50ms"
  # 使用 tsi1 索引（建议磁盘类型为 SSD）
  index-version = "tsi1"
  # 分片允许最大内存，当超过最大内存时会拒绝写入
  # 内存越大，多个新老分片会占用更多的堆空间
  cache-max-memory-size = "2g"
```

```
  # 当 cache 超过 128m 时，会进行快照落盘
  cache-snapshot-memory-size = "128m"
  #cache 冷冻写入时间
  cache-snapshot-write-cold-duration = "30m"
  # 进行全量压缩时间，若要降低 CPU 压缩计算和磁盘 I/O，则适当调大，例如 24h
  compact-full-write-cold-duration = "24h"
  # 并行压缩处理器
  max-concurrent-compactions = 8
  # 压缩每秒落盘数据量
  compact-throughput = "16m"
  # 压缩每秒最大落盘数据量
  compact-throughput-burst = "16m"
  #wal 日志超过 128m 时会被压缩为索引文件，并删除
  max-index-log-file-size = "128m"
[monitor]
  # 关闭监控
  store-enabled = false
```

9.3.4 VictoriaMetrics 时序数据库介绍

在生产环境中，随着监控节点的持续增加，内存及磁盘压力持续增大，尤其是磁盘 I/O 压力，经常会出现 InfluxDB 机器由于 iowait 过高而负载很高的情况，最后导致时序数据库的机器处于宕机（Pending 假死）状态。鉴于业务方出于成本考虑不愿采用昂贵的 NVMe 磁盘，这里可以推荐使用 VictoriaMetrics 时序数据库。

在做替代的技术选型方案时，这里也将 VictoriaMetrics 与 InfluxDB 和 Thanos 等时序数据库做了综合对比，对比其他一些主流的时序数据库，VictoriaMetrics 具有如下优势：

- 指标数据的收集和查询具有极高的性能和良好的垂直和水平伸缩性，比 InfluxDB 和 TimesscaleDB 的性能高出 20 倍。
- 在处理高技术时间序列时，内存方面做出了优化，比 InfluxDB 少 10x 倍，比 Prometheus、Thanos 或 Cortex 少 7 倍。
- 数据存储的压缩方式更加高效。比 TimescaleDB 少 70 倍，与 Prometheus、Thanos、Cortex 相比，所需存储空间也少 7 倍。
- 针对高延迟 IO 和低 IOPS 存储进行了优化。
- 单节点的 VictoriaMetrics 即可替代 Thanos、M3DB、Cortex、InfluxDB 或 TimescaleDB 等竞品中等规模的集群。
- 对于 Prometheus 具有良好的兼容性，能够支持 Prometheus 的配置文件、PromQL、各类 API、数据格式，并有一些独有的增强 API。
- VictoriaMetrics 针对 HDD 做了优化，所以基本上没必要使用昂贵的 SSD。VictoriaMetrics

采用高性能的数据压缩方式，存储的数据量只有 Thanos 的 1/10。这就意味着与 Thanos 相比，VictoriaMetrics 需要更少的磁盘空间，存储相同容量数据的成本更低。

VictoriaMetrics 的安装及部署非常简单，这里推荐二进制的部署方式，可以做成服务自启动，其详细步骤如下。

编辑 /etc/system/system/victoria-metrics-prod.service 文件，文件内容如下：

```
[Unit]
Description=For Victoria-metrics-prod Service
After=network.target

[Service]
ExecStart=/usr/local/bin/victoria-metrics-prod -httpListenAddr=0.0.0.0:8428
-storageDataPath=/data/victoria -retentionPeriod=180d

[Install]
WantedBy=multi-user.target
```

可以用如下命令启动 VictoriaMetrics 服务并做成自启动服务，命令如下：

```
systemctl start victoria-metrics-prod.service
systemctl enable victoria-metrics-prod.service
```

启动 victoria-metrics-prod.service 服务以后，服务就监听 8428 端口了；配置 Prometheus 的本地保存数据时间为 1 天。Prometheus 配置命令如下：

```
./prometheus --config.file="prometheus.yml" --storage.tsdb.retention.time=1d
```

Prometheus 配置需要新增以下内容：

```
# 10.1.0.203 是部署了 VictoriaMetrics 服务的地址
remote_write:
     - url: "http://10.1.0.203:8428/api/v1/write"
```

2000 个左右的 HPC 计算节点的业务监控场景下，相同的硬件环境下，部署 VictoriaMetrics 的时序数据库比 InfluxDB 的磁盘占用率确实要小很多，大约为十分之一，而且再也没发生因为磁盘 iowait 过高而出现机器宕机的现象了。

9.4　Alertmanager 告警系统

在整个 Prometheus 告警系统中，要理解两个核心组件：Prometheus 负责数据采集，Alertmanager 负责处理告警。这两个组件都是独立的，后面的内容会讲到。另外，需要引入 alertmanager-webhook 组件，它负责处理 Alertmanager 的告警内容并发送到对应的介质中（如钉钉或企业微信等）。

9.4.1 Alertmanager 的基础概念

图 9.1 所示为 Prometheus 系统架构图，从中可以了解 Prometheus 的警报工作机制，其中 Prometheus 与 Alertmanager 是两个分离的独立组件。

使用 Prometheus Server 端通过静态或者动态配置，主动拉取（pull）部署在 Kubernetes 或云主机上的各种类别的监控指标数据。Prometheus 会根据配置的参数周期性地对警报规则进行计算，如果满足警报条件，产生一条警报信息，则将其推送到 Alertmanager 组件，Alertmanager 收到警报信息之后，会对警告信息进行处理，进行分组 Group 并将它们通过定义好的路由 Routing 规则转到正确的接收器 receiver，如 Email、钉钉、企业微信等，最终异常事件 Warning、Error 通知给定义好的接收人。其工作流程如图 9.2 所示。

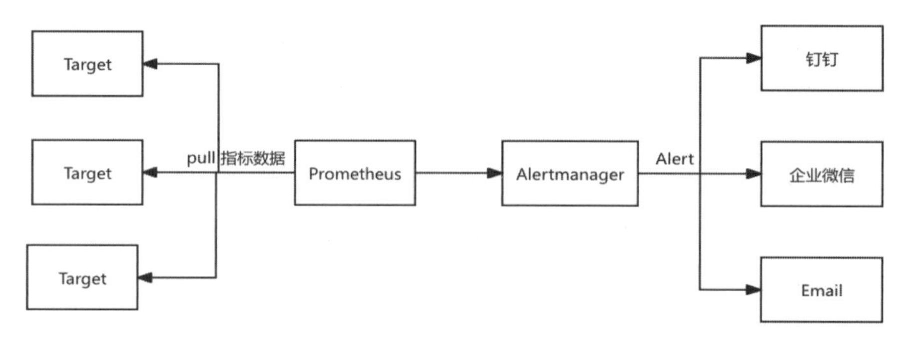

图 9.2　Alertmanager 与 Promtheus 结合工作流程

在 Prometheus 中，不仅可以对单条警报定义规则，更多的时候是对相关的多条警报进行分组后统一定义。这些定义会在后面说明其管理方法。下面介绍 Alertmanager 中的分组 Grouping、抑制 Inhibition、静默 Silences 核心特性，便于系统性地学习与理解。

1. Alertmanager 的三个核心概念

（1）分组（Grouping）。Grouping 是指 Alertmanager 把同类型的警报进行分组，合并多条警报到一个通知中。在生产环境中，特别是云环境下的业务之间密集耦合时，若出现多台 Instance 故障，可能会导致成百上千条警报触发。在这种情况下使用分组机制，可以把这些被触发的警报合并为一个警报进行通知，从而避免瞬间突发性地接收大量警报通知，使得管理员无法对问题进行快速定位。

例如，在 Kubernetes 集群中运行重量级规模的实例时，即便是集群中持续很短一段时间的网络延迟或者延迟导致网络抖动，也会引发大量类似服务应用无法连接数据库的故障。如果在警报规则中定义每一个应用实例都发送警报，那么最后会有大量的警报信息发送给 Alertmanager 。

作为运维组或者相关业务组的开发人员，可能更想在一个通知中就可以快速查看到哪些服务实例被本次故障影响了。为此，对服务所在集群或者服务警报名称的维度进行分组配置，把警报汇总成一条通知，就不会受到警报信息的频繁发送影响了。

（2）抑制（Inhibition）。Inhibition 是指当某条警报已经发送时，停止重复发送由此警报引发的其他异常或故障的警报机制。在生产环境下的 IDC 托管机柜中，若每个机柜的接入层仅仅是单台交换机，那么该机柜接入交换机故障会造成机柜中服务器非 up 状态警报。另外，服务器上部署的应用服务不可访问也会触发警报。

这时，可以通过配置 Alertmanager，忽略因交换机故障而造成此机柜中的所有服务器及应用不可达而产生的警报。

在灾备体系中，当原有集群故障宕机而业务彻底无法访问时，会把用户流量切换到备份集群中，这样为故障集群及其提供的各个微服务状态发送警报机会失去意义，此时，Alertmanager 的抑制特性就可以在一定程度上避免管理员收到过多无用的警报通知。

（3）静默（Silences）。Silences 提供了一个简单的机制，根据标签快速对警报进行静默处理；对传进来的警报进行匹配检查，如果接收到的警报符合静默的配置，Alertmanager 则不会发送警报通知。

除了分组、抑制是在 Alertmanager 配置文件中配置外，静默需要在 WEB UI 界面中设置临时屏蔽指定的警报通知，如图 9.3 所示。

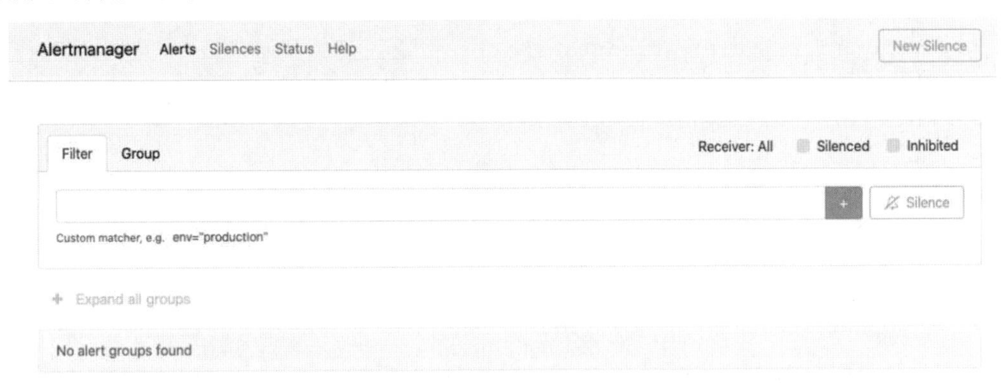

图 9.3　Alertmanager 工作界面 UI 展示

参考官方网站，其配置模板文件如下：

```
global:
  # The default SMTP From header field
  [ smtp_from: <tmpl_string> ]
  # The default SMTP smarthost used for sending emails, including port number
  # Port number usually is 25, or 587 for SMTP over TLS (sometimes referred
  # to as STARTTLS)
  # Example: smtp.example.org:587
  [ smtp_smarthost: <string> ]
  # The default hostname to identify to the SMTP server
  [ smtp_hello: <string> | default = "localhost" ]
  # SMTP Auth using CRAM-MD5, LOGIN and PLAIN. If empty, Alertmanager doesn't
  # authenticate to the SMTP server
  [ smtp_auth_username: <string> ]
  # SMTP Auth using LOGIN and PLAIN
```

```
    [ smtp_auth_password: <secret> ]
    # SMTP Auth using LOGIN and PLAIN
    [ smtp_auth_password_file: <string> ]
    # SMTP Auth using PLAIN
    [ smtp_auth_identity: <string> ]
    # SMTP Auth using CRAM-MD5
    [ smtp_auth_secret: <secret> ]
    # The default SMTP TLS requirement
    # Note that Go does not support unencrypted connections to remote SMTP
    # endpoints
    [ smtp_require_tls: <bool> | default = true ]

    # The API URL to use for Slack notifications
    [ slack_api_url: <secret> ]
    [ slack_api_url_file: <filepath> ]
    [ victorops_api_key: <secret> ]
    [ victorops_api_key_file: <filepath> ]
    [ victorops_api_url: <string> | default =
"https://alert.victorops.com/integrations/generic/20131114/alert/" ]
    [ pagerduty_url: <string> | default =
"https://events.pagerduty.com/v2/enqueue" ]
    [ opsgenie_api_key: <secret> ]
    [ opsgenie_api_key_file: <filepath> ]
    [ opsgenie_api_url: <string> | default = "https://api.opsgenie.com/" ]
    [ wechat_api_url: <string> | default = "https://qyapi.weixin.qq.com/cgi-bin/" ]
    [ wechat_api_secret: <secret> ]
    [ wechat_api_corp_id: <string> ]
    [ telegram_api_url: <string> | default = "https://api.telegram.org" ]
    [ webex_api_url: <string> | default = "https://webexapis.com/v1/messages" ]
    # The default HTTP client configuration
    [ http_config: <http_config> ]

    # ResolveTimeout is the default value used by alertmanager if the alert does
    # not include EndsAt, after this time passes it can declare the alert as
    # resolved if it has not been updated
    # This has no impact on alerts from Prometheus, as they always include EndsAt.
    [ resolve_timeout: <duration> | default = 5m ]

  # Files from which custom notification template definitions are read
  # The last component may use a wildcard matcher, e.g. 'templates/*.tmpl'
  templates:
```

```
  [ - <filepath> ... ]

# The root node of the routing tree
route: <route>

# A list of notification receivers
receivers:
  - <receiver> ...

# A list of inhibition rules
inhibit_rules:
  [ - <inhibit_rule> ... ]

# DEPRECATED: use time_intervals below
# A list of mute time intervals for muting routes
mute_time_intervals:
  [ - <mute_time_interval> ... ]

# A list of time intervals for muting/activating routes
time_intervals:
  [ - <time_interval> ... ]
```

参考官方的配置文件，Alertmanager 配置主要分为 5 个部分，分别是全局设置（global）、警报模板（templates）、警报路由（route）、接收器（receivers）和抑制规则（inhibit_rules），详细内容如下：

（1）global，即全局设置。在 Alertmanager 配置文件中，只要全局设置的选项，全部为公共设置，还可以配置继承，作为默认值，可以在参数中覆盖其设置。其中，resolve_timeout 用于设置处理超时时间，也是生命警报状态未解决的时间，这个时间会直接影响警报恢复的通知时间，需要自行结合实际生产场景来设置主机的恢复时间，默认是 5 分钟。在全局设置中可以设置 SMTP 服务，同时也支持 slack、victorops、pagerduty 等方式，这里的内容只涉及常用的 Email、钉钉和企业微信等，当然了，可以自己使用 Go 语言进行二次开发，对接自定义 webhook 通知源。

（2）templates。警报模板可以自定义通知的信息格式，以及其包含的对应警报指标数据，可以自定义 Email、企业微信的模板，配置指定的存放位置，对于钉钉的模板会单独讲如何配置，这里的模板是指发送的通知源信息格式模板，如 Email 和企业微信。

（3）route。警报路由模块描述了在收到 Prometheus 生成的警报后，将警报信息发送给接收器 receiver 指定的目标地址规则。Alertmanager 对传入的警报信息进行处理，根据所定义的规则与配置进行匹配。对于路由可以理解为树状结构，设置的第一个 route 是根节点，其下是包含的子节点，每个警报传进来以后，会从配置的根节点路由进入路由树，按照深度优先从左向右遍历匹配，当匹配到节点后停止，进行警报处理。

分组就是将多条告警信息聚合成一条发送，这样就不会收到连续的报警了，从而达到降噪的目的。将传入的告警按标签分组（标签在 Prometheus 的 rules 中定义）。例如，接收到的告警信息中有许多具有 cluster=A 和 alertname=LatencyHigh 的标签，这些告警将被分为一个组。

（4）receivers 接收器。接收器是一个统称，每个 receivers 都需要设置一个全局唯一的名称，并且对应一个或者多个通知方式，包括 Email、微信、Slack 和钉钉等。

（5）inhibit_rules 。在模块中设置警报抑制功能，可以指定在特定条件下需要忽略的警报条件，还可以使用此选项设置首选。例如，优先处理某些警报，如果同一组中的警报同时发生，则忽略其他警报。因此，使用 inhibit_rules 可以减少频繁发送没有意义的警报。

这里较复杂的是 route 分组，可以参考下面的 Alertmanager 配置案例文件。其文件的详细内容如下：

```
## Alertmanager 配置案例文件
global:
  resolve_timeout: 5m
  # smtp 配置
  smtp_from: "123456789@qq.com"
  smtp_smarthost: 'smtp.qq.com:465'
  smtp_auth_username: "123456789@qq.com"
  smtp_auth_password: "auth_pass"
  smtp_require_tls: true
# Email、企业微信的模板配置存放位置，钉钉的模板会单独讲如何配置
templates:
  - '/data/alertmanager/templates/*.tmpl'
# 路由分组
route:
  receiver: ops                  # 默认的接收器名称
  group_wait: 30s                # 如果 30 秒内出现的组内规则，则统一归纳成一组进行报警

  group_interval: 5m             # 已发送初始通知的一组告警接收到新告警后，再次发送通
                                 # 知前等待的时间
  repeat_interval: 24h           # 发送报警间隔，如果指定时间内没有修复，则重新发送报警
  group_by: [alertname]          # 报警分组，根据 prometheus 的 labels 进行报警分组，
                                 # 这些警报会合并为一个通知发送给接收器，也就是警报分组
  routes:
      - match:
          team: operations
        group_by: [env,dc]
        receiver: 'ops'
      - match_re:
          service: nginx|apache
        receiver: 'web'
      - match_re:
```

```
          service: hbase|spark
        receiver: 'hadoop'
      - match_re:
          service: mysql|mongodb
        receiver: 'db'
# 接收器
# 抑制测试配置
      - receiver: ops
        group_wait: 10s
        match:
          status: 'High'
# ops
      - receiver: ops  # 路由和标签，根据match来指定发送目标，如果 rule的label 包
                       # 含 alertname，使用ops来发送
        group_wait: 10s
        match:
          team: operations
# web
      - receiver: db   # 路由和标签，根据match来指定发送目标，如果 rule的label 包含
                       # alertname，使用 db 来发送
        group_wait: 10s
        match:
          team: db
# 接收器指定发送人以及发送渠道
receivers:
# ops分组的定义
- name: ops
  email_configs:
  - to: '9935226@qq.com,10000@qq.com'
    send_resolved: true
    headers:
      subject: "[operations] 报警邮件"
      from: "警报中心"
      to: "yuhongchun027@163.com "
  # 钉钉配置
  webhook_configs:
  - url: http://localhost:8070/dingtalk/ops/send
    # 企业微信配置
  wechat_configs:
  - corp_id: 'ww5421dksajhdasjkhj'
    api_url: 'https://qyapi.weixin.qq.com/cgi-bin/'
```

```
      send_resolved: true
      to_party: '2'
      agent_id: '1000002'
      api_secret: 'Tm1kkEE3RGqVhv5hO-khdakjsdkjsahjkdksahjkdsahkj'

  # web
  - name: web
    email_configs:
    - to: '9935226@qq.com'
      send_resolved: true
      headers: { Subject: "[web] 报警邮件 "}            # 接收邮件的标题
    webhook_configs:
    - url: http://localhost:8070/dingtalk/web/send
    - url: http://localhost:8070/dingtalk/ops/send
  # db
  - name: db
    email_configs:
    - to: '9935226@qq.com'
      send_resolved: true
      headers: { Subject: "[db] 报警邮件 "}             # 接收邮件的标题
    webhook_configs:
    - url: http://localhost:8070/dingtalk/db/send
    - url: http://localhost:8070/dingtalk/ops/send
  # hadoop
  - name: hadoop
    email_configs:
    - to: '9935226@qq.com'
      send_resolved: true
      headers: { Subject: "[hadoop] 报警邮件 "}         # 接收邮件的标题
    webhook_configs:
    - url: http://localhost:8070/dingtalk/hadoop/send
    - url: http://localhost:8070/dingtalk/ops/send

# 抑制器配置
inhibit_rules:     # 抑制规则
  - source_match: # 源标签警报触发时抑制含有目标标签的警报，在当前警报匹配 status:
                  # 'High'
      status: 'High'  # 此处的抑制匹配一定要在最上面的 route 中配置，否则会提示找不到 key
    target_match:
      status: 'Warning' # 目标标签值正则匹配，可以是正则表达式，如 ".*MySQL.*"
```

```
    equal: ['alertname','operations', 'instance'] # 确保这个配置下的标签内容相同
                                    # 才会抑制，也就是警报中必须有这三个标签值才会被抑制
```

这里以某个生产环境下 Alertmanager 的配置文件来说明，其详细内容如下：

```
global:
  ## 持续多长时间没有触发告警，则认为处于告警问题已经解决状态的时间
  resolve_timeout: 5m
  ## 配置邮件发送信息
  smtp_smarthost: 'smtp.qiye.163.com:465'
  smtp_from: 'robot@example.com'
  smtp_auth_username: 'robot@example.com'
  smtp_auth_password: '123456'
  smtp_require_tls: false
# 所有报警信息进入后的根路由，用来设置报警的分发策略
templates:
  - '/etc/alertmanager/*.tmpl'
route:
  ## 这里的标签列表是接收到报警信息后的重新分组标签。例如，接收到的报警信息中有许多具有
  # status=XX 这样的标签，可以根据这些标签将告警信息批量聚合到一个分组中
  group_by: ['alertname','status']
  ## 当一个新的报警分组被创建后，需要等待至少 group_wait 时间来初始化通知，这种方式可以
  # 确保有足够的时间为同一分组汇入尽可能多的告警信息，然后将这些汇集的告警信息一次性触发
  group_wait: 60s
  ## 当第一个报警发送后，等待 group_interval 时间来发送新的一组报警信息
  group_interval: 60m
  ## 如果一个报警信息已经发送成功了，则需要等待 repeat_interval 时间才能重新发送
  repeat_interval: 10m
  ## 配置默认的路由规则
  receiver: 'webhook'

  routes:
    - receiver: 'business'
      group_wait: 10s
      match:
        status: jobruntime


receivers:
  - name: 'webhook'
    webhook_configs:
      - url: http://10.1.0.199:8088/adapter/wx
```

```
    - name: 'business'
      email_configs:
        - to: 'support@example.com'
```

在以上配置中，所有满足 Label 中包含 status 为 jobruntime 的实例会发送告警到 suppport@example.com，其他报警则是默认的告警，即发送到企业微信群。

2. 告警规则的定义

Prometheus 中配置有告警规则，告警规则也是 YAML 格式的文件。告警规则允许用户基于 PromQL 定义告警条件，并在触发告警时发送通知给外部接收者。

为了能够让 Prometheus 启用定义的告警规则，需要在 Prometheus 全局配置文件中通过 rule_files 指定一组告警规则文件的访问路径，其相关配置如下：

```
rule_files:
  - "first_rules.yml"
```

告警规则的定义遵循下面的风格：

```
ALERT <alert name>
   IF <expression>
     [  FOR <duration>  ]
       [  LABELS <label set>  ]
         [  ANNOTATIONS <label set> ]
```

以上规则中的参数说明如下。

（1）alert：告警规则的名称。

（2）expression：用于报警规则的 PromQL 查询语句。

（3）for：评估等待时间（Pending Duration），用于表示只有当触发条件持续一段时间后才发送告警，在等待期间新产生的告警状态为 pending。这个参数主要用于降噪，很多类似响应时间这样的指标都是有抖动的，通过指定 Pending Duration，可以过滤掉这些瞬时抖动，将注意力都放在真正有持续影响的问题上。

需要注意的是，如 for 后面是 10 分钟，表示如果 10 分钟以内有 10 个采集指标，即使有 50% 的指标超出规则条件，也不会触发告警；只有采集到的指标 100% 满足条件，且持续 10 分钟，才会进行告警。

（4）labels：自定义标签，允许用户指定额外的标签列表并将其附加在告警上。

（5）annotations：指定另一组标签，它们不作为告警实例的身份标识，而常用于存储一些额外的信息，用于报警信息的展示。

一个报警信息在 Alertmanager 生命周期内有下面 3 种状态。

（1）pending：表示在设置的阈值时间范围内被激活。

（2）firing：表示超过设置的阈值时间被激活。

（3）inactive：报警规则没有得到满足或者已经过期（还没触发或者已经修复）。

参考官方文档，一个较简单的 alert.yml 配置规则如下（表示定义的 Instance，要么是主机要么是服务，如果持续5分钟up值为0，则表示处于down状态）：

```
groups:
- name: Instances
  rules:
  - alert: InstanceDown
    expr: up == 0
    for: 5m
    labels:
      severity: page
    # Prometheus templates apply here in the annotation and label fields of
    # the alert
    annotations:
      description: '{{ $labels.instance }} of job {{ $labels.job }} has been
down for more than 5 minutes.'
      summary: 'Instance {{ $labels.instance }} down'
```

在被监控主机上安装node-exporter以后，常用的监控配置文件rules.yml内容如下：

```
groups:
    - name: 主机状态 - 监控告警
      rules:
      - alert: CPU 使用状况
          expr: 100-(avg(irate(node_cpu_seconds_total{mode="idle"}[5m]))
by(instance)* 100) > 90
        for: 1m
        labels:
          status: 通常告警
        annotations:
          summary: " CPU 使用率太高! "
          description: " CPU 使用大于 90%( 目前使用 :%)"
      - alert: 内存使用
        expr: 100 -(node_memory_MemTotal_bytes
-node_memory_MemFree_bytes+node_memory_Buffers_bytes+node_memory_Cached_bytes
) / node_memory_MemTotal_bytes * 100> 80
        for: 1m
        labels:
          status: 严重告警
        annotations:
          summary: " 内存使用率太高! "
          description: " 内存使用大于 80%( 目前使用 :%)"
      - alert: IO 性能
          expr: 100-(avg(irate(node_disk_io_time_seconds_total[1m]))
```

```
by(instance)* 100) < 60
            for: 1m
            labels:
              status: 严重告警
            annotations:
              summary: " 流入磁盘 IO 使用率太高! "
              description: " 流入磁盘 IO 大于 60%(目前使用:)"
          - alert: 网络
            expr: ((sum(rate (node_network_receive_bytes_total{device!~'tap.*|
veth.*|br.*|docker.*|virbr*|lo*'}[5m])) by (instance)) / 100) > 102400
            for: 1m
            labels:
              status: 严重告警
            annotations:
              summary: " 流入网络带宽太高! "
              description: "流入网络带宽持续 2 分钟高于 100MB. RX 带宽使用率"
          - alert: 网络
            expr: ((sum(rate
(node_network_transmit_bytes_total{device!~'tap.*|veth.*|br.*|docker.*|virbr*
|lo*'}[5m])) by (instance)) / 100) > 102400
            for: 1m
            labels:
              status: 严重告警
            annotations:
              summary: " 流出网络带宽太高! "
              description: "流出网络带宽持续 2 分钟高于 100MB. RX 带宽使用率"
          - alert: TCP 会话
            expr: node_netstat_Tcp_CurrEstab > 1000
            for: 1m
            labels:
              status: 严重告警
            annotations:
              summary: " TCP_ESTABLISHED 太高! "
              description: " TCP_ESTABLISHED 大于 1000%(目前使用:%)"
          - alert: 磁盘容量
            expr:
100-(node_filesystem_free_bytes{fstype=~"ext4|xfs"}/node_filesystem_size_bytes
{fstype=~"ext4|xfs"}*100) > 90
            for: 1m
            labels:
              status: 严重告警
```

```
        annotations:
          summary: " 磁盘分区使用率太高！ "
          description: " 磁盘分区使用大于 90%（目前使用：%)"
```

以上都是官方推荐的常见 PromQL 告警例子，读者可以结合 PromQL 的官方文档学习和熟悉 PromQL 基础语法，然后结合自己的实际业务来写符合要求的 PromQL 告警规则。

9.4.2　Alertmanager 系统告警时间

相信很多读者朋友使用 Alertmanager 时有这种困惑，为什么自己的告警没有及时发出来，或者不应该告警的时候出现告警了？下面先了解一下 Prometheus 系统和 Alertmanager 系统的默认时间，见表 9.1。

表 9.1　Prometheus 系统和 Alertmanager 系统的默认时间

系　统	选　项	作　用	默认值
Prometheus	scrape_interval	服务端抓取数据的时间间隔	60 秒
	scrape_timeout	服务端抓取数据的超时时间	10 秒
	evaluation_interval	评估告警规则的时间间隔	60 秒
Alertmanager	group_wait	发送一件新的告警的初始等待时间，也就是初次发告警的延时	30 秒
	group_interval	如果初次告警已经发送，需要等待多长时间再次发送同组新产生的告警	300 秒
	repeat_interval	如果告警已经成功发送，间隔多长时间再重复发送	4 小时

Alertmanager 遵循以下流程来正确发送告警：

（1）Prometheus 系统以 scrape_interval 为一个采集周期，然后以 evaluation_interval 为平均评估周期定期对告警规则进行评估；如果采集对象出现问题，Prometheus 会持续获取数据，直到 scrape_timeout 时间后停止尝试。

（2）Prometheus 会进行告警规则判断，如果表达式为真，那么告警状态会切换到 pending；如果持续时间超过 for 语句指定的时间（如 10 分钟），告警状态变更为 active，并将告警从 Prometheus 发送给 Alertmanager。

（3）如果下一个计算周期表达式仍然为真，且持续时间超过 for 语句指定的时间（如 10 分钟），则持续发警告给 Alertmanager；直到某个计算周期表达式为假，告警状态变更为 inactive，发送一个 resolve 给 Alertmanager，说明此告警已解决。

（4）Alertmanager 收到告警数据以后，会将告警信息进行分组，然后根据 Alertmanager 配置的 group_wait 时间先等待，在 group_wait 时间过后再发送告警信息。

（5）Alertmanager 会判断同一个 group 组的告警，如果之前的告警已经成功发出，那么等待 group_interval 时间后再重新发送新的告警信息；如果告警组的告警一直没发生变化并且已经成功发送，则等待 repeat_interval+group_wait 时间发送相同的告警；如果之前的告警没有成功发送，需要等待 group_interval 时间后发送。

9.5 用 Docker 搭建完整的监控告警系统

考虑到各种复杂环境，包括客户的内网环境（无公网环境），需要一个通用的监控告警系统的交付方案，所以这里用 Docker 来部署这套完整的监控告警系统。

主机为 192.168.1.199，已提前配置好 Docker 服务并且启动。

前面主要介绍了 Prometheus 下如何进行白盒监控，监控主机的资源用量、容器的运行状态、数据库中间件的运行数据等，这些都是支持业务和服务的基础设施，通过白盒能够了解其内部的实际运行状态，通过对监控指标的观察能够预判可能出现的问题，从而对潜在的不确定因素进行优化。而从完整的监控角度来看，除了大量使用白盒监控以外，还应该添加适当的 Blackbox（黑盒）监控，黑盒监控即以用户的身份测试服务的外部可见性，常见的黑盒监控包括 HTTP 探针、TCP 探针等，用于检测站点或者服务的可访问性，以及访问效率等。

黑盒监控与白盒监控最大的不同在于黑盒监控以故障为导向，当故障发生时，黑盒监控能快速发现故障，而白盒监控则侧重于主动发现或者预测潜在的问题。一个完善的监控目标既要能够从白盒的角度发现潜在问题，也能够从黑盒的角度快速发现已经发生的问题。

Blackbox Exporter 是 Prometheus 社区提供的官方黑盒监控解决方案，允许用户通过 HTTP、HTTPS、DNS、TCP 以及 ICMP 的方式对网络进行探测。

InfluxDB 的安装步骤如下：

（1）InfluxDB 和 Prometheus 的安装过程前面已经介绍过，这里不再介绍。

（2）通过 Docker 安装 Blackbox。

Blackbox 的配置文件 /data/blackbox/conf/blackbox.yml 的内容如下：

```
modules:
  http_2xx:   # http 检测模块  Blockbox-Exporter 中所有的探针均是以 Module 的信息
              # 进行配置
    prober: http
    timeout: 10s
    http:
      valid_http_versions: ["HTTP/1.1", "HTTP/2"]
      valid_status_codes: [200] # 这里最好做一个返回状态码，方便后续 Grafana 出图
      method: GET
      preferred_ip_protocol: "ip4"
  http_post_2xx: # http post 监测模块
    prober: http
    timeout: 10s
    http:
      valid_http_versions: ["HTTP/1.1", "HTTP/2"]
      method: POST
      preferred_ip_protocol: "ip4"
```

```
    tcp_connect:  # TCP 检测模块
      prober: tcp
      timeout: 10s
    dns:  # DNS 检测模块
      prober: dns
      dns:
        transport_protocol: "tcp"            # 默认是 udp
        preferred_ip_protocol: "ip4"         # 默认是 ip6
        query_name: "kubernetes.default.svc.cluster.local"
```

其运行脚本如下：

```
    docker run -d  -p 9115:9115 --name blackbox -v /etc/timezone:/etc/
timezone:ro -v /etc/localtime:/etc/localtime:ro -v /data/blackbox/conf:/config
10.1.0.201:8083/blackbox-exporter:0.16.0 --config.file=/config/blackbox.yml
--log.level=debug
```

Prometheus 增加相应监控选项，具体如下：

```
    - job_name: '线上官网 HTTPS 检测'
      metrics_path: /probe
      params:
        module: [http_2xx]
      static_configs:
        - targets:
            - https://www.example.com:1443    # 探测的目标
      relabel_configs:
        - source_labels: [__address__]
          target_label: __param_target
        - source_labels: [__param_target]
          target_label: instance
        - target_label: __address__
          replacement: 10.1.0.199:9115
```

（3）安装 alertmanager-webhook 程序，用于接收 Alertmanager 的通知并可以通过 Email、钉钉或企业微信发送，这里推荐 alertmanager-webhook-adapter，其 git 地址如下：

```
    https://github.com/bougou/alertmanager-webhook-adapter.git
```

git clone 下载代码库，其命令如下：

```
    git clone https://github.com/bougou/alertmanager-webhook-adapter.git /
Users/yuhongchun/repo/workspace/cloudnative/monitoring_center/
```

此代码库是一个通用的 webhook 服务器，用于接收 Prometheus Alertmanager 的通知并通过不同的渠道类型发送；使用它的主要原因是其模板功能丰富且强大，支持钉钉、微信和企业微信等，效果如图 9.4 和图 9.5 所示。

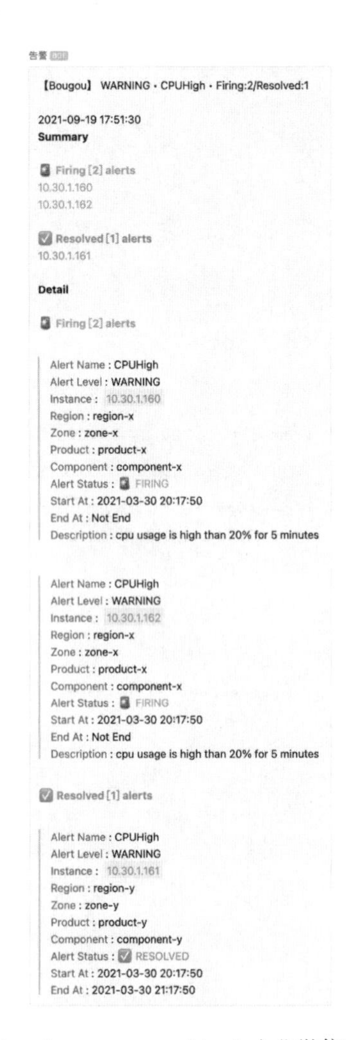

图 9.4 alertmanager-webhook 企业微信效果

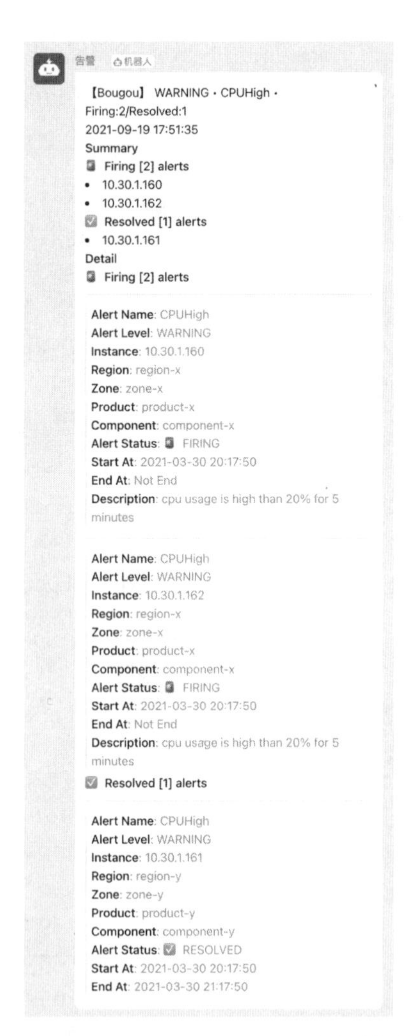

图 9.5 alertmanager-webhook 钉钉效果

以企业微信为例，这里为了更好的告警效果，所以采取中文模板，并且时间改为 UTC+8（东八区时间，也就是北京时间）。更新后的模板文件为（/Users/yuhongchun/repo/workspace/cloudnative 为作者工作目录，下面简称 /working）：

```
/working/monitoring_center/alertmanager-webhook/pkg/models/templates/weixin.zh.tmpl
```

其内容如下：

```
{{ define "__subject" -}}
【{{ .Signature }}】

{{- if eq (index .Alerts 0).Labels.severity "ok" }} OK{{ end }}
{{- if eq (index .Alerts 0).Labels.severity "info" }} INFO{{ end }}
```

```
{{- if eq (index .Alerts 0).Labels.severity "warning" }} WARNING{{ end }}
{{- if eq (index .Alerts 0).Labels.severity "critical" }} CRITICAL{{ end }}

{{- ` • ` }}

{{- if .CommonLabels.alertname_cn }}{{ .CommonLabels.alertname_cn }}{{
else if .CommonLabels.alertname_custom }}{{ .CommonLabels.alertname_custom }}
{{ else if .CommonAnnotations.alertname }}{{ .CommonAnnotations.alertname }}{{
else }}{{ .GroupLabels.alertname }}{{ end }}

{{- ` • ` }}

{{- if gt (.Alerts.Firing|len) 0 }}告警中:{{ .Alerts.Firing|len }}{{ end }}
{{- if and (gt (.Alerts.Firing|len) 0) (gt (.Alerts.Resolved|len) 0) }}/{{ end }}
{{- if gt (.Alerts.Resolved|len) 0 }}已恢复:{{ .Alerts.Resolved|len }}{{ end }}

{{ end }}


{{ define "__externalURL" -}}
{{ .ExternalURL }}/#/alerts?receiver={{ .Receiver }}
{{- end }}

{{ define "__alertinstance" -}}
{{- if ne .Labels.alertinstance nil -}}{{ .Labels.alertinstance }}
{{- else if ne .Labels.instance nil -}}{{ .Labels.instance }}
{{- else if ne .Labels.node nil -}}{{ .Labels.node }}
{{- else if ne .Labels.nodename nil -}}{{ .Labels.nodename }}
{{- else if ne .Labels.host nil -}}{{ .Labels.host }}
{{- else if ne .Labels.hostname nil -}}{{ .Labels.hostname }}
{{- else if ne .Labels.ip nil -}}{{ .Labels.ip }}
{{- end -}}
{{- end }}

{{ define "__alert_list" }}
{{ range . }}
> <font color="comment"> 告警名称 </font>:
{{ if .Labels.alertname_cn  }}{{ .Labels.alertname_cn }}{{ else
if .Labels.alertname_custom  }}{{ .Labels.alertname_custom }}{{ else
if .Annotations.alertname  }}{{ .Annotations.alertname }}{{ else }}{{ .Labels
.alertname }}{{ end }}
>
> <font color="comment"> 告警级别 </font>:{{ ` ` }}
```

```
    {{- if eq .Labels.severity "ok" }}OK{{ end -}}
    {{- if eq .Labels.severity "info" }}INFO{{ end -}}
    {{- if eq .Labels.severity "warning" }}WARNING{{ end -}}
    {{- if eq .Labels.severity "critical" }}CRITICAL{{ end }}
    >
    > <font color="comment"> 实例 </font>: `{{ template "__alertinstance" . }}`
    >
    {{- if .Labels.region }}
    > <font color="comment"> 地域 </font>: {{ .Labels.region }}
    >
    {{- end }}
    {{- if .Labels.zone }}
    > <font color="comment"> 可用区 </font>: {{ .Labels.zone }}
    >
    {{- end }}
    {{- if .Labels.product }}
    > <font color="comment"> 产品 </font>: {{ .Labels.product }}
    >
    {{- end }}
    {{- if .Labels.component  }}
    > <font color="comment"> 组件 </font>: {{ .Labels.component }}
    >
    {{- end }}
    > <font color="comment"> 告警状态 </font>: {{ if eq .Status
"firing" }}🚨{{ else }}✅{{ end }} <font color="{{ if eq .Status
"firing" }}warning{{ else }}info{{ end }}">{{ .Status | toUpper }}</font>
    >
    > <font color="comment"> 开始时间 </font>: {{ (.StartsAt.Add 28800e9).
Format "2006-01-02 15:04:05" }}
    >
    > <font color="comment"> 结束时间 </font>: {{ if .EndsAt.After .StartsAt }}{{
(.EndsAt.Add 28800e9).Format "2006-01-02 15:04:05" }}{{ else }}Not End{{ end }}
    {{- if eq .Status "firing" }}
    >
    > <font color="comment"> 告警描述 </font>: {{ if .Annotations.description_
cn }}{{ .Annotations.description_cn }}{{ else }}{{ .Annotations.description }}
{{ end }}
    {{- end }}

    {{ end }}
    {{ end }}
```

```
{{ define "__alert_summary" -}}
{{ range . }}
<font color="{{ if eq .Status "firing" }}warning{{ else }}info{{ end }}">{{
template "__alertinstance" . }}</font>
{{ end }}
{{ end }}

{{ define "prom.title" -}}
{{ template "__subject" . }}
{{ end }}


{{ define "prom.markdown" }}
{{ (.MessageAt.Add 28800e9).Format "2006-01-02 15:04:05" }}

{{ if gt (.Alerts.Resolved|len) 0 }}
#####  <font color="info">✅ 已恢复告警
[{{ .Alerts.Resolved|len }}]alerts</font>
{{ template "__alert_summary" .Alerts.Resolved }}
{{ end }}

#### 详请

{{ if gt (.Alerts.Firing|len ) 0 }}
##### <font color="warning">🚨 触发中告警
[{{ .Alerts.Firing|len }}]alerts</font>
{{ template "__alert_list" .Alerts.Firing }}
{{ end }}


{{ if gt (.Alerts.Resolved|len) 0 }}
#####  <font color="info">✅ 已恢复告警
[{{ .Alerts.Resolved|len }}]alerts</font>
{{ template "__alert_list" .Alerts.Resolved }}
{{ end }}
{{ end }}

{{ define "prom.text" }}
{{ template "prom.markdown" . }}
{{ end }}
```

然后进入 /working/monitoring_center/alertmanager-webhook/ 目录打包镜像，命令如下：

```
docker buid -t alertmanager-webhook-adapter:20230615.
```

这里通过脚本可以生成 wechat-webhook 应用：

```
docker run -d --name wechat-webhook --restart always -p 8090:8090 \
alertmanager-webhook-adapter:20230615 \
/alertmanager-webhook-adapter --signature=ZCloud --tmpl-lang=zh
```

（4）使用 Docker 安装 Alertmanager，在前面提到的 webhook 中增加 Alertmanager 配置项：

```
receivers:
## 这个是alertmanager默认的通知选项
  - name: 'webhook-default'
    webhook_configs:
      - url: http://10.1.0.199:8088/adapter/wx
        send_resolved: true
## 增加alertmanager-webhook选项
  - name: 'business'
    webhook_configs:
      - url:
"http://10.1.0.199:8090/webhook/send?channel_type=weixin&token=xxxxxx "
        send_resolved: false
```

启动 Alertmanager 的命令如下：

```
docker run -d --restart=always \
--name=alertmanager \
-p 9093:9093 \
-v /data/alertmanager/conf:/etc/alertmanager \
-v /etc/localtime:/etc/localtime \
10.1.0.201:8083/altermanager:0.25.0
```

最终，Alertmanager 告警效果（其时间为北京时间）如图 9.6 所示。

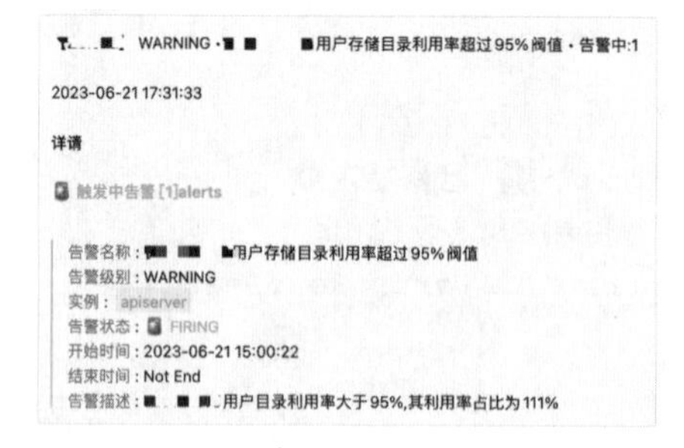

图 9.6　自定义 alertmanager-webhook 企业微信中文模板效果

另外，像 Grafana 这种监控面板也可以作为可选项考虑安装，由于项目或业务场景更看重的是自动化预警 + 告警功能，而且也有专业前端可以对接后端的 API，因此这里没考虑安装 Grafana，用户可以自行选择。

9.6　用 Go 定制开发 Exporter 组件

Exporter 在 9.2 节中已经提到过，它与 Job 的作用一样，是 Prometheus 的监控对象；独立的 Exporter 类似于一个代理层，主要采取零侵入式、独立部署的方式在目标系统上运行。Exporter 会从目标系统获取监控数据，并且转成 Prometheus 支持的格式，最后 Prometheus 会以 HTTP 轮询的形式从 Exporter 代理层获取加工后的符合 Prometheus 格式的数据，其工作流程如图 9.7 所示。

图 9.7　Exporter 在 Pormetheus 监控体系中的工作流程

对于 Exporter 而言，它的功能主要是将数据周期性地从监控对象中取出来进行加工，然后将数据规范化后通过端点暴露给 Prometheus，所以主要包含以下 3 个功能：

（1）封装功能模块获取监控系统内部的统计信息。

（2）将返回数据进行规范化映射，使其成为符合 Prometheus 要求的格式化数据。

（3）Collect 模块负责存储规范化后的数据，最后当 Prometheus 定时从 Exporter 提取数据时，Exporter 就将 Collect 收集的数据通过 HTTP 方式在 /metrics 端点进行暴露。

常见的 Exporter 数据采集方式如下：

（1）HTTP/HTTPS 方式。例如，RabbitMQ exporter 通过 RabbitMQ 的 HTTPS 接口获取监控数据。

（2）TCP 方式。例如，Redis exporter 通过 Redis 提供的系统监控相关命令获取监控指标，MySQL server exporter 通过 MySQL 开放的与监控相关的表获取监控指标。

（3）本地文件方式。例如，Node exporter 通过读取 proc 文件系统下的文件，计算得出整个操作系统的状态。

（4）标准协议方式。例如，SNMP 或 IPMI 协议。

基于 Exporter 的数据采集方式，Prometheus 提供了各种语言的依赖库以支持开发者进行 Exporter 的研发工作，其中推荐使用 Go、Python、Java 这 3 种语言进行 Exporter 的开发工作。下面主要以 Go 语言为主来介绍。

下面用 Go 语言编写一个小程序，用于实时统计 /tmp、/bin、/etc、/sbin 下有多少文件夹。代码如下：

```
package main

import (
	"fmt"
	"github.com/prometheus/client_golang/prometheus"
	"github.com/prometheus/client_golang/prometheus/promhttp"
```

```
        "github.com/siddontang/go/log"
        "net/http"
        "os/exec"
        "strconv"
        "strings"
)

// define a struct   from prometheus's struct  named Desc
type ovsCollector struct {
        ovsMetric *prometheus.Desc
}

func (collector *ovsCollector) Describe(ch chan<- *prometheus.Desc) {
        ch <- collector.ovsMetric
}

var constLabel = prometheus.Labels{"component": "ovs"}

//  get the value of the metric from a function who would execute a command
// and return a float64 value
func (collector *ovsCollector) Collect(ch chan<- prometheus.Metric) {
        str := []string{"/tmp", "/bin", "/etc", "/sbin"}
        for _, dirName := range str{
                result := getDirectoryCount(dirName)
                // 去除空格
                result = strings.Replace(result, "", "", -1)
                // 去除换行符
                result = strings.Replace(result, "\n", "", -1)
                metricValue, err := strconv.ParseFloat(result, 64)
                if err != nil{
                        fmt.Println("metricValue", metricValue)
                }
                //vV := []string{"yhc"}
                vV := []string{dirName}
                ch <- prometheus.MustNewConstMetric(collector.ovsMetric,
prometheus.CounterValue, metricValue, vV...)
                yhc := ch
               log.Infof("ch:%v", yhc)
        }
}
```

```
// define metric's name、help
func newOvsCollector() *ovsCollector {
	// 将字符串转换成字符串切片，否则 Prometheus.NewDesc 不能识别
	var vValue = []string{"directoryName"}
	return &ovsCollector{
		ovsMetric: prometheus.NewDesc("ovs_directory_count",
			"Show ovs directory status", vValue,
			constLabel),
	}
}

func getDirectoryCount(dir string) (m string) {
	postcmd := "ls -lR" + " " + dir + "|wc -l"
	//fmt.Println(postcmd)
	result, err := exec.Command("bash", "-c", postcmd).Output()
	if err != nil{
		log.Error("result:", string(result))
		log.Error("command failed:", err.Error())
	}
	res := string(result)
	return res
}

func main() {

	ovs := newOvsCollector()
	prometheus.MustRegister(ovs)

	http.Handle("/metrics", promhttp.Handler())

	log.Info("begin to server on port 8080")
	// listen on port 8080
	log.Fatal(http.ListenAndServe(":8080", nil))
}
```

然后可以用下面的地址来访问，命令如下：

```
curl http://127.0.0.1:8080/metrics
```

其结果如下：

```
# HELP ovs_directory_count Show ovs directory status
# TYPE ovs_directory_count counter
ovs_directory_count{component="ovs",directoryName="/bin"} 37
```

```
ovs_directory_count{component="ovs",directoryName="/etc"} 1
ovs_directory_count{component="ovs",directoryName="/sbin"} 64
ovs_directory_count{component="ovs",directoryName="/tmp"} 1
```

其开发流程遵循如下规律：

```
定义集群指标采集器 --> 数据采集工作 --> 实现 describe 接口，写入描述信息 --> 将收集的数据导入 collect--> 结构体实例化赋值 --> 定义注册表 --> 注入自定义指标 --> 暴露 metric
```

这里再用一个更复杂的程序来演示，代码如下：

```
package main

import (
	"github.com/prometheus/client_golang/prometheus"
	"github.com/prometheus/client_golang/prometheus/promhttp"
	"github.com/siddontang/go/log"
	"net/http"
)

// 定义结构体，这是一个集群指标采集器
type HostMonitor struct {
	cpuDesc    *prometheus.Desc
	memDesc    *prometheus.Desc
	ioDesc     *prometheus.Desc
	labelVaues []string
}

// 创建结构体及对应的指标信息
func NewHostMonitor() *HostMonitor {
	return &HostMonitor{
		cpuDesc: prometheus.NewDesc(
			"host_cpu",
			"get host cpu",
			// 动态标签 key 列表
			[]string{"instance_id", "instance_name"},
			// 静态标签
			prometheus.Labels{"module": "cpu"},
		),
		memDesc: prometheus.NewDesc(
			"host_mem",
			"get host mem",
			// 动态标签 key 列表
			[]string{"instance_id", "instance_name"},
```

```
                //静态标签
                prometheus.Labels{"module": "mem"},
        ),
        ioDesc:prometheus.NewDesc(
                "host_io",
                "get host io",
                //动态标签key列表
                []string{"instance_id", "instance_name"},
                //静态标签
                prometheus.Labels{"module": "io"},
        ),
        labelVaues: []string{"myhost", "yunwei"},
    }
}

//实现Describe接口，传递指标描述符到channel
func (h *HostMonitor) Describe(ch chan<- *prometheus.Desc) {
    ch <- h.cpuDesc
    ch <- h.memDesc
    ch <- h.ioDesc
}

//实现Collect接口，将执行抓取函数并返回数据
func (h *HostMonitor) Collect(ch chan<- prometheus.Metric) {
    ch <- prometheus.MustNewConstMetric(h.cpuDesc, prometheus.GaugeValue,
70, h.labelVaues...)
    ch <- prometheus.MustNewConstMetric(h.memDesc, prometheus.GaugeValue,
30, h.labelVaues...)
     ch <- prometheus.MustNewConstMetric(h.ioDesc, prometheus.GaugeValue,
90, h.labelVaues...)
}

func main() {

    ovs := NewHostMonitor()
    prometheus.MustRegister(ovs)

    http.Handle("/metrics", promhttp.Handler())

    log.Info("begin to server on port 8081")
    // listen on port 8081
```

```
    log.Fatal(http.ListenAndServe(":8081", nil))
}
```

然后可以用下面的命令来访问：

```
curl127.0.0.1:8081/metrics
```

其结果如下：

```
# HELP host_cpu get host cpu
# TYPE host_cpu gauge
host_cpu{instance_id="myhost",instance_name="yunwei",module="cpu"} 70
# HELP host_io get host io
# TYPE host_io gauge
host_io{instance_id="myhost",instance_name="yunwei",module="io"} 90
# HELP host_mem get host mem
# TYPE host_mem gauge
host_mem{instance_id="myhost",instance_name="yunwei",module="mem"} 30
```

事实上，如作者目前维护的HPC超算系统，除了Node Exporter、Infiniband Exporter、Lustre Exporter这些开源组件能够基本完成HPC集群的监控，还有很多业务指标的采集数据，需要用Go语言来另行开发。

9.7 小　　结

本章首先详细介绍了整个Prometheus监控体系的安装及部署流程，包括核心组件，如InfluxDB、Prometheus和Alertmanager的详细配置和优化，以及DevOps监控开发工作中涉及的Prometheus API等；然后介绍了用Go语言进行Exporter二次开发的相关步骤，这些都与云原生监控DevOps生产实践相关，希望读者能熟练掌握本章内容，针对自己的系统和业务平台，能够搭建监控系统，并且针对不同的业务线，能够利用Go语言开发Exporter组件。

附录 A　Go 语言开发中的常见错误

下面介绍 Go 语言开发中的常见错误。

1. 大括号问题

在大多数编程语言中，都可以将大括号放在任意位置，但是 Go 语言不能将大括号中的左括号放到新的一行，并且与 Python 语言一样都不需要分号（即使含有分号，也不会报错）。示例如下：

```
package main
import "fmt"
func main()
{
fmt.Println("hello there!")
}
```

执行报错如下：

```
./test.go:4:1: syntax error: unexpected semicolon or newline before {
```

正确语句如下：

```
package main
import "fmt"
//正确语句
func main()  {
fmt.Println("hello there!")
}
```

2. 变量未使用

如果在 Go 语言中出现没有使用的变量，则程序无法完成编译。如果在函数中声明了变量，则必须要使用，但是全局变量除外。如果将一个新的值分配给一个未使用的变量，不算使用该变量。示例如下：

```
package main

import "fmt"
// 因为six是全局变量，即使未使用，程序也不会报错
var six  int
```

```
//Go 代码中不能出现未使用已定义变量的情况
func main() {
      var one int
      _ = one

      two := 2
      fmt.Println(two)

      var three int
      three = 3
      one = three

      var four int
      four = four

      // var five int
      // 如果定义了 five 变量又没有使用，则有如下报错
      // five declared and not used
}
```

3. 导入的包未使用

在使用集成开发环境时，一般包都是自动导入和删除的，所以导入的包未使用这个问题一般不会出现。Go 语言也不会允许出现未使用的包，一般都会将不使用的包删除或者注释掉。但是在一些特殊情况需要只导入包而不使用（如连接数据库的包），这时就要用到 _（在导入路线前添加下划线，表示只导入该库的初始化函数，而不对其他导出的对象进行使用。这样可以避免编译时的未使用包错误，因为 Go 语言的数据库驱动都会在 init 函数中注册）。示例如下：

```
package main

import (
"fmt"
      "log"
)

func main() {
      fmt.Println("hello.yhc")
}
```

执行上面的程序会报错：

```
./test.go:5:2: "log" imported and not used
```

4. 误用 nil 初始化没有显式类型的变量

例如，接口、函数、指针、哈希表、切片和通道的默认值是 nil，但是如果没有指明一个变量的类型而为其赋值 nil 或将其用于初始化则是行不通的，因为 Go 语法无法推断出变量的类型。示例如下：

```
    package main

func main() {
    var x = nil //error

    _ = x
}
```

执行上面的程序会报错，报错信息如下：

```
./test.go:4:10: use of untyped nil in variable declaration
```

5. 在多行表示的切片、数组和字典中缺少逗号

在编写代码时，单行列表或元组的最后一个元素后面可以不用加逗号（有逗号也不会出现错误），但在多行列表或元组中，每行最后都需要加逗号。示例如下：

```
package main
func main() {
    x := []int{
    1,
    2 // 此处应该有“,”号，如果没有则会报错
    }
    _ = x
}
```

报错信息如下：

```
./test.go:5:56: syntax error: unexpected newline in composite literal;
possibly missing comma or }
```

6. Go 语言中字典的线程安全

Go 语言中的 map 不是并发安全的，当有多个并发的 goroutine 读写同一个 map 时，会产生 panic 报错，报错信息如下：

```
concurrent map read and map write
```

示例如下：

```
var mMap map[int]int

func TestMyMap(t *testing.T) {
    mMap = make(map[int]int)
```

```
    for i := 0; i < 5000; i++ {
        go func() {
            mMap[i] = i
        }()
        go readMap(i)
    }
}
func readMap(i int) int {
    return mMap[i]
}
```

解决办法有以下两种：

（1）sync.Mutex 或 sync.RWMutex。

（2）sync.Map。

7. 超出索引范围

在 Go 语言中，当尝试访问数组、切片、字符串或者 map 的索引超出它们的范围时，程序会 panic，并抛出 index out of range 的运行时错误。

例如，当尝试访问一个长度为 0 的数组、切片、字符串或者 map 的第 0 个元素时，程序就会 panic。这是因为在这些数据结构中，第 0 个元素的索引是有效的，但是此时数据结构的长度为 0，表示它并没有任何元素，所以尝试访问第 0 个元素就会导致程序出现 panic 错误，程序直接退出。示例如下：

```
package main

import "fmt"

func main() {
    arr := [3]int{1, 2, 3}
    fmt.Println(arr[3])
}
```

运行程序，直接抛出 panic 错误，结果如下：

```
./test.go:7:18: invalid argument: index 3 out of bounds [0:3]
```

那么，如何解决这个问题呢？首先使用 len() 函数获取数组 arr 的长度，然后判断该长度是否大于 3。如果是，则访问 arr[3]，否则输出“索引超出范围”，示例如下：

```
package main

import "fmt"
```

```
func main() {
    arr := [4]int{1, 2, 3,4}
    fmt.Println(len(arr))
    if len(arr) > 3{
        fmt.Println(arr[3])
    } else {
        fmt.Println(" 索引超出范围 ")
    }
}
```

8. map 字典需要用 make 函数赋值

Go 语言中的 map 值是 nil，这个 map 是不能赋值的；刚开始编写 Go 语言代码时很容易犯这个错误，示例如下：

```
package main

import "fmt"

func main() {
    jLabels := make(map[string]map[string]string)
   // 用make 函数创建一个非 nil 的 map，nil map 不能赋值
    jLabels["job_name"] = make(map[string]string)
    jLabels["job_user"] = make(map[string]string)
    jLabels["job_queue"] = make(map[string]string)
    jobid := "45282"

    jLabels["job_name"][jobid] = "45281"
    jLabels["job_user"][jobid]= "yhc"
    jLabels["job_queue"][jobid] = "gpu"

    fmt.Println(jLabels)
}
```

9. 可变参数是空接口类型

当可变参数是空接口类型时，传入空接口的切片需要注意参数展开的问题，但输出的结果是完全不同的。示例如下：

```
package main

import "fmt"

func main() {
```

```
        var a = []interface{}{1, 2, 3}

        fmt.Println(a)
        fmt.Println(a...)
    }
```

不管参数是否展开，编译器都无法发现其错误，输出如下：

```
[1 2 3]
1 2 3
```

10. Go 语言中是显式转换没有隐式转换

Go 语言的静态性体现在类型上，需要显式定义与显式转换，是一种强类型语言，这一点与 Java 和 Python 语言截然不同，需要特别注意。示例如下：

```
package main

import "fmt"

func main() {

    var str string = "hello world"
    fmt.Println("str 的长度: %d" + len(str))
}
```

报错如下：

```
./test.go:8:14: invalid operation: "str 的长度: %d" + len(str) (mismatched types untyped string and int)
```

解决方法其实也很简单，其他类型的转换可以使用 strconv 包中的方法。例如，这里需要转换的话，可以用如下代码实现：

```
strconv.itoa(len(str))
```

11. 读取数据库中的 Null 值问题

如果直接从数据库中读取 Null 值到 string/int，则会发生以下错误：

```
sql: Scan error on column index 1: unsupported Scan, storing driver.Value type <nil> into type *string
```

更新下面的结构体配置：

```
type User struct {
    ID        int
    ProjectID    int
    ref_rep string
    channel string
```

```
    nsp_name string
    docker string
    //create_time sql.NullTime      //防止Null空值
    //update_time sql.NullTime      //防止Null空值
    create_time sql.NullString
    update_time sql.NullString
    cluster_id string
    enabled string
}
```

附录 B　Go 语言中的关键字

关键字（也称为保留字）是被编程语言保留而不让编程人员作为标识符使用的字符序列。

关键字是被 Go 语言赋予了特殊含义的单词，也可以称为保留字。

Go 语言中的关键字一共有 25 个，之所以刻意将 Go 语言中的关键字保持这么少，是为了简化在编译过程中的代码解析。和其他语言一样，关键字不能用作标识符。

Go 语言的关键字主要分成 3 类，即包管理、程序实体声明和定义、程序流程控制，如表 B.1 所示。

表B.1　Go语言的关键字及其作用

关键字分类	关键字名称	作　用
包管理	import	用于导入包，这样就可以使用包中被导出的标识符
	package	用于声明包的名称，需放在 go 文件所有代码的最前面。一个包由一个或多个 go 源文件组成，需放在同一个目录下
程序实体声明和定义	chan	用于声明 channel（信道）。信道提供一种机制使两个并发执行的函数实现同步，并通过传递具体元素类型的值来通信
	const	用于定义常量，一旦创建，不可赋值修改
	interface	用于定义接口，一个接口一个方法集合
	map	用于声明映射变量
	struct	用于定义结构体
	type	用于定义类型
	var	用于定义变量
	func	用于定义函数。Go 函数支持变参且返回值支持多个，但不支持默认参数
程序流程控制	go	用于创建 goroutine，实现并发编程
	select	用来选择哪个 case 中的发送或接收操作可以被立即执行。它类似于 switch 语句，但是其 case 涉及 channel 有关的 I/O 操作。也就是说，select 用于监听与 channel 有关的 I/O 操作

续表

关键字分类	关键字名称	作　用
程序流程控制	switch	提供多路执行，表达式或类型说明符与 switch 中的 case 相比较从而决定执行哪一分支
	case	与 switch 搭配使用，决定是哪一个分支
	fallthrough	与 switch 搭配使用，用于继续后面的 case 或 default 子句
	default	与 switch 搭配使用，用于默认分支
	defer	用于预设一个函数调用，即推迟函数的执行
	else	if 与 else 实现条件控制
	break	用于终止最内层的 for switch 或 select 语句的执行
	for	for 与 range 搭配使用，用于循环 Go 语言中的数据结构
	goto	用于将程序的执行转移到与其标签相应的语句
	if	if 与 else 实现条件控制
	range	for 与 range 搭配使用，用于循环 Go 语言中的数据结构
	return	用于函数执行的终止并可选地提供一个或多个返回值
	continue	通常用于结束当前循环，提前进入下一轮循环

附录 C　Go 语言中如何处理结构复杂的 JSON 文件

在工作中，我们经常会与各种系统的 HTTP API 打交道，这些 API 通常采用 RESTful 风格，并返回 JSON 格式的结果。Go 语言自带的 encoding/json 库能够方便地在 JSON 字符串与 JSON 对象之间进行转换。对于结构简单的 JSON 字符串，这个库非常好用。然而，当 JSON 字符串变得复杂或嵌套层次较多时，使用 encoding/json 为每个嵌套字段定义 struct 类型的方式就显得烦琐且不切实际。


```
{
    "rc": 0,
    "error": "Success",
    "type": "stats",
    "progress": 100,
    "job_status": "COMPLETED",
    "result": {
        "total_hits": 803254,
        "starttime": 1528434707000,
        "endtime": 1528434767000,
        "fields": [],
        "timeline": {
            "interval": 1000,
            "start_ts": 1528434707000,
            "end_ts": 1528434767000,
            "rows": [{
                "start_ts": 1528434707000,
                "end_ts": 1528434708000,
                "number": "x12887"
            }, {
                "start_ts": 1528434720000,
                "end_ts": 1528434721000,
                "number": "x13028"
            }, {
                "start_ts": 1528434721000,
                "end_ts": 1528434722000,
                "number": "x12975"
```

```
                }, {
                    "start_ts": 1528434722000,
                    "end_ts": 1528434723000,
                    "number": "x12879"
                }, {
                    "start_ts": 1528434723000,
                    "end_ts": 1528434724000,
                    "number": "x13989"
                }],
                "total": 803254
            },
            "total": 8
        }
    }
```

上面的 JSON 字符串嵌套很深，如果要逐个定义对应的 struct 结构，无疑是一件比较费力的事情，可以用原始官方的 JSON 解析库 encoding/json 来查看，其代码如下：

```
package main

import (
    "encoding/json"
    "fmt"
)

func main() {
    // 定义名为 JSONData 的复杂结构体
    type JSONData struct {
        Rc        int    `json:"rc"`
        Error     string `json:"error"`
        Type      string `json:"type"`
        Progress  int    `json:"progress"`
        JobStatus string `json:"job_status"`
        Result    struct {
                  TotalHits int           `json:"total_hits"`
                  Starttime int           `json:"starttime"`
                  Endtime   int           `json:"endtime"`
                  Fields    []interface{} `json:"fields"`
                  Timeline  struct {
                            Interval int `json:"interval"`
                            StartTs  int `json:"start_ts"`
                            EndTs    int `json:"end_ts"`
```

```
                        Rows      []struct {
                        StartTs int    `json:"start_ts"`
                        EndTs   int    `json:"end_ts"`
                        Number  string `json:"number"`
                    } `json:"rows"`
                    Total int `json:"total"`
                } `json:"timeline"`
                Total int `json:"total"`
            } `json:"result"`
    }

    jsonData:= []byte(`
{
    "rc": 0,
    "error": "Success",
    "type": "stats",
    "progress": 100,
    "job_status": "COMPLETED",
    "result": {
        "total_hits": 803254,
        "starttime": 1527000,
        "endtime": 1527000,
        "fields": [],
        "timeline": {
            "interval": 1000,
            "start_ts": 1527000,
            "end_ts": 1527000,
            "rows": [{
                "start_ts": 1527001,
                "end_ts": 1527003,
                "number": "x12887"
            },
              {
                "start_ts": 1527002,
                "end_ts": 1527004,
                "number": "x13028"
            },
              {
                "start_ts": 1527003,
                "end_ts": 1527005,
                "number": "x12975"
```

```
            },
                        {
                "start_ts": 1527004,
                "end_ts": 1527005,
                "number": "x12879"
            },
                        {
                "start_ts": 1527005,
                "end_ts": 1527006,
                "number": "x13989"
            }],
            "total": 803254
        },
        "total": 8
    }
}`)


    var jsondata JSONData
    err := json.Unmarshal(jsonData, &jsondata)
    if err != nil{
        fmt.Println(err)
    }

    // 获取 JSON 数据中的 Timeline.Rows
    rowslist := jsondata.Result.Timeline.Rows
    // 遍历 Timeline.Rows 中的每一项
    for _, vv := range rowslist{
    // 打印每一项的 Number、StartTs 和 EndTs
        fmt.Println(vv.Number, vv.StartTs, vv.EndTs)
    }
}
```

这里如果使用其他库，如使用 bitly 公司开源的 simplejson 库，实现就简单了。示例如下：

```
package main

import (
    "fmt"
    simplejson "github.com/bitly/go-simplejson"
    "log"
)
```

```
var json_str string = `{
    "rc": 0,
    "error": "Success",
    "type": "stats",
    "progress": 100,
    "job_status": "COMPLETED",
    "result": {
        "total_hits": 803254,
        "starttime": 1527000,
        "endtime": 1527000,
        "fields": [],
        "timeline": {
            "interval": 1000,
            "start_ts": 1527000,
            "end_ts": 1527000,
            "rows": [{
                "start_ts": 1527001,
                "end_ts": 1527003,
                "number": "x12887"
            },
                        {
                "start_ts": 1527002,
                "end_ts": 1527004,
                "number": "x13028"
            },
                        {
                "start_ts": 1527003,
                "end_ts": 1527005,
                "number": "x12975"
            },
                        {
                "start_ts": 1527004,
                "end_ts": 1527005,
                "number": "x12879"
            },
                        {
                "start_ts": 1527005,
                "end_ts": 1527006,
```

```
                "number": "x13989"
            }],
            "total": 803254
        },
        "total": 8
    }
}`

func main() {
    js, err := simplejson.NewJson([]byte(json_str))
    if err != nil || js == nil{
        log.Fatal("something wrong when call NewFromReader")
    }
    //获取json字符串中result的 timeline 下的 rows 数组并且打印
    arr, err := js.Get("result").Get("timeline").Get("rows").Array()
        fmt.Println(arr) //&{map[test:map[array:[1 2 3]
arraywithsubs:[map[subkeyone:1] map[subkeytwo:2 subkeythree:3]]
bignum:8000000000]]}
    // 利用 for 循环遍历
    for i, _ := range arr{
        // 依次取真实的值
        end_time_index :=
js.Get("result").Get("timeline").Get("rows").GetIndex(i)
        endtime := end_time_index.Get("end_ts").MustInt()


        number_index :=
js.Get("result").Get("timeline").Get("rows").GetIndex(i)
        number :=  number_index.Get("number").MustString()
        start_time_index :=
js.Get("result").Get("timeline").Get("rows").GetIndex(i)
        starttime := start_time_index.Get("start_ts").MustInt()
        fmt.Println(endtime,number,starttime)
    }
}
```

最终获取结果如下：

```
map[end_ts:1527003 number:x12887 start_ts:1527001] map[end_ts:1527004
number:x13028 start_ts:1527002] map[end_ts:1527005 number:x12975 start_
ts:1527003] map[end_ts:1527005 number:x12879 start_ts:1527004] map[end_
```

```
ts:1527006 number:x13989 start_ts:1527005]]
    1527003 x12887 1527001
    1527004 x13028 1527002
    1527005 x12975 1527003
    1527005 x12879 1527004
    1527006 x13989 1527005
```

更多 simplejson 库中函数的具体用法，可以参考 simplejson 库的官方文档。